AMBUSHES AND SURPRISES.

GENERAL LORD MARK KERR, K.C.B.

From a Photograph by Messrs. Maull & Fox, 187A, Piccadilly.

AMBUSHES AND SURPRISES:

BEING A DESCRIPTION OF SOME OF THE MOST FAMOUS INSTANCES OF THE LEADING INTO AMBUSH AND THE SURPRISE OF ARMIES, FROM THE TIME OF HANNIBAL TO THE PERIOD OF THE INDIAN MUTINY.

WITH A PORTRAIT OF GENERAL LORD MARK KERR, K.C.B.

BY

COLONEL G. B. MALLESON, C.S.I.,

AUTHOR OF "THE DECISIVE BATTLES OF INDIA," ETC.

"BUT, MY LADS, MY LADS, TO-MORROW MORNING, BY FOUR O'CLOCK, EARLY AT GADSHILL. THERE ARE PILGRIMS GOING TO CANTERBURY WITH RICH OFFERINGS, AND TRADERS RIDING TO LONDON WITH FAT PURSES. I HAVE VISORS FOR YOU ALL, YOU HAVE HORSES FOR YOURSELVES. GADSHILL LIES TO-NIGHT IN ROCHESTER; I HAVE BESPOKE SUPPER TO-MORROW NIGHT IN EASTCHEAP; WE MAY DO IT AS SECURE AS SLEEP."—*Henry IV.*

LONDON:

W. H. ALLEN & CO., 13 WATERLOO PLACE. S.W.

PUBLISHERS TO THE INDIA OFFICE.

1885.

DEDICATION.

DEAR LORD MARK KERR,

I VENTURE TO ASK YOU TO ACCEPT THE DEDICATION OF A WORK, WHICH, UNDER THE TITLE "AMBUSHES AND SURPRISES," CLOSES WITH A DESCRIPTION OF THE SOLDIERLY MANNER IN WHICH YOU, IN 1858, NOT ONLY EXTRICATED THE FORCE UNDER YOUR COMMAND FROM A DANGEROUS POSITION, BUT LED IT TO THE DEFEAT OF THE ENEMY WHO HAD DRAWN YOU INTO AN AMBUSH AND SURPRISED YOU. UNDER SIMILAR CIRCUMSTANCES FLAMINIUS SUCCUMBED AT LAKE TRASIMENUS; VARUS, IN THE FOREST OF TEUTOBURG; ROLAND, AT RONCESVALLES; BRADDOCK, AT FORT DUQUESNE. YOU OCCUPY, THEN, A POSITION UNIQUE IN MILITARY HISTORY; FOR LECOURBE, AND THE DEFENDERS OF INKERMAN—WHO ALONE, BESIDES YOURSELF, OF ALL THE SOLDIERS WHO HAVE BEEN SURPRISED, REPULSED THEIR ENEMIES—WERE ATTACKED IN POSITION, WHEREAS THE ENEMY WERE LYING IN WAIT FOR YOU IN THE JUNGLE.

TRUSTING THAT AN OPPORTUNITY MAY YET BE AFFORDED YOU OF PROVING THAT AN OFFICER OF THE OLD AND UNREFORMED SCHOOL CAN AGAIN LEAD HIS MEN TO VICTORY,

I REMAIN,

DEAR LORD MARK KERR,

YOURS VERY TRULY,

1 JULY 1885. G. B. MALLESON.

GENERAL LORD MARK KERR, K.C.B.

PREFACE.

WHILST, in this book, I have endeavoured to illustrate, by a series of striking examples, a special and very interesting phase of the art of war, I have striven so to select the examples that each, in its turn, should illustrate a distinct epoch of history. Thus, whilst the citizens of consular Rome, after the surprise of Flaminius by Hannibal, pose before us as men worthy of their high renown, undaunted, daring, still rushing to the front to meet the destroyer of their armies, the descendants of the same men, occupying the same city, but just become Imperial, appear, after Teutoburg, enervated, timid, shrinking from danger, each of them caring only for his own life. Had it been possible that the same man should have witnessed both epochs, he could not surely, contrasting the ignominy of the second with the gallantry of the first period, have withheld the words expressive of every thought of his soul: *Quantum mutata ab illâ Româ!* Gibbon dates, indeed, the history of the decline and fall from the age of Trajan and the Antonines; but the utter prostration of the population of the Mistress of the World, from the Emperor downwards, on hearing of the defeat at TEUTOBURG, proves that she was, at all events, quite ready to start on the fatal incline when Octavius wore the purple.

In the romantic story of RONCESVALLES I have endeavoured to give a glimpse of Europe emerging for a moment from the chaos of the dark ages under the guidance of the three great Carlovingians ; whilst in leading, in the fourth chapter, to the surprise which gave Constantinople to the Osmánli, I have given in, I trust, not unnecessary detail, the earlier history of the race which fled from the tyranny of Chengiz Khán in Khorásán to found an empire on the shores of the Bosphorus. A great part of that chapter was written in close vicinity to the events which it records.

The story of the attempted surprise of Charles V. at INNSBRUCK shows us Central Europe just affected by the spirit of the Reformation ; how the interests, or supposed interests, of the rulers and the ruled came gradually to diverge under its influence until a war of thirty years' duration became necessary to enable those interests to harmonise. In this chapter I have endeavoured to show Charles V. as the man he really was, and not as the man his admirers have represented him to be.

The account of the surprise of FORT DUQUESNE, again, gives the reader a glimpse of the American colonies as those colonies were constituted and governed, and of the motives which actuated the colonists, before a fatal policy drove them to sever their connection with the mother country. The almost contemporaneous surprise of MAXEN will be interesting as showing at his best the principal opponent of Frederic II. of Prussia during the Seven Years' War, and as giving a striking example of the manner in which that King, one of the greatest of generals, and yet one of the meanest of men, never hesitated to make a scapegoat of a general who was unsuccessful only because he obeyed the King's positive directions.

Of all the campaigns of the Revolution and the Empire there is scarcely one the details of which are less generally known than the campaign, in Italy and Switzerland, of 1799. In the eighth chapter I have given an outline of the leading military incidents in both countries in that year. Those incidents formed a prelude to the crushing defeat which the genius of Masséna and the splendid daring of Lecourbe combined to inflict upon Souvoroff and his generals at ZURICH and on ST. GOTHARD.

The SURPRISES recorded in the last two chapters occurred in our own time. Their right to a place in this volume will be denied by no one. I will add that the subjects of both have been studied by me with the greatest care. If I have not, as I am conscious I have not, brought into the prominence they deserved the names of many gallant soldiers, it is because in so short a sketch it has been impossible to individualise the actions of all. I hope, however, I may be allowed to cherish the conviction that, at all events, I have been unjust to no one.

If I have omitted from the list of surprises those of Von Rantzau at Tuttlingen, and of the illustrious Turenne at Mergentheim, it is because I have so recently told the story in another work, *The Battle-fields of Germany.* I would gladly have added the account of the surprise of Aroyo Molino, but that it is impossible to equal the description given of it by William Napier in his classic history of the Peninsular War. Equally was I debarred from describing a more modern surprise, that of Fontenoy-sur-Moselle, by the implied conditions in which a translation of the authorised German account, by Captain Trotter, R.A., was inserted in the *Journal* of the Royal United Service Institution.

Of all the soldiers who have led the surprised or surprising expeditions recorded in this volume General Lord Mark Kerr, K.C.B., is the sole survivor. It seemed to me, then, that the work would be incomplete if it were to withhold from the public the living likeness of one who not only represents in his own person a race of leaders who enjoyed experiences in warfare not shared by the crowd, but who is a successful type of that race. It is for this reason that I begged and obtained the permission which has enabled me to adorn the book with the portrait of the soldier who conquered at A'zamgarh.

G. B. MALLESON.

27, West Cromwell Road.
1st *July* 1885.

CONTENTS.

DIRECTION TO BINDER

AMBUSHES AND SURPRISES.

CHAPTER I.

LAKE TRASIMENUS.*

THE name of Hannibal will ever retain its fascination for mankind. Of all the great captains of the world, he was the greatest. Napoleon, who had studied his campaigns with the greatest care, could not detect in his conduct of them a single mistake. "Hannibal," he says, "who was the most daring, the most wonderful (*étonnant*) of all; so bold, so sure, so grand in all things; who, at the age of twenty-six, conceives a plan which is scarcely conceivable, and executes that which one was bound to regard as impossible; who, abandoning all communication with his own country, traverses territories hostile or unknown, attacks and conquers the inhabitants, then escalades the Pyrenees and the Alps—believed to be insurmountable—and pays for the right to descend into Italy, to occupy a field of battle and to fight there, by the sacrifice of one-half of his army; who occupies, traverses, and governs that same Italy for sixteen years, brings several times

* For this chapter I have studied the works of Livy, of Polybius, of Appian, and other ancient writers: I have read also Zander, Rollin, de Vaudoncourt, Macdougall, and Bosworth Smith; but I desire to express my special obligations to the excellent work of M. E. Hennebert, *Histoire d'Annibal*, most exhaustive in its completeness, for upon it I have indented largely and often.

the terrible and formidable Rome to within two inches of destruction, and only quits his prey when her children profit by the lesson which he has given them to transfer the war to his own soil!"

Such, in brief, was Hannibal: the greatest of all captains; the master in the art of war to all who were to come after him; the conceiver and executor of campaigns which were to exist as models for all time. To his unwritten maxims the greatest of his successors in the field of glory, the Scipios and Cæsars before the Christian era, and Alexander Farnese, Maurice of Nassau, Spinola, Condé, Turenne, Eugène, Marlborough, Frederic, and Napoleon, in the more modern times, have appealed to justify their own actions. Such was he, a terror and an example to his enemies, winning their esteem whilst he beat their armies and occupied their provinces; a wonder, full of the deepest interest, to the world for ever!

The civil wars of Greece, the wars of Greece with Persia, the stupendous conquests of Alexander, gave undoubtedly earlier examples of the successful exercise of the art of war. For the most perfect demonstration of the strategic science in all its branches, however, there is no necessity to travel back beyond Hannibal. The question of ambushes and surprises forms no exception to this rule. One of the earliest, and certainly the most instructive, well-authenticated instance of the destruction of a large army by enticing it into a position whence it could not escape is that which forms the heading of this chapter. It will, I think, be convenient if, before entering upon that subject, I trace as briefly as the subject will permit, the career of the great Carthaginian up to the time when he successfully drew the Consul Flaminius into the position which proved fatal to him.

Hannibal was born at Carthage, 247 B.C., one of six children, and the eldest of four brothers.* His father, who placed in him all his hopes, educated him with great care, and, whilst giving him

* The other brothers were Hasdrubal, Mago, and Hanno. One sister married a Numidian prince; the other, Hasdrubal, surnamed "the handsome."

a physical training which developed all the muscles of his body, excited his mind by constantly reciting to him the story of the campaigns made in Sicily and in Africa. When he was nine years old, Hamilcar resolved to take him with him into Spain to familiarise his youthful mind with scenes of war. Prior to his departure the boy solemnly swore, in the temple of Jupiter, that he would be the eternal enemy of the nation which was the great rival of his own country. Arrived in Spain, Hannibal devoted himself with ardour to the duties which devolved upon him. He could not have learned in a better school. His father was a man of transcendent genius. He had revolutionised the art of war as it had existed before him. Up to his time, military operations had been conducted on a principle from which enterprise was absolutely wanting. The general commanding had been accustomed to mass his troops on a fortified town, which served as a pivot whence timidly and hesitatingly to feel his way towards the enemy. Hamilcar had altered all that. He was the inventor of the science of marching with boldness and confidence against an enemy wherever he was to be found. By his unwonted daring he had astonished his enemies, and had gained a name and a prestige which had given him great influence amongst his countrymen. He was now, at the period at which we have arrived, B.C. 237, carrying out that principle in Spain; and it was in watching the working of it that young Hannibal imbibed those daring ideas which he knew so well how to apply in the years that were to follow.

The Carthaginians had, in preceding years, attempted to colonise some parts of the coasts of Spain. Their endeavours, however, to cultivate friendly relations with the children of the soil—a people half savage, resolute, patient, industrious, and fierce—had proved unavailing. The Iberians always regarded the settlers as strangers, and, if the record is to be believed, they lost no opportunity of harassing, even of attacking them. The necessity of punishing the indigenous tribes for their unfriendly behaviour formed, at all events, the pretext of the war which Hamilcar undertook in the year 237.

That war was conducted by Hamilcar during the eleven years which followed. He had operated first in Alemtejo, had defeated in the first battle the army of the two brothers who commanded there, and had achieved a second victory over the forces of a powerful chief who had endeavoured to avenge the death of his kinsmen in the first encounter. With this second victory opposition to the progress of the Carthaginian leader ceased. Hamilcar then put in practice a policy of conciliation. He restored to liberty, without conditions, the ten thousand prisoners he had made. Ascending then the western coast, he forced to submit the various tribes which occupied Portugal and Gallicia. Their territories subdued, he gave a new opening to the industrious energies of his countrymen. Under the protection of his arms, the silver mines especially, the rumoured existence of which had excited the cupidity of his countrymen, were worked and developed; that which had been a fable soon became a reality, and the officers of the army of Spain were soon able to keep their wine in silver casks, and to drink it from silver goblets.*

Turning then westward, Hamilcar reached a point on the coast whence he could gaze in the direction of Italy—ever the object of his thoughts. Between that point and the Italian peninsula were the Balearic Isles. At first he thought of constituting those islands as a base for a further advance; but he abandoned the idea to create, on the coast itself, one more solid. He selected for this purpose a position not far from the town of Saguntum. It was at this spot that he died, whilst rallying his troops to oppose a sudden incursion of hostile tribes, his arms in his hands, in the year 227.†

In the command of the army in Spain Hamilcar was succeeded by his son-in-law Hasdrubal, a man whose manly beauty gave him the name of "the handsome," and whose military talents were very considerable. But the young Hannibal returned to Carthage, his

* *Histoire d'Annibal.* Hennebert, vol. i. p. 151. Consult likewise Strabo and Polybius.

† *Ibid.*, pp. 152 and 238, and notes.

presence there being required by the position which had devolved upon him as the head of the family. At Carthage Hannibal employed his time in welding that political connection without which, he felt certain, his actions in the field would be paralysed. It required three years of incessant labour to obtain in the senate the necessary majority—a majority which should be permanent. When that result had been accomplished Hannibal returned to Spain to serve under his brother-in-law (223 B.C.)

That brother-in-law, meanwhile, had proved himself a worthy successor to the great Hamilcar. Nominated Governor-General of the Peninsula and General-in-Chief of the sea and land forces, he signalized his assumption of these titles by gaining a great victory over an influential chief named Orisso. This victory tranquillized the country. The greater part of the chiefs of the several districts submitted at once; and Hasdrubal, having established peace and order in the interior, found leisure to devote himself to the development of a plan which his late father-in-law had long secretly nourished but had not been able to execute—the creation of a new Carthage in Spain.

To the genius, the energy, the enterprise of Hasdrubal Carthage owed, and the Spain of the present day owes, the founding of Carthagena, then the principal seat of the colonists from the mother-country, now the chief arsenal in the peninsula. In a few years the new town became, so to speak, the capital of Carthaginian Spain, the principal store-house of the mother-country in the Peninsula. The locality was peculiarly adapted to become the seat of a first-class maritime establishment, and it speedily supplied a want which had been long felt by the enterprising people on the opposite shore of the continent of Africa. This is not the place to describe it. It must suffice to state that, whilst its harbour was capacious and defensible, the town itself was built on the declivity of a hill and was covered towards the interior by five considerable heights, two of them lofty and abrupt, the other three rugged and difficult of access. Its value to the mother country as a depôt for her magazines, for her armies, for her commerce, cannot be overrated.

But Hasdrubal was more than the founder of this new Carthage. Prudent, far-sighted, and conciliatory, he knew how to bring the native populations to recognise the importance to themselves of the connection with the mother-country; to feel that their interests were bound up with the maintenance of her power and of her authority. It was the news of his increasing influence in Spain which caused the patricians of Rome to feel that it was necessary to come to some understanding to prevent the further development of a dominion which might soon embrace the entire peninsula.

"The Romans," says Polybius,* "dared not, however, dictate to them" (the Carthaginians) "new laws nor arm against them. They had enough to do to hold themselves in a state of preparation against the Gauls, by whom they were menaced and who might any day attack them. It seemed to them preferable to act in a conciliatory spirit with respect to Hasdrubal. They sent him then ambassadors, who, without referring to the rest of Spain, required from him a stipulation that he would not carry the war beyond the Ebro." A treaty on somewhat wider terms was agreed to,† and, for the moment, the apprehensions of Rome were stayed.

A very short time afterwards those apprehensions, which had been based principally on the knowledge Rome possessed of the genius of the great Carthaginian Governor-General, were entirely dissipated by the intelligence that Hasdrubal was no more.

The intelligence was too true. After a tenure of office lasting only seven years, during which he had rendered inestimable services to his country, Hasdrubal was assassinated (220 B.C.). Whilst some writers attributed the murder to the vengeance of a Gaul whose master Hasdrubal had put to death, there are not wanting others who attribute the deed to the fears of the Roman senate. This is a matter which must ever remain doubtful.‡

* *Polybius II.*, xiii.

† "It was expressly stipulated," writes Appian, "that the Ebro should form the limit of the Carthaginian empire in Spain; that the Carthaginians were not to carry on war beyond that river; that the Saguntians and other Greek colonies established in the Peninsula should preserve their independence and their autonomy."

‡ Hennebert, vol. i. p. 156.

Four years prior to the death of Hasdrubal, Hannibal had, as already stated, returned to Spain to serve under his brother-in-law (223 B.C.) His arrival had caused the greatest excitement. The soldiers who had served under his father saw in him again their old leader. There was the same energetic face, the same bold carriage, the same eye sparkling with the fire of genius. Hannibal soon became the idol of the soldiers. Hasdrubal confided to him important military commands, and the manner in which he executed his instructions gained for him alike the approval of his chief and the enthusiastic confidence of the men.

When, then, Hasdrubal was assassinated, the voice of the army was unanimous in calling upon his young kinsman to fill the post he had vacated. The election of the army was ratified at once by the Carthaginian Senate. A description of the personal appearance of the new Commander-in-Chief, derived from the best sources,* may here be interesting.

An exception to all, or almost all, of the greatest of his successors in the art of war, Hannibal was tall. The breadth of his shoulders gave evidence of the strength of his body. His head was well set on, was well shaped, and well carried. His features were handsome—it was the handsomeness of intelligence—and his glance was sufficient to bring the boldest to his senses. It was his habit to leave his head uncovered, warmed only by his abundant hair, the thick locks of which were kept in their place by a golden band. His habits were temperate. He had trained himself to bear privations, and when it was necessary could make long fasts. He drank but little wine. He could, on occasions, dispense with sleep, and was always ready for action. When he did sleep it was his wont, during his campaigns, to lay down on a lion's skin in the middle of his camp, or with the advance guard, wherever he deemed his presence most necessary. Like Napoleon, he acquired, after a time, the faculty of dropping off to sleep in a moment, and of awakening at his will. Neither heat nor cold seemed to affect him. His example in all things was contagious; his men strove to emulate

* Hennebert, vol. i. pp. 247–8 and notes.

him. The admiration which he evoked in their hearts was undoubtedly one of the secrets of his success.

Such was the man who early in the year 220 B.C. assumed command of the Carthaginian army in Spain. Events were to test his powers at the very outset. The change in the leadership, the transfer from a man who, like Hasdrubal, had made himself beloved by the peaceable and feared by the discontented, to a young chief untried in the exercise of unfettered authority, induced several of the indigenous tribes to rise and strike a blow for liberty. The first to give the signal was the tribe of the Olcades. The Olcades occupied the most rugged districts of Spain—the most difficult part of the districts which in the Peninsular War were the scene of the operations of Marshal Suchet, and the subjugation of which, after five years of intense labour, gained for him his title and his renown. They signalled their insurrection by occupying in force the important town of Carteia.* There Hannibal attacked them, took Carteia, and by his vigorous action stamped out the rebellion. The manner in which the task was accomplished impressed the people of that part of the Peninsula that the death of Hasdrubal had brought them no advantage. They bent their heads then under the yoke.

But with the Vaccæi, the turbulent people who inhabited old Castille, Leon, and the Basque provinces, it was different. They, too, seized the opportunity to raise the standard of independence. Against them, Hannibal, after devoting the winter to the carrying out the plans of his predecessor for the development of Carthagena, marched in the summer of the year 220 B.C. The march was long, the traversing of the eastern slopes of the great Iberian plateau was difficult; it was necessary to undertake the siege of two strong cities, Tordesillas and Salamanca. But all these obstacles were overcome. Tordesillas was taken; Salamanca, after a vigorous resistance, was forced to open her gates, and the Vaccæi submitted. Scarcely, however, had their submission been assured

* Called by Polybius "Althaia": supposed to be represented by the modern Ocaña.

than the young general was called upon to encounter a third rising, fomented, if not designed, by discontented members of the two tribes which had been subdued, in the vicinity of Toledo. This rising was the most serious of all. In a country but just conquered, and therefore still secretly hostile, Hannibal calculated that he had to deal with a mass of a hundred thousand armed men. Should they mass their forces and compel him to accept battle, his defeat was certain. From this dilemma he extricated himself by the skill with which he knew how to draw every advantage from the difficult ground on which he was acting; he did it in the following manner.

He was at some distance from the right bank of the Tagus: between him and that bank were the enemy's masses: it was necessary that he should cross over to the left bank. As he approached the right bank towards the evening of the day preceding that which he had fixed for decisive action, the entire army of the enemy moved to attack him. Hannibal contented himself with merely repulsing the attacks, and marched on till he had reached a position near the right bank. By this time night had set in, and the enemy, confident that on the morrow they would destroy the Carthaginian army, desisted, as Hannibal had anticipated, from the attack. But that night was spent by Hannibal in a manner of which they had not thought. Having ranged his army in sight of the enemy as though it were in permanence on the right bank, he turned to receive the reports of the trusted explorers whom he had sent out during the day to sound the fords of the Tagus. One of these reported that he had discovered an easy ford at no great distance. At the darkest hour of the night, then, Hannibal broke up his camp, crossed by the ford indicated, and, ranging his army in battle array on the left bank, stood to watch the movements of the enemy. Soon after break of day the Spaniards, noticing that the Carthaginians had crossed the river, rushed after them with all the disordered fury of wild animals baulked of their prey. Strong in their numbers they rushed into the river pell-mell, without order, each man feeling that he was the most fortunate who should first reach the fleeing enemy. Hannibal, who had foreseen all this, and who had posted his army in such a manner as to make every variation of ground subserve his purpose, selected the proper moment to order a charge of cavalry. The effect was electric. The enemy's infantry, unable to reach the left bank, were driven by the very weight of their masses over the line of the ford, and were drowned by hundreds. Blindly and madly, however, the masses behind followed on the same track. Then Hannibal struck his decisive blow. He suddenly ordered his cavalry to fall back, uncovered his infantry, and charging with them across the river, swept the disordered Spaniards from his path. Most of them were driven into the current: those—and they were few—who reached the left bank were at once crushed by the elephants; those who gained the right were cut down by the cavalry who had recrossed the ford in the track of the infantry. By his skill, his daring, and the happy employment of his reserves, Hannibal had caused to disappear an enemy outnumbering his own army in the proportion of at least four to one.

The victory of Toledo caused the absolute submission of the entire Cis-Iberian peninsula. The authority of Hannibal, for a moment disputed, was re-established from the roadstead which bore his name to the east of Cape St. Vincent to the ford of Amposta on the Ebro. This end attained, Hannibal returned to Carthagena, to meditate there, during the winter of 220–19, on the plans the pursuance of which led to the remarkable action which forms the main subject of this chapter.

In the course of his meditations during that winter Hannibal had recognised the fact that the progress of the Republic of Rome was fraught with danger to the very existence of his own country. Already the influence of Rome preponderated, if it did not actually predominate, in the western world. The Greek and Massilian colonists had placed themselves either under her dependence or her protection, and, by means of Saguntum and Massilia (Marseilles), her hand reached to the banks of the Rhone and of the Guadalaviar.

Her commerce, carried in her own ships or by the smaller states of Italy under her flag, was making enormous strides to the detriment of that of Carthage. The first Punic war had revealed to her the strength of her military marine. Hannibal realised in fact that the Roman Republic was making sure and steady strides; that each year brought nearer of accomplishment the project she secretly nurtured of making of the Mediterranean a Roman lake; of imposing her overlordship everywhere, over every other people; of becoming, in a word, the Queen of the West.

For Carthage under these circumstances there was, he felt, only one course possible. Unless she were prepared to agree to disavow all her manliness, and to die, bit by bit, at the bidding of her rival, she must seize the very earliest opportunity to meet that rival face to face, to cast herself upon her, and, in a supreme struggle of life or death, to endeavour to strangle her There was no possible middle course. The Roman eagle, though but just emerging from its shell, already threatened the world. It must be attacked before its pinions had become matured, its strength developed, or it would dominate the world!

It is remarkable that the thoughts which passed through the brain of the great Carthaginian during that winter, represented the thoughts, the feelings, the convictions of the leading statesmen of Rome. Rome knew that the co-existence of two rival republics was, in the then state of the world, impossible. The famous expression, uttered at a later period, *delenda est Carthago*, was but the echo of the feelings of every Roman who turned his mind to the subject. Rome had watched with anxiety the progress of the Carthaginians arm under Hamilcar: the seven years of progress under Hasdrubal had increased the feeling to one almost of dread. With the death of Hasdrubal Rome breathed again. Two men, unexceptionally great, had passed away; it was scarcely possible that a third equal to them in ability should take their place. She watched, then, the first movements of Hannibal with intense interest; and it was when she discovered that he not only combined in his own person the abilities of his two predecessors, but that he had recognised all the points of the situation, that she felt that the supreme hour had arrived when, if not victorious, she must succumb for ever.

Given, then, the inevitable conflict, the question which Hannibal had to resolve was how to bring it to a successful issue. His meditations during that long winter at Carthagena had forced upon him the conviction that the Romans could only be conquered in Italy. A Carthaginian invasion of Italy would drive all the Roman citizens to take refuge under the walls of Rome. The blood would flow to the heart. The allies and tributaries of Rome would, on the other hand, profit by the success he might obtain to recover their freedom. An invasion, in fact, would rouse Italy against Rome.

Should he proceed by land or by sea? The sea route was easy and it was short. But there was one great objection to it. Rome was mistress of the seas.* A check at sea would derange the whole scheme. On the other hand the land route, though long and bristling with obstacles, was feasible. He had tried his soldiers in Spain. Numerous as were the enemies he might encounter, there were none who could stop them!

He decided, then, on the land route. But of land routes there were many; which one should he take? The decision at which he arrived reveals more than any other act of his life the extraordinary genius of Hannibal. It is a decision to contemplate which the greatest men who have ever lived have paused in admiration. It forms the theme of every writer of antiquity upon the subject. The more modern writers are not one whit behind them. "When one considers," wrote Montesquieu, "the numerous obstacles which presented themselves to Hannibal, and how that extraordinary man conquered them all, we have before us the grandest subject with which antiquity has furnished us." "When I think," wrote Saint-Évremond, "that Hannibal set out from Spain where his

* The student of history will notice how the same cause preserved England from an invasion in the early days of the present century. "England was mistress of the seas." Unfortunate will it be for England should the time arrive when she shall be unable to make the same proud assertion.

affairs were not on a very firm basis; that he traversed the country of the Gauls whom he was bound to regard as enemies; that he crossed the Alps to make war upon the Romans, who had just driven the Carthaginians from Sicily; when I think that, in Italy, he had neither magazines, nor strong places, nor assured aid, nor the least hope of retreat in case of disaster, I am astonished at the boldness of his plan. But when I consider his greatness and his conduct, I can only admire Hannibal, and I value him far beyond his work." The opinion of Napoleon, who in his earlier career was his most apt pupil, has been recorded in the first page of this chapter. "Of all the finest human exploits," says Montaigne, "can I not point with certainty to that of Hannibal?" "It was," writes Michelet, "an extraordinary boldness to penetrate into Italy, across so many barbarous nations, so many rapid rivers, and those Pyrenees, and those Alps, of which no regular army had ever seen the eternal snows." With respect to no man who ever lived is the consensus of opinion that is worth having so complete.

Yes! In that winter at Carthagena Hannibal had thought out his great plan. He would force the mountain barrier which separated Spain from Gaul, and that other which divided the last-named country from Italy. The Pyrenees and the Alps should lead him into the fertile plains over which Rome reigned supreme, and whence she aspired to issue her mandates to the rest of the known world!

Bold as the plan was, startling even in its audacity, it was far from being chimerical. The genius of the young commander had devised every sort of arrangement to minimise the more obvious dangers which attended it. His base, far behind him though it might be, would be the Cis-Iberian peninsula and the shores of Africa. With the latter the communication would always be sure. Thence there would reach him without impediment, from Carthage to Mers-el-Kebir (Mazalquiver) and Tangiers, and thence to Carthagena or Cadiz, men, munitions, and supplies. The Roman sailors were brave, but not even they would venture to affront their enemy in the waters of Gibraltar. Africa, then, would be solidly united to Spain. In Spain the Carthaginians were undisputed masters of the country, as far as the line of the Ebro. In advance of that line, Catalonia, which was indispensable to them, would be an impregnable fortress, which they would use as a depôt. That province gained, the last barriers were the mountains!

The treaty of 223 B.C. had not merely prohibited to the Carthaginians the passage of the line of the Ebro: it had stipulated for the independence of the people of Saguntum and other Greek colonists. Hannibal waited, then, at Carthagena, busily engaged in making his preparations, in endeavouring likewise to assure his position in Carthage during an absence which must necessarily be long, until he was quite ready. When everything for the carrying out of his plans had been completed, he made use of a quarrel, in which the citizens of the doomed city were embroiled with the citizens of Torbola, to espouse the cause of the latter, to cross the Guadalaviar, and to lay siege to Saguntum (219 B.C.).

Several reasons combined to induce Hannibal to select Saguntum as the point of attack. In the first place that city was in alliance with Rome, and an attack upon it would not fail to produce the declaration of war which would give his hand free course. Then, the city was an ancient city, a place of renown, the first in rank among the cities beyond the Guadalaviar. Whilst, on the one side, he could not leave such a place behind him, ready to serve as a base for the Roman armies or their allies, its capture would ensure the submission of the districts surrounding it. Finally, the place was rich, and he would find in it abundant supplies and the means of provisioning others. In fact, were he master of Saguntum, he would be in a position to venture the terrible duel with Rome!

Hannibal appeared before Saguntum, reconnoitred it, saw that its capture would require a regular siege, and promptly invested it. The siege lasted eight months: the place resisted three assaults, and succumbed only to the fourth. Before one even had been delivered the young general was wounded in the thigh; and it deserves to be recorded, as a sign of the supreme influence of his presence, that during the short period he was compelled to lay up for his recovery, the operations were limited to a strict blockade. At last Saguntum fell; the survivors of the garrison had to submit to the rigorous laws of war of the period; and the place was rased to the ground. But the winter (219–8) had set in, and Hannibal could undertake nothing further until the following spring. He drew back his army, laden with booty, to Carthagena, to make thence the advance the last obstacle to which had been removed.

Rome had not observed without great disquietude the action of Hannibal. Before even the Carthaginian general had crossed the Guadalaviar, her Senate, suspicious of his designs, had sent ambassadors to Carthagena, to remind him that Saguntum was under her protection. Hannibal had replied by declaring that if he were to interfere in the affairs of the citizens of that city, it would only be at the request of the best and most influential amongst them. The ambassadors, failing at Carthagena, had proceeded to Carthage, where, too, they were forced to be satisfied with generalities. Later, when the siege had begun, and during the interval between the first and second assaults, other ambassadors had arrived within sight of the place. Warned, however, that they could not enter it in safety, they had continued their journey to Carthage, to demand there, and to demand in vain, the delivery of the person of Hannibal. Following this abortive effort, a third Roman embassy proceeded direct to Carthage to declare war, if the Carthaginian Senate should refuse to disavow the action of their general, and, according to Polybius, should decline to deliver him to the Romans. The Carthaginian Senate refused, and the orator of the war party, Gestar, carried away by excitement, cried out, "Talk no more of Saguntum, or of the Ebro; the policy which has been long hatching in your hearts is now declared in open day." Scarcely had he ceased to speak when one of the Roman ambassadors, Quintus Fabius, rose, and, folding his toga, said, "We bring you peace or war; choose." "Choose yourself," replied Gestar, fiercely. "It shall be war, then," replied Fabius, shaking out his toga. "War be it," replied all the senators; "we will sustain it with the same enthusiasm with which we accept it." Just after this announcement the news reached Carthage that Saguntum had fallen.

His hands thus set free by the declaration of Rome, Hannibal proceeded to conquer the province beyond the Ebro, from whose base he would have to traverse the Pyrenees—the province of Catalonia. The conquest of that province cost him two months' time and twenty-one thousand men of the one hundred and two thousand whom he had raised for his expedition. But it was completed, and the province was organised under the administration of his brother Hanno, at whose disposal he placed a force of eleven thousand men. Leaving, then, behind him his brother Hasdrubal, with an effective of fifteen thousand men, as Governor-General of the Peninsula, and taking with him his third brother, Mago, Hannibal, in the spring of 218, broke up his camp at Ampurias, and set out for the Pyrenees. He had under him fifty thousand footmen, nine thousand cavalry, and thirty-seven elephants.

The town of Ampurias lies on the shores of the Gulf of Rosas, twenty-four miles to the north-east of Gerona. Marching as near to the sea-coast as possible, in sight almost of the Carthaginian ships which in a manner escorted him, Hannibal, forming his army into three columns, crossed in succession the Fluvia and Muga rivers, and reached Castillion. There the columns separated. The right one wheeled brusquely to its right, and turning the crest of the great chain by Rosas, crossed heights and threaded defiles till it reached Port Vendres by the pass of Las Portas. Filing then by the base of the heights, it arrived, by way of Collioure, at the mouth of the Massana.

Whilst the right was executing this turning movement, the centre and left columns marched due northwards by way of Perelada, Espolla, and Saint Genis. At Saint Genis they separated. The centre column proceeded to cross the slopes of Banyuls, and descended thence by way of La Tuilerie and d'Amont; there it wheeled to the left, turned the mamelon of the Tour de Madeloch, and, following the valley of the Ravenel, joined the first column at the mouth of the Massana. To the same point marched the left column, by the way, after quitting Saint Genis, of the defile of Carbassera, by Sorède, and by the pass of Pourné. The three columns, re-united, crossed then the Tech, and encamped for the first time on Gaulish soil at Elne.*

The passage of the Pyrenees was not effected without some difficulty. Hannibal had to contend against the ill-will of the mountaineers, and he had more than once to give evidence of the prowess of his troops. But so judicious were his operations in this respect, that his losses, up to the moment of his reaching the plains of Roussillon, were small.

The first act of Hannibal after concentrating his army at Elne was to open negotiations with the native chiefs, already almost terrified into hostility by rumours, industriously spread, of the intentions of the Carthaginian general. The result of these negotiations, conducted ultimately by those chiefs in person in the Carthaginian camp, was the concession of a general permission to traverse their country. Hannibal, impatient to enter Italy, marched then with all speed, by way of Castel-Roussillon,† to the right bank of the Rhone. The exact point at which he touched that famous river is doubtful. He established his camp, says Polybius, at four days from the sea. The weight of authority seems to indicate a point three miles above Roquemaure, known as the Ardoise, almost opposite to Caderousse.‡ A straight line drawn on the map from Narbonne to Roquemaure will represent with sufficient clearness and correctness the exact line of march followed by the Carthaginian columns.

The degree of reception meted to those columns by the inhabitants of the countries between the Pyrenees and the Rhone had not

* The Illiberis of Livy: "Pyrenæum transgreditur, et ad oppidum Illiberis castra locat."

† About three miles north-east of Perpignan.

‡ Hennebert, vol. i 430–35. The conclusion arrived at is in general concurrence with the dictum of Napoleon. "Hannibal," wrote that great soldier, "must have crossed above the conflux of the Durance, because he had no intention to march on the Var: he must have crossed below the conflux of the Ardèche, because at that spot commences the chain of mountains almost perpendicular which dominates the right bank of the Rhone."

2

always been the same. Fear often provokes hostility; and it thus happened that whilst the chiefs whom Hannibal had been able personally to influence showed a friendly disposition, others had manifested hostile sentiments. As far as Béziers, then, the march had been easy; but at Béziers the Carthaginians entered the country of the Arecomici, and the members of this tribe, terrified at the unaccustomed sight of the hardy warriors, had fled to the further bank of the Rhone to give their hand to their brethren who they doubted not would successfully defend the passage of that river.

And, in fact, when Hannibal reached Roquemaure he beheld the left bank of the Rhone lined with multitudes of armed men, fierce in aspect and menacing in gesture. His first care was to establish a friendly communication with the peoples who dwelt on the right bank. These men for the most part lived by trade with the Massilian colonies, and were accustomed to strangers. With them, then, the task was easy, more especially as the strict discipline maintained by Hannibal, and the practice he followed of paying cash for supplies, inspired confidence. From them Hannibal obtained the boats and the labour necessary for the purpose he contemplated.

The labour so obtained he employed to construct a flotilla of a peculiar character. So perfect was the general organisation, and so well regulated the division of labour, that in two days this flotilla, formed of monoxyles, or canoes, each formed from a single tree,* was ready for use.

But meanwhile the populations on the left bank had not been idle. The Arecomici who had fled thither from the vicinity of Béziers, became more threatening; whilst another tribe, the Volcæ, made it clear that on the first movement in advance they would move on the Carthaginian flank. To the ordinary mind the position seemed full of danger.

Not so, however, to Hannibal. Well aware that the uneducated and the bigoted are always dominated by the unforeseen, he de-

* "Pirogues monoxyles," or canoes made of a single tree. *Vide* Polybius, who visited the spot and interrogated eye-witnesses of the fact.

spatched one of his lieutenants, Hanno by name, a man whom he had proved in many an expedition requiring delicacy of touch, boldness, coolness, and dexterity, in command of a strong detachment to effect a diversion. Hanno quitted the camp at Roquemaure at nightfall, and marched silently up the bank, until, after he had covered a distance of twenty-five miles, he reached a point where several islands are formed in the bed of the river, where its current was less rapid and the depth of water was inconsiderable.* On the opposite bank, moreover, there were no signs of an enemy. Here, then, Hanno crossed, and gave his men a repose of twenty-four hours. The following night he descended the left bank and marched as far as Orange. On a mamelon which was visible from the camp of Hannibal, and which had been determined upon beforehand, he caused a fire to be lighted to announce his presence to his chief.

Hannibal had awaited only that signal to set his troops in motion. On the instant he embarked his men, the infantry on board the monoxyles, the cavalry, dismounted, in the large boats lent by the people on the left bank, and which, placed in front, broke the force of the current; most of the horses, saddled and bridled, swimming. At this sight the Arecomici chanted their war-songs and struck their javelins against their shields, whilst their light troops discharged thousands of darts. It seemed as though their opposition, combined with the force of the current, would baffle the plan of the Carthaginian leader, when, suddenly, Hanno, who had meanwhile set fire to the camp of the Volcæ, dashed upon the Arecomici and threw them into the wildest disorder. The success of the passage was thenceforth assured.

Once solidly upon the left bank, Hannibal mustered his army. He found that of the fifty-nine thousand troops with whom he had commenced the march from Ampurias he had lost thirteen thousand.† With the forty-six thousand who remained to him, he

* This point has been recognised by modern enquirers as "Pont Saint Esprit." Hennebert, vol. i. pp. 447–8, *and notes.*

† Twelve thousand infantry and one thousand cavalry.

resolved to put to profit the topographical studies which had occupied him during that long winter at Carthagena, and ascended the Rhone as far as the Isara (l'Isère), four miles above Valence.* Just before he set out, and before his elephants had traversed the river, the Roman Consul, Scipio, who had just arrived near Marseilles with an army, missed, in a manner presently to be related, a chance which offered of making upon his enemy an attack which, even had it not succeeded, would have seriously inconvenienced him. But before Scipio could arrive, Hannibal had set out, and making a complete turn then to the right, had entered the country of a chieftain with whom he had long previously negotiated an alliance, and in whose capital, Calaro (the modern Grenoble), he had caused to be stored immense supplies of provisions, of clothing, and of stores.

Grenoble reached, there remained the passage of the Alps. Having received the reports of the several agents he had sent to examine the various routes, held many consultations with his engineers, re-clothed and re-fitted his army, Hannibal addressed his soldiers in language of encouragement, language to be reproduced, in its general sense, some two thousand years later by the great modern leader who most nearly rivalled his exploits. He quitted Grenoble the 18th October, with the intention of marching direct upon the Durance by one of the defiles which cut by the shortest route the Alps of the Dauphiné.

Quitting Grenoble, as I have said, the 18th October,† Hannibal began to ascend the slopes of the long crest which, sharply pointed, runs from east to west between the course of the Drac and that of the torrent of Ancelle. Ascending this crest, he soon reached an altitude of 3,916 feet. Still mounting, he arrived at Saint-Hilaire, 4,241 feet, and then at Ancelle, about two hundred feet higher. At

* For the details of the march it is interesting to study the exhaustive arguments contained in Chapters II. and III., book v. vol. ii. of Hennebert's great work.

† The most learned of astronomers, amongst them the English astronomer, Maskelyne, have derived, from the histories of Livy and Polybius, the fact that Hannibal reached the crest of the Alps the 26th October. He took eight days to reach that crest from the valley of Champsaur. *Vide* Hennebert, vol. ii. pp. 205-6.

this village, the centre-point of a valley shut in by mountains, he halted to rest his troops.

Up to that point all had gone well. But the future seemed to promise difficulties. Encamped upon the borders of the torrent of Ancelle, a confluent of the Drac, the Carthaginians had to spring thence into the valley of the Panerasse, a confluent of the Avance, itself a confluent of the Durance, and, consequently, to manage one of the steep descents which debouch into the valley of Ancelle. Of these descents there are four, those of Comheous, of Rouanette, of La Couppa, of La Pioly. It seems tolerably certain that Hannibal chose the last-named.*

He had but just engaged in the difficulties of the desert when he was threatened by the Katoriges, a mountain tribe, established in force at La Tour, Saint Philippe, and on the plateau of Chategré. Their position was strong, and Hannibal soon ascertained by means of his spies that not only were they resolved to bar to him the road, but that every other descent was similarly defended. He would have to force his way or to renounce his enterprise.

To a man such as Hannibal renunciation was impossible. But a regular front attack, made in the full light of day, against warriors greatly superior in numbers, and of whose capabilities his troops had had no experience, would have been dangerous. His genius then suggested another method, a method in which his intellectual superiority should more than atone for the lack of numbers.

Having ascertained that the mountaineers were in the habit of withdrawing from their strong positions and retiring under the walls of their "castellum" as soon as darkness set in, he led his troops forward towards evening to the entry of one of the gorges which led to the crests beyond his position, as though he were about to force the passage; then, suddenly, as if discouraged, he renounced the attack, planted palisades, lighted fires, and acted as though he were about to rest for the night. The mountaineers, satisfied on this point, retired as was their custom,

* Hennebert, vol. ii. p. 207.

leaving not a single guard behind them. Till midnight silence reigned around. But then Hannibal rose, and, leading a chosen band of infantry, entered the defile of Pioly. It was unoccupied! The entire army then received the order to follow. Before morning broke the bulk of the Carthaginian army held all the strong positions of the mountain.

On discovering that they had been thus outwitted the rude warriors seemed at first overcome with despair. But, rushing from one extreme to another, they threw themselves suddenly upon the flank of a column which had not quite completed its movement. The attack was so sudden, the position occupied by the assailed was so little capable of defence, that the column was in great danger of being cut off from the main body and destroyed. But, again, Hannibal was equal to the occasion. At the very critical moment his light troops threw upon the advancing enemy a storm of projectiles, hurled with so true an aim that the attack was diverted, and the mountainers fell back, baffled, leaving the road open to the invaders!

The young general was able, then, to push on without opposition to the valley of the Panerasse. Thence he proceeded to and occupied Chorges, the capital of the tribe he had discomfited on the crest, and the key of the communications between the valleys of the Drac and the Durance.

At Chorges, which the enemy had abandoned, Hannibal obtained a number of horses and mules with their conductors, as well as several days' supplies of food. He halted there the 20th October, and proceeding thence south-westward along the right bank of the Marasse, reached Saint-Michel in the valley of the Upper Durance.

Thence the valleys seemed at first to offer an easy route to the advancing army; but they gradually closed in; the streams became torrents; and the passage of these torrents was difficult and perilous. At last the dangerous pass of Réalon was cleared and the columns reached Savines. From Savines the road became easier, and they arrived at the strong town of Embrun, a place which a few men might have held against them. The spirit of the mountaineers had been broken, however, by their defeat on the Pioly.

On the 22nd October, when descending from Châteauroux, Hannibal was met by the chiefs of a tribe, the Brigiani, who had lived under the protection of the more powerful Katoriges. They came unarmed, and with branches in their hands, to offer the Carthaginian who had vanquished their patrons, peace, a free passage, supplies, and guides. Without too much trusting their professions Hannibal treated the chiefs with courtesy, accepted their provisions, and received their guides. He took care, however, to keep these apart, to allow them no opportunity of cultivating friendly relations with his own men. He then pushed on to Phazy, and passed there the night of the 23–24 October.

From Phazy Hannibal set out, the morning of the 24th, ascending the Durance, to effect the passage of Mont Genèvre. The way was difficult, the paths were rugged, the falling-in of the earth was frequent, the necessity to cross and re-cross the river was incessant: many men were drowned in the torrents, more were hurt by the falling of the rocks and earth, some slid down the precipices: the march was, in a word, one of those marches in which difficulty strives with danger to impede the progress of an army. No wonder, then, that when the surviving soldiers, tired out, their nerves unstrung, faint with hunger, reached the summit, the view which then opened before their eyes completed the moral exhaustion of their natures!

They were, in fact, in an unknown region, a region which the superstition in which they had been brought up had taught them to be the abode of malevolent deities. It seemed as though they had overcome the difficulties behind them only to encounter others more insuperable in advance. Fortunate was it for Hannibal that at that moment no discordant shouts of barbarous warriors pierced the air. He was able, though with difficulty, to re-assure his troops. A little rest aided his efforts. Pushing on after a time, they reached the confluent of the Biasse, in sight of the plateau of the

Pallon. Again they advanced. The fort (*oppidum*) of Arama, which a few warriors might have held against them, offered no resistance: they defiled without danger under its walls, and pressed on to l'Abessée, a military position of great importance, for there, too, a small number might have checked a host.

Hitherto the promises given by the native chiefs had been kept; and the soldiers, who had seen one strong position after another given as it were into their hands, passed from an excess of doubt and despair to an excess of confidence. Hannibal alone maintained the coolness of judgment which was one of his remarkable qualities. The Brigiani were known to him by report as famous for their astuteness. It was true, he said, that he had been allowed to master strong places which might have been held against him:—but he was not yet at Briançon. Possibly there might be before him a position which would lend itself more easily than any which he had passed to the destruction of his army! Never, then, did Hannibal march more warily, keep his troops more in hand, or maintain a better look-out than after he had cleared the defile of l'Abessée.

It happened that the road from that place to Queyrières was singularly adapted for defensive warfare. It resembled nothing so much as a dark lane between two walls of perpendicular rock, whose sharp edges seemed to rend the sky—one of those wild passes the soil of which is torn by the rushing waters of a torrent. It was, in fact, a gorge with vertical walls, walls whose discordant stratifications bulge out in sharp and pointed projections, and over whose floor there rushed tumultuously the wild waters of the Durance.

At the moment of penetrating into this gorge Hannibal hesitated. The danger of allowing his whole army to be entangled in such a place was apparent to him at a glance. He then halted his troops, sends out parties to reconnoitre, to climb the crests, to search everywhere for any menace of danger. Whilst they were searching he remained pensive, his brow, usually so clear, clouded and anxious. At last they returned and reported that no living being was to be seen anywhere, and that the heights were clear. Then, though with doubt and anxiety, Hannibal moved forward.

Hitherto, in the plains, the order of advance had been in this wise: the infantry had formed the head of the column; then had followed the elephants; then the supplies; the cavalry had brought up the rear. This order was now modified. The elephants, preceded by a small detachment of footmen, formed the advance; then came the cavalry; then the supplies; the infantry formed the rear.

The column entered the pass, traversed it; its heads emerged from the other end. Scarcely, however, did they emerge when, in the very centre of the gorge, a tremendous boulder detaches itself from the wall of rock, and rolling down smashes ten of the leading files. It is followed by a second, by a third, and finally by a tenth! The officers, in the excitement of the moment, turn to look for the Brigiani guides. They have disappeared! The fall of the boulders is followed by showers of stones. Suddenly the crests around the column are alive with mountaineers; the grottoes and the ravines vomit them forth. Attacked in front, in rear, on the flanks, the column, surprised, harassed, disordered, is in danger of being destroyed to the last man.

In such a situation a really great man is divine. Preserving all his coolness, his presence of mind, his quick intelligence, in an instant Hannibal comprehended the situation and how to meet it. His front was secure, because the natives, terrified by the elephants, would not dare to attack it; his rear was secure, because his infantry could hold their own; his centre was weak because the cavalry was harassed by having to defend the convoy, and the convoy, he felt sure, would soon be the main object of attack. He allowed, then, the barbarians to pierce his centre and to rush upon his convoy. Waiting, then, till they were busily engaged in pillage, he led half his infantry through a ravine to a commanding position which he had noted, and which was unapproachable except through that ravine, and from it poured showers of darts upon the enemy. The discharge was so well directed and so continuous that, after a short hesitation, the Brigiani yielded their hold on the convoy and took refuge in flight.

With the disappearance of the enemy order was restored, but morning had set in before the last man emerged from the terrible gorge. The army marched that day along the Durance, and encamped that evening, 25th October, in sight of Briançon, 4,333 feet above the sea.

The following morning Hannibal, still following the course of the Durance, marched to the Alberts, an altitude of 4,668 feet. Thence, suffering in its progress from the bad water and the rarification of the air, the army ascended to Druentium,* 6,100 feet. From that summit there was but one step to take to behold a vast plateau covered with fields of rye, and bordered by forests of larch. The Carthaginians accomplished that step, and the plains of Italy were beneath them!

Bright and hopeful as was the prospect, it produced no effect on the jaded soldiers of Hannibal. They invoked not, as in after years invoked with shouts of joy the soldiers of Napoleon, the Italy which lay at their feet. Their past labours, fatigues, losses, and discouragement made them still doubt the future. Each countenance bore an expression of desolation, of discouragement, of despair.

Again did Hannibal show himself a king of men. Pointing out to them the fertile country which lay before them, and extending his hands in the direction of Rome, "We hold," he said, "in our hands the keys of Italy. You, soldiers, have just conquered the acropolis; the rest of the journey will be easy to accomplish. The Cisalpine Gauls are only awaiting the propitious moment to join us. One or two successful combats will place the entire peninsula in our power." These and similar words, followed by a halt of forty-eight hours, re-animated the troops. With renewed confidence they commenced, the 29th October, the descent of the valley of the Dora. A march of five miles, involving a descent of a little over fifteen

* A station supposed to be near the present village of Mont-Genèvre

hundred feet, brought them to Scincomagus—the modern Césanne.* Thence, with a view to gain the pass of Sestrières, an altitude of 6,580 feet, the column has to turn and ascend 2,100 feet. From this pass of Sestrières the Carthaginians behold, on the one side, the summits of the Pelvoux, 13,442 feet above the sea; on the other, the valley of the Chisone. The march into that valley is the last of their descents. The torrent, of which they behold the meanderings, leads them surely into the plains of Piedmont!

Yet their terrors are not quite passed. The gorge of Pragelas, which they have to traverse, brings to their minds some sinister reminiscences. A narrow defile, shut in by walls of abrupt rocks, at the base of which the river tears its channel! Yet they traverse it, pass below the village of Sestrières, and, at length, see opening to their right the ravine made by the impetuous Chisone, of which they had till then followed an affluent. The difficulties of the descent increase from this point every moment—so much so that at last a vast chasm opens in the very path of the advance-guard: the troops can proceed no further: the minds of the men are filled with despair.

At the moment Hannibal was with the rear-guard. He had not long to wait to be sensible of the sudden block. Inquiring the cause he is told that there is no longer any road; that the column is in a blind alley; that even if the infantry might be able to emerge from it, such a course would be impossible for the elephants and the cavalry.

Once again does the will of the one man extricate thousands from a difficulty. Rushing to the front, Hannibal recognised that immediately in front of the spot where his advance-guard had halted there had occurred a crumbling away of the ground sufficient to form a precipice.† It was nature, not man, which had caused this last obstacle: the work of disintegration, proceeding for years, had

* Smith's *Dictionary of Greek and Roman Geography*, vol. ii. p. 931; also Hennebert, vol. ii. p. 235–9.

† Livy calls it, borrowing from Polybius, *lapsus terræ*. Michelet thus describes it: "Tout à coup, on se trouve arrêté par un éboulement de terre qui avait formé un précipice." See the excellent commentaries of Hennebert, vol. ii. pp. 240–5.

displaced the friable rocks. These had fallen into the valley below, leaving a chasm not less than nine hundred feet in width. How was this gulf to be passed?

Hannibal gazed for a few moments at the obstacle. It is too true: to advance is impossible. To add to his anxieties he perceives that marauders are collecting upon his flanks.

However, in his mind are other feelings than despair. Summoning to his side his engineers, he indicated to them the mode in which he thought the difficulty might be surmounted. The engineers set to work at once. Retiring a few yards they trace out a path over the summit of the rocks. They climb up, examine the ground, and plant poles to guide the men who are to follow. But as soon as the forward movement begins new dangers present themselves. The fresh snow which had supported a few detached men, crumbles and disappears under the weight of whole columns. Soon the bare surface, sharp and slippery, has to be encountered. The situation becomes cruel. To slip, or to stumble, is to be carried down into the depths below. Many do slip or stumble. The pack-mules share the danger. The fate of many of these is strange. They fall; they seek to rise; but, in many cases, they succeed only in breaking the crust of the hardened snow. In it they become entangled; their feet are caught as it were in a trap; the weight of their burdens deprives them of the power to extricate themselves.

In the midst of these difficulties night fell. To proceed further was impossible. On that slippery crest, then, Hannibal halted till morning. When day broke he again summoned his engineers. It was as evident to them as to their leader, that to proceed by way of the summit would be impossible. After a long consultation they resolved to try, now, another plan; they would hew a path along the wall of the chasm. Hannibal at once ordered his African pioneers to set themselves to the task.

Nobly and vigorously did these hardy men respond to the call. Each of them toiled as though upon his exertions alone depended the safety of the army, The engineers arranged that they should work by relays, so that the energy might ever be fresh and the labour continuous. Constant supplies of pickaxes were brought up to replace those bent or broken by the hard rock. At last, after twenty-four hours' incessant labour the path was declared practicable. The infantry then crossed: the cavalry and the convoy followed. But for the elephants it was still too narrow. To enlarge it the engineers made use of a material employed in those days for the purposes of demolition, which was known to the Greeks as ὄξος; and which possessed all the explosive and destructive powers of the modern dynamite.* The application of this substance accomplished the purpose, and the elephants crossed the chasm in safety.

This was the last difficulty. It had cost three precious days, and entailed considerable losses. But it was finally accomplished. Then Hannibal, who had remained on the further side of the chasm till the last of the men and animals had crossed over, continued the descent of the valley of the Chisone, and passing by Fenestrelle, Perosa, and Pignerol, debouched unopposed into the plains of Piedmont. He had marched, from Ampurias, three hundred and fifty miles: he had crossed two chains of mountains, traversed broad plains, and passed many rapid rivers. Setting out from Carthagena, the 30th May, he had crossed the Ebro the 15th July, had spent nearly two months in the conquest of Catalonia, had quitted Ampurias the 10th September. On the 1st November (218) he was under the walls of Turin. When he set out from Ampurias he had, we have seen, under his command, fifty-nine thousand men, of whom nine thousand were cavalry. This total number had been reduced, after the passage of the Rhône, to forty-six thousand. The passage of the Alps had diminished these by nearly one-half. When he entered the plains of Piedmont, his army was composed of twenty thousand infantry,† six thousand cavalry, and thirty-seven elephants. "He had paid," wrote Napoleon in admi-

* For the erudite argument on the subject of the material used by Hannibal on this occasion, I must refer the reader to Hennebert's *Vie d'Annibal* vol. ii. pp. 252–267.

† Of the infantry eight thousand were Spaniards.

ration, "one half of his army to acquire his field of battle." True! But, with an army of twenty-six thousand men, the horses of which were utterly worn out; the men so exhausted and limping, that they could scarcely bear the weight of their arms*; the elephants the shadow of their former selves, how was Hannibal to meet the fresh soldiers of Rome? Before answering that question, it is necessary to inquire into the measures which the Roman Senate had taken to meet the invasion—the certain consequence, they were well aware, of the war which their own ambassador had declared.

The Consuls for the year 218 B.C. were Publius Cornelius Scipio and Sempronius Longus. It was Scipio who had introduced in the Roman Senate the deputies who had come from Saguntum to implore prompt assistance against Hannibal. Instead of sending an army, the Romans had, as we have seen, deputed an embassy to Carthage. As soon, however, as the news of the fall of Saguntum reached Rome, the Senate ordered the levy of three armies, and, retaining one, under the pretors Manlius and Attilius, in Italy, despatched the second under Sempronius to Sicily, and the third under Publius Scipio to Spain, to hinder the egress of Hannibal from that country. Scipio was too late to effect this purpose. Embarking at Pisa, he had reached the port of Bouc, not far from Marseilles, when he learned that Hannibal had crossed the Pyrenees and was already on the Rhone. Scarcely believing this news, he despatched three hundred chosen horsemen, supported by some Gaulish auxiliaries, to reconnoitre. Hannibal, on his side, having been told that some Roman ships had anchored at the mouths of the Rhone, had sent five hundred light cavalry to examine and report. It happened that the two reconnoitring parties met not far from the village of Védènes. They at once came to blows. After a sharp contest the Romans prevailed, and the Carthaginians, having lost at least two hundred of their number,†

* "Ταῖς ἐπιφανείαις, καὶ τῇ λοιπῇ διαθέσει . . . οἷον ἀποτεθηριωμένοι πάντες ἦσαν (Polybius iii. lx.) . . . illuvie, squalore enecti . . . effigies, imo umbræ hominum . . . squalida et prope efferata corpora . . ." (Livy xxi.).

† *Victi amplius ducenti ceciderunt.*—Livy. Polybius, who, as M. Hennebert truly remarks, never exaggerates, states that the greater number of them were placed *hors de combat.*—Hennebert, vol. i. p. 444; vol. ii. 33–4.

fled for their lives. The pursuers followed their enemy so closely as to come within the range of the Carthaginian camp. It happened at the time that Hannibal had crossed his entire army, the elephants alone excepted, to the left bank. If the leader of the Roman horsemen had had his wits about him he would have recognised that the passage had been virtually accomplished. But he had an eye only for the elephants. His gaze was riveted upon the huge animals guarded by a few infantry, and these to his disordered mind represented the Carthaginian army. He returned, then, to inform Scipio that Hannibal was still encamped on the right bank. A report of the true state of the case would have caused the Roman Consul to proceed with all haste to the plains of Piedmont to be beforehand there with his enemy. The false report induced in his mind the belief that he might surprise Hannibal at the critical moment when he should attempt the passage.

Scipio, then, disembarked his troops, and marched up the left bank of the Rhone towards the place indicated by his reconnoitring officer. On reaching the Durance he changed the order of march for the order of battle, and proceeded with great caution. As he approached the goal of his journey he continued to cherish the hope that he would yet catch his enemy in the very act. The bitterness of the awakening from the delusion may be imagined when, on reaching the ground which the Carthaginian troops had occupied, he learned that they had quitted it three days before!

The Scipios were a great family, and although Publius Cornelius was not the most illustrious of its members, he was yet a man to bear the test of fortune. Keenly disappointed though he was at the departure of his enemy, he did not allow it to discourage his roops. Representing to them that the flight of the Carthaginians had been the consequence of the news of their advance, he returned to his ships, despatched the greater number of his men under his brother, Cneius Cornelius, to Spain, and embarking, with a few followers, on board a galley, set sail for Genoa. Learning, at Genoa, that the third army, raised by order of the Senate, was under the command of the pretors Manlius and Attilius, at Taneto, some ten miles to the east of Parma, he re-embarked, proceeded to Pisa, and landed there.

It had been the hope of Scipio to be able to meet Hannibal as he descended, worn and exhausted, on the plains of Piedmont. But already the fatal "too late!" sounded in his ears. His army was not yet in his hand; the distance to be traversed was great. He was forced, then, to content himself with the formation of a plan to cover Rome, and, as the most effectual means for that purpose, to occupy and defend the great river arteries which flowed between that city and the base of the Apennines.

Whither, in the first instance, should he direct his steps? In deciding upon this question he could not but be influenced by a consideration of the strategic reasons which would naturally guide the invader. It was clear to Scipio, as it must be clear to every-one who studies the topographical conditions of Northern Italy, that an invading army descending the Alps is inevitably drawn to the river Po. To an invader that river constitutes the first obstacle. To the defender it is the great base of offensive operations. To maintain himself as long as possible *à cheval* of the river which traverses all the roads of invasion, to preserve his entire liberty of action on the plains on both sides of it, to prevent it from being crossed, such must be the pre-occupation of the defender of the Italian soil. Now, to arrive at such an end, the defender has at his disposition a central position between the Ligurian Sea and the Adriatic, most suitable for every kind of manœuvring. This position is that comprised by the three cities,—Pavia, Stradella, and Piacenza. It is the base indicated for all defensive military operations against an invader from the north in the plains of North Italy.*

These considerations decided Scipio to march upon Stradella, the central point of the position. The route which he adopted was probably that by way of Lucca, Pietra-Santa, Massa, Carrara, Sarzana, Aulla, Pontremoli, Berceto, Cassio, Fornovo, and Parma.

* For the exhaustive argument which proves this thesis I must refer the reader to Hennebert, vol. ii. pp. 386–90.

As he advanced he enlisted some ten thousand troops from among the inhabitants of the country, many of them old cavalry soldiers who had seen service. It was not till he approached Taneto, however, that he came upon the regular army. In the vicinity of that place the pretors Manlius and Attilius joined him with at least twenty thousand* men. This junction raised the number of the Consular army to thirty thousand. Serving in its ranks were the poet Ennius, Cato the Elder, and the heroic Sergius. A few days later Scipio reached Piacenza. I propose to leave him in that strong position whilst I return to the camp of Hannibal.

We left Hannibal just descended into the plains of Piedmont. Arrived on those plains he, too, had to decide the direction he should take. The decision was not easy. The Romans occupied, in the west of Italy, the line from Pisa to Stradella; to the east, they were concentrated under the walls of Ariminum (Rimini). Their legions barred absolutely the entrance into the peninsula. They could, at will, manœuvre in the Emilian plain or behind the curtain of the Apennines; they thus held command of all the bases of the lines of operations which an invader might attempt.

On what point, then, was Hannibal to move. As an objective base, he had simply the entrance—quite closed behind him—into the plains of Piedmont. How was he to force the strong barrier in front of him? Were he to try his adversary by the right bank of the Po, he would be forced to carry by a front attack the western opening to that Stradella which alone could give him Piacenza, and with Piacenza the means of passing the Apennines. Were he to operate on the left bank, it would be scarcely possible to advance very far beyond the Ticino. Were he to push forward into the Milanese, his flank would be exposed to attacks from Piacenza and Cremona. Were he to follow the Po, its winding course would remove him too far from his objective. He would be forced, sooner or later, to march on Piacenza.

* The contingent of Manlius had consisted, a few months earlier, of nineteen thousand six hundred men; that of Attilius of nine thousand three hundred. The numbers in the text are certainly not too high a computation.

Recognising, then, Piacenza as the true key of the Peninsula, as the point which, whether he should act by the right or by the left bank of the Po, it was absolutely necessary that he should either occupy or render untenable by the enemy's army, Hannibal resolved to move rapidly upon that place and to operate by both banks of the Po. With this view he chose for the front of his operations, on the left, the Sesia, a confluent of the Po, which, springing from Monte Rosa, augments its volume in its course, and joining a bend of the greater river at Frassinetto, constitutes thus, from the base of the mountain to Valenza, a continuous line; on the right, the Tanaro, which, though a river quite of second importance, forms a good flanking defence to the Sesia-Po; in the centre, the Po, from Valenza to the confluence with it of the Tanaro.

To assure this position, this front of Sesia-Po-Tanaro, Hannibal began by mastering the three important places, Verceil (Vercella), Valenza (Valentinum), and Asti (Asta). It was then of importance, with a view to assure to himself the free use of the valley of the Upper Po, to secure the forts which guarded the line behind him, running perpendicularly to the line he occupied, that of Verceil-Frassinetto-Valenza. With this object in view he reduced Ivrea, Chivasso, Bodenkmag,* and Carbantia.* He then marched on Turin.

The relations which Hannibal had been careful to cultivate with the Cisalpine Gauls had rendered the acquisition of the places I have mentioned comparatively easy. It was not so with Turin. That city closed her gates to his army. A civil war was raging within her walls, a war produced by the uprising of the democratic element. Hannibal had failed in two attempts to open negotiations with the dominant party; he was obliged to have recourse, then, to a regular siege. Of this siege, the following account, compressed from the conclusions drawn by M. Hennebert, may be accepted as that best supported by the scanty evidence we can command.

About sixty yards (δύο πλέθρα) from the west salient angle of

* Names of ancient places which have not been traced.

the *enceinte* the Carthaginians organised their parallel (στοὰν παράλληλον), the trace of which cuts the approaches to the modern citadel. Under cover of this they sank a well, at the bottom of which was constructed the ὑπόνομος, curriculum or gallery, much wider than the ordinary modern gallery. The work at this was incessant. Divided into brigades, to each of which a service of six hours was allotted, the men toiled without a moment's relaxation. The earth displaced was thrown into (cophini) baskets, which were then conveyed to the base of the first descending point by means of sidings, and thence drawn up. To obtain a constant supply of fresh air, vertical wells were sunk at certain distances, and these were made to communicate with each other by means of oblique branches in the form of syphons.

The mine thus opened by the Carthaginians was pushed by them with great vigour, during three days and three nights, under the rampart. It was then extended laterally to a width equal to its length—sixty yards—and filled with combustibles and explosives. Then it was fired, and, an enormous breach having been thus effected, the troops rushed to the assault, and Turin was carried. The civil dissensions within her walls had prevented the citizens from either interrupting the progress of the mine or from organising a sufficient resistance.

The very day that Hannibal had stormed Turin, Scipio had occupied Piacenza. The Carthaginian was still in the city which he had conquered when the news reached him that the position which he had already recognised as that best adapted for the successful defence of the line of the Po had been seized by his enemy. His first thought was how to dislodge him before he should establish himself there too solidly to be displaced. There was not, he felt, an instant to lose: he must, without a day's unnecessary delay, march to attack him.

No one can study the history of the greatest generals whom the world has known without being struck by the bond of sympathy which has existed between themselves and the men they led to victory. It was so with Cæsar and his legions, with Charles the

Great and his armies, with Gustavus, with Cromwell, with Turenne, with Eugene, with Villars, with Frederic, with Loudon, and most especially with Napoleon, and their followings. These illustrious warriors had won the confidence of their soldiers not less by their genius than by the manner in which they were in the habit of taking them into their confidence as to the objects of the manœuvres they were about to execute. On the eve of Austerlitz Napoleon visited his several corps and confided to his soldiers the plan which he should send them to execute on the morrow. The men thus trusted went into action with a double certainty of victory. They were animated not alone by a faith in their leader, but also by the conviction that his plan was a plan which could not fail if they only were to do their part. The precedent for such a line of conduct in those great leaders can be traced to Hannibal. That great commander had led his army across the Pyrenees and the Alps; his men had suffered terribly; they had not yet recovered from the unwonted dangers and fatigues they had undergone; he was now about to lead them against a new enemy, the most formidable enemy in the known world, a severely disciplined enemy, led by a captain who had known how to occupy at a critical moment a very strong position. At that supreme hour Hannibal took his soldiers into his confidence. Summoning them to attend a spectacle in which the captives taken on the march were arrayed to fight in pairs to the death against each other—the dying being thus delivered from the agony of suspense, the victors being granted their liberty and loaded with rich presents —he thus, after the combat, addressed them:

"You can see for yourselves, soldiers, the necessities of your situation: you have had under your eyes a living image of the fate which awaits you. We are enclosed between two seas, without possessing a single ship which could take us back to our country. Before us is the Po, wider than the Rhone; behind us are the Alps, the crossing of which was accomplished with so much difficulty. Can we take refuge in flight? It is not to be thought of; there are no means of issue from our position; we are prisoners in

Upper Italy. What, then, have we to do? If we beat the Romans we shall gain their lands and their riches; but if we are beaten by them, shall we fall alive into their hands? Rather, a thousand times, let us die: let us imitate these Gauls, and let us say with them, 'We will conquer, or die!'"

Softening then his tone, he spoke to those about him of his own happy convictions; of how he counted on success, and hoped with his own hands to destroy the gates of Rome. His words caused the greatest enthusiasm. Shouts of applause followed the conclusion of his speech. The next day he took the road to Piacenza.

Meanwhile Scipio was making preparations to meet the advance, which he foresaw, of his great antagonist. After due consideration he had come to the conclusion that it would be more advantageous to meet him on the Ticino than on the Po. The section of the former river between Lago Maggiore and the Po measures a distance of about fifty-nine miles, and presents along its slightly winding course an imposing front. Throughout this distance the width of the river varies from sixty to a hundred and thirty yards; the depth from two and a half inches to thirteen feet in the driest season, from ten to forty-six feet when the water is at its height. The current is extremely rapid.

The comparative shortness of its length, fifty-nine miles, the strength afforded to that length by the supports at both its ends—the Lago Maggoire and the Po—gives to the line of the Lower Ticino considerable strategical value. But to obtain the fullest advantage from the occupation of this line it is necessary to occupy more than one point of it between its extreme points, Pavia and Sesto Calende; to command both banks at all points where a passage is practicable, so as, when necessary, to be able to operate on both.

As to whether Scipio, holding both banks, extended his front from Pavia to St. Martino, and whether he threw across the river many bridges to ensure to himself freedom of action, the records of the time make no mention; but this is certain: he did throw across the river a bridge of boats or rafts, and he did erect a bridge-head on the right bank. It may be accepted that this bridge was constructed near Pavia, for the necessity of maintaining the Pavia-Stradella-Piacenza position on the Po would not allow the Roman General to expose himself to be cut off from that city. By its construction Scipio had the command of entry into the Lomellina.* He did not himself await on the Ticino the approach of his enemy, but crossed the river with his cavalry, and proceeded in the direction of Vercelli.

It was on that place Hannibal had marched from Turin. Thence, having ascertained that Scipio had thrown a bridge over the lower Ticino, near Pavia, he despatched a regiment of cavalry commanded by Maharbal, to reconnoitre. Scipio, meanwhile, creeping forward with his cavalry only, had arrived within about five miles of Vercelli. From his position he carefully examined the Carthaginian position, and, believing that he had divined the intention of its leader, sent orders to his entire army to cross the Ticino. Hannibal, on learning of this, crossed the Sesia. This happened on the afternoon of the 30th November.

These movements brought the cavalry of the two armies, the following morning, within striking distance of each other. Hannibal, having reconnoitred Scipio's position, recalled Maharbal's men and assigned them a place in his line. He had resolved to receive the enemy's attack and then to overwhelm him with superior numbers.

Just before the combat engaged, the two generals, in accordance with the custom then general, addressed a few spirit-stirring words to their men. Then Scipio gave the order to advance.

His front was covered with a swarm of foot skirmishers mixed with irregular horsemen of Gaul; the Romish cavalry and the cavalry of the Italian allies of Rome formed his main body. The skirmishers moved at first at an ordinary pace, but they quickened it as they approached the Carthaginian line.

Hannibal had watched calmly the advance across the Lomellina of these armed masses. As they quickened their pace he

* The broad plain of the Po, south-west from Pavia.

passed the word to his cavalry to be in readiness; the instant they came within charging distance he gave the order to charge.

The Carthaginian warriors, inciting their horses to the utmost, charged with a force so irresistible that the skirmishers were swept from the field and the Roman cavalry was broken by the shock. Not in the centre only. Availing himself of his superiority in numbers, Hannibal had taken care that the flanks of the Roman line should be almost simultaneously assailed, and Scipio had not recovered from the effects of the first charge when he beheld the swarthy troops of Africa dashing upon his flanks and his rear. He was enclosed, indeed, in an iron circle.

That charge decided the action. The Romans, surrounded, became completely demoralised. Thenceforth it was not a fight, but a slaughter; each man striving only to save himself. A great number were killed and wounded. Amongst the latter was Scipio himself. Fighting bravely, he had received a severe hurt, and he owed his life to the gallantry of his son, also called Publius Cornelius, then seventeen years old, and who was in after years to avenge at Zama the many defeats which Hannibal inflicted upon his countrymen in Italy.

The wounded general, whilst retreating as best he could, despatched hurried orders to his infantry to recross the Ticino. He then moved rapidly with the remnant of his cavalry in the same direction, crossed the Ticino in advance of his infantry, and gained Piacenza in safety. There he occupied a strong position, a central point whence he could issue with both arms to smite any enemy who should attempt the passage of the river.

Why did Hannibal allow Scipio to recross the Ticino? It would seem that in the very sharp cavalry action his men, too, had suffered considerably; that, on its termination, he, expecting to be attacked by the Roman infantry, had retired, taking with him his wounded, into his palisaded camp. His dead he then caused to be buried, and it was only twelve hours after the Roman infantry had begun its retreat that the Carthaginian received information of the movement. Then, in all haste, he traversed the plain of the

Lomellina; and although he was not successful in capturing a single loiterer, he yet reached the bridge-head on the Ticino before the men who guarded it, six hundred in number, had crossed to the opposite bank. He made prisoners of them all.

Hannibal had no desire to cross the Ticino. His aim was Rome; and, to reach Rome, the position held by Scipio on the Po, the Pavia-Stradella-Piacenza position, must be forced or turned. Instead, then, of crossing the Ticino, a movement which would have led him from his true line of advance, Hannibal made a half-turn and ascended the Po by its left bank, with a view to effect the passage of that river at a point where he would not be hindered by the Romans.

The exact point he selected for this purpose has been much disputed, but the weight of probabilities inclines in favour of Cambio.* It is certain that the passage took place at a point two days' march up the river from the confluence with it of the Ticino; and where the throwing of a bridge was easy. Cambio fulfils both these conditions. It possessed, moreover, the advantage of being beyond immediate striking distance from Pavia, and of communicating with a country, on the opposite bank, inhabited by a clan of the Ligurians who were friendly to Carthage.

Just beyond the town of Cambio, then, Hannibal threw a bridge of boats across the Po. It seems probable that to cast this bridge he ranged his elephants in the river, just above the point on which he had fixed, to break the force of the current; certain it is that the passage was effected in good order and without opposition. By crossing near Cambio Hannibal had gained a position between the Tanaro and the Scrivia—a position which uncovered to him the western opening to Stradella, left unguarded by the enemy. He marched then in that direction, and, crossing the Staffora below Voghera, was pressing on, when suddenly he found his road barred by a Roman *oppidum*, guarded by a strong and well supplied garrison. That *oppidum* was Casteggio, the principal depôt of the stores of the Roman army!

* Hennebert, vol. ii. pp. 466–8.

To attack the place would have occupied precious time: to mask it would have deprived him of precious troops. Hannibal adopted a third course. By means of his native allies he managed to induce the commander of the garrison to close his eyes. He then continued his march; his light cavalry, led by his brother Mago, leading the way. Still pressing on, the Carthaginians, the third day after the passage of the Po, came within sight of Piacenza, and of the Roman army ranged under its walls.

Scipio, meanwhile, quite demoralised by his cavalry defeat on the Ticino, had written to the Roman Senate to represent the impossibility of his being able to meet Hannibal single-handed, and, indicating that it was the evident intention of the Carthaginian leader to march upon Parma, and, turning thence eastward, to force his way through the Lunigiana,* had presented an urgent request that he might be reinforced by Sempronius, then lying at Rimini. The united consular armies, he represented, would take up on the Trebia a position, which, covering the Apennines, would at the same time threaten the rear of the Carthaginians and force their leader to suspend his forward movements.

Pending the arrival of Sempronius, Scipio resolved to remain immovable in his lines. He did remain immovable, whilst Hannibal, defiling within sight of Piacenza, pitched his camp at a point where the Via Emilia is crossed by the river Nure, on both banks of that river, between five and six miles from that town. This position was very strong against attack, and it gave the Carthaginian leader facilities of opening communications with the Boii, a people with whom the Romans were at war, and whose friendship had been secured.

The task before Hannibal was such as to task all the abilities, all the resources, all the skill of a great general.† He had to

* The district of which Sarzana is the capital, so called from the Roman town of Luna, destroyed by the Arabs A.D. 1016.

† Napoleon, commenting on the earlier phases of his marvellous campaign of 1796, said: "I was in a position more favourable than that of Hannibal. The two consuls had but one common interest—to cover Rome; the two generals whom I

keep separate from each other two consular armies, which, occupying the two points of Rimini and Piacenza, were eagerly seeking the opportunity to unite in order to attack him. His clear brain enabled him easily to detect the plan that would occupy all their thoughts. He had traversed the Pyrenees and the Alps that he might attack Rome; but between him and the Po arose another barrier, the formidable chain of the Apennines. Scipio had missed him after he had crossed the Pyrenees; he had been baffled when, after Hannibal had descended from the Alps, he had tried to check his progress; but, united with Sempronius, he hoped to crush him at the foot of the Apennines. Believing this to be the plan of the consuls, Hannibal had to defeat it—to prevent, if possible, a junction between the two armies; if that should prove impossible, to fight them when combined; and to cross the Apennines in spite of them. It was with that object that he had crossed the Po, and had thrown himself between the two armies, in that position on the Nure, which, whilst it gave him a safe communication with the Boii, cut the line which connected the army of Scipio with the Lunigiana and Pisa on the one side and the army of Sempronius on the other.

To Scipio the position seemed, after a short reflection, full of danger. A discovery which he made of the indisposition of the Gaulish auxiliaries serving under his orders to fight against the new liberator of Gaul and Italy, brought to him further the conviction that to stay in his actual position was to court destruction. He resolved, then, to quit Piacenza, and, leaving there a garrison which might be victualled and, if need were, strengthened by means of the Po, to take up a strong position on the Trebia. On the banks of that river, in the territory of the Anamani, faithful allies of Rome, he would be within reach of his magazines at Casteggio; could maintain his communications with Rome and Genoa; and, finally, could await in a good defensive position the arrival of Sempronius.

attacked had separate interests, Beaulieu to cover the Milanese, Colli to cover Piedmont. It sufficed for me to throw myself between the two armies and threaten both countries at once to separate them for ever."

Scipio executed this plan. Quitting Piacenza under favour of a very dark night he marched to the Trebia, and crossed that river near the point where, in the present day, there is the railway viaduct. He then ascended the left bank by Casaliggio, and took up a position at Rivalta, a rising ground forming almost the last and lowest of the heights which constitute the extension towards the lower country of the slopes of the Apennines.* There he fortified himself strongly.

But a few hours elapsed before Hannibal made himself acquainted with the movement of his enemy. He immediately broke up his camp on the Nure and took a new position on the Trebiola, not far from the existing village of Settima. Thence, whilst within watching distance, four or five miles, of the Roman camp, he could keep an eye on Piacenza, and, he believed, still prevent the junction of Sempronius with his colleague.

Sempronius, meanwhile, recalled from Sicily, had marched upon Rome by the Via Appia, and from Rome to Rimini by the famous Via Flaminia, which had been completed only two years previously by the Censor Flaminius, now on the eve of becoming consul for the second time.†

The exact route taken by Sempronius to join his colleague on the Trebia is not known. Considering, however, that he was a man of action, and that he had to combine two objects—to march as speedily as possible and yet to make such a detour as would take him beyond the reach of Hannibal—and that he accomplished those objects, it seems fair to surmise that he advanced into the Emilian plain beyond the point which is now known as Forli; that he pushed on thence to Faenza (Faventia); that there he threw himself on the Apennines by the Via Faventina, which descended with tolerable directness on Florentia (Florence); and that he accomplished this march in four or five days.‡

* Hennebert, vol. ii. pp. 477–8. Rivalta is about sixty feet above the bed of the river. Desaix occupied it on the eve of the battle of Marengo.

† He had marched, halts inclusive, at the rate of twenty-five kilometres—nearly fifteen English miles—per day. Hennebert, vol. ii. pp. 480–1.

‡ Hennebert, vol. ii. pp. 481–2.

Once at Florence the junction with Scipio was assured. Marching by the Via Clodia, Sempronius reached Lucca in three days. Thence he pushed forward by the Via Aurelia as far as the mouth of the Magra, two miles to the east of the gulf of Spezzia. Here he had to carefully consider his next movement. Were he, turning inland at a right angle, to march on Parma, he could assuredly threaten the rear of Hannibal, firmly fixed at Settima; but he would not the less expose himself to the attacks of the Boii, the allies of the Carthaginian. He rejected, then, that idea, and preferred to continue his march along the coast—six days—as far as Genoa. Thence he pushed forward by mountain paths—for there was no road—by Torriglia and Ottone as far as Bobbio. At Bobbio he was on the Trebia, in the valley of that name, not very far—twenty miles—from Rivalta. The next day he virtually effected the junction, encamping at Statto, within an easy distance from the camp of his colleague.

The march of Sempronius had been so well conceived and so ably executed that Hannibal had been unable to prevent it. During the greater part of the movement the Apennines had shielded Sempronius. To harass that consul Hannibal would have had to quit his watch-tower over Scipio. He resolved, then, to stay at Settima, to cement there his alliance with the Gauls, already impressed by the stupendous grandeur of his passage over the Alps, and eager to welcome in him their champion against the all-devouring Rome. The discipline exercised by the Carthaginian, the prompt reparation he always made for any acts of violence on the part of his troops, the promises he made to the Gauls to recover their independence, worked wonderfully in his favour. By the time that Sempronius joined Scipio all the Cisalpine Gauls, the Anamani and some of the Cenomani excepted, were prepared to rise in his support. He had kept in reserve one great *coup* to give them still greater confidence in his star. The day following that on which Sempronius had arrived at Bobbio, Hannibal had surprised Casteggio, and become master of the magazines of the Roman army!

How did Hannibal obtain possession of Casteggio? According

to Livy, he bought it from the prefect of the garrison, Dasius of Brundusium (Brindisi), with whom it will be recollected he had come in contact previously, for four hundred golden crowns.* However that may have been, he had it, and its possession was of inestimable value to him; for it gave him the power, should his enemy display any enterprise, to force upon him an action whenever—almost wherever—he might choose.

The effect upon the Roman consuls was not less striking. Sempronius, a man of energy and vigour, full of military pride, was at first anxious to march at once and crush the audacious Carthaginian in his very camp. He gradually calmed down, however, and, seeking an interview with Scipio, consulted with him regarding the military position and the course to be pursued. He listened with attention to the arguments of his colleague that, notwithstanding the loss of Casteggio, the Roman position on the Trebia was impregnable; that they possessed a very safe line of retreat into the valley of the same name, part of which Sempronius had already traversed, and which led, by paths from which Hannibal could not cut them off, to Genoa and thence to Rome. Scipio proceeded, then, to urge the advisability of halting where they were; of continuing to occupy the camp even throughout the coming winter. They could only gain, Hannibal could only lose, by delay. His Gaulish allies, seeing him inactive, would tire of furnishing him with supplies and would gradually abandon him. "Finally," urged Scipio, hoping to touch his colleague by a personal allusion to himself, "I am still far from well; my wounds are not yet healed; give me time to be cured. As soon as I shall be able to join my efforts to yours, we shall both be able to render splendid service to our country."

Sempronius could not deny the justice of this reasoning: he could not help being convinced by it. But he was an ambitious man; the time of the consular elections was approaching, and he was unwilling to leave for his successor the glory of crushing Hannibal. The very reference made by Scipio with the hope of

* Equivalent to a little over £325 sterling.

working upon his feelings, the reference to his unhealed wounds, produced, therefore, an effect the opposite to that which had been intended. If Scipio could take no part in the battle, the entire glory of the victory would accrue to Sempronius.* Making use, then, of the state of feeling at Rome—a feeling which represented ignorance of the state of affairs in the camp and a desire for victory at any price—he told his colleague that public opinion already reproached him for failing to seize two opportunities, on the Rhone and on the Ticino, and that he was regarded as wanting in energy and enterprise. He concluded by declaring himself opposed to the policy of delay, and desirous of immediate action.

The matter was not settled at the moment. Two parties, however, formed in the Roman camp : the party of delay, and the party of action. Hannibal, who was perfectly informed of his adversaries' proceedings, and who longed, more ardently even than Sempronius, to see the Roman army in motion, resolved to excite still more the sensitiveness of that susceptible consul. With that object he despatched two thousand infantry and a thousand horsemen to ravage the lands of some Gaulish tribes which still held for the Romans. This action produced the desired result. Sempronius sent at once a detachment composed of the greater part of his cavalry, supported by a thousand skirmishers, to drive back the Africans. In the combat which followed the Romans were repulsed. Still more furious, Sempronius marched to support them with all his infantry and the remainder of his cavalry. But Hannibal had attained his end. He suddenly called in his troops, already well in hand, and drew them off before the fresh troops of Rome could arrive in line. The sight of the escape of his prey—as he had deemed it—augmented to a still greater extent the fury of Sempronius.

Hannibal now felt that he had him; that a battle was simply a

* History abounds with instances of the same jealousy. It was the same feeling displayed by Marshal Victor with respect to Soult, which saved Wellington at Talavera; the same evinced by Marmont towards King Joseph and Marshal Jourdain, which caused him to make the movement which lost for him the battle of Salamanca. These are but two instances amid hundreds

question of days, and that he might even choose the ground. Where should he encounter the famed legions of Rome? The plain, close to the position he occupied at Settima, seemed to offer conditions admirably adapted to the special qualities of his troops. Flat, generally bare, and exposed, it was traversed by a stream, the Trebiola, which meandered some distance below the steep banks which rose abruptly from its bed. These banks were so profusely covered with thick bushes, reeds, and brushwood, that the ordinary observer would not notice the long but narrow gap which formed the two sides of the rivulet. It was the very place, then, for an ambuscade.

Confident that Sempronius would cross the Trebia, Hannibal issued from his camp at Settima the evening of the 25th December* and marched nearly a mile southward in the direction of the present town of Niviano. There he halted, and drawing up his main body in a compact form facing the Trebia, despatched the corps of his brother, Mago, to occupy the ravine formed by the Trebiola, at a point near to where that rivulet runs into the river. Whilst Mago's corps was thus hidden from view by the brushwood and reeds which covered abundantly the steep banks of the Trebiola, it occupied a position which would bring it on the left rear of an army which, crossing the Trebia, should move straight forward to attack Hannibal.

Thus prepared for attack, the great Carthaginian resolved to provoke it. Early on the morning of the 26th December he despatched a strong body of horsemen across the river to insult the Roman camp. Crossing the Trebia, partly by the ford, partly swimming, these hardy soldiers galloped to the palisades which covered that camp; then, by their cries, their gestures, and the hurling of javelins, they roused the Roman soldiers from their morning slumbers. Sempronius was one of the first to mount his horse. Mad with rage, he summoned to his side officers, tribunes, and centurions, to witness the outrage, and to demand of them if

* The arguments regarding the exact localities and the date are to be found in Hennebert's admirable work, vol. ii. pp. 499–504.

it did not call for instant vengeance. Almost with the same breath he gave orders for the men to turn out and prepare for battle. His orders were obeyed; the cavalry, forming the head of his columns, issued first from the palisades, the light infantry followed; the heavy infantry formed the rear. Long before they had formed up the insulting Africans had vanished.

However, the die was cast. Sempronius, quitting the positions the two consular armies had occupied at Statto-Rivolta, marched to the river. The ford of Mirafiore offered them a passage, but the waters, swollen by rain which had fallen heavily during the previous night, reached up to the waists of his men. Having crossed, he halted to form his men in order of battle. He had under him from forty-six to fifty thousand men, of whom one-tenth were cavalry. He formed his infantry in three parallel lines of equal strength, and placed his cavalry on the flanks. His line occupied a frontage of about five miles.*

* "The infantry," says Hennebert, alluding to the general composition of the consular armies, "were divided into light infantry (*levis armatura*) and infantry of the line. The *levis armatura* comprised only *chasseurs* or *tirailleurs*, carrying only javelins; the *gravis armatura* was composed of three bodies, destined to fight in line, and provided for this purpose with defensive armour and weapons which they could wield. The three bodies were the *hastati*, the *principes*, and the *triarii*. These four component parts of the legionary infantry were under all circumstances essentially distinct from each other.

"Theoretically, the legion consisted of 600 *triarii*, 1,200 *principes*, 1,200 *hastati*, and an indefinite number of *tirailleurs*; but whilst the effective of the *triarii* always remained at 600, the number of the others varied according to time and circumstances. At the Trebia, the consuls disposed of 16,000 men of infantry of the line, and of 6,000 light infantry, distributed in two armies, that is to say, in four national legions. We may, therefore, affirm that each of these Roman legions counted 1,700 *hastati*, 1,700 *principes*, 600 *triarii*, and 1,500 men of *levis armatura*, or a total effective of 4,500 men. The Latins, allied to Rome by ancient treaties, had besides supplied her, for the duration of this war, with a contingent of 20,000 infantry, divided into four legions of 5,000 men each; these allied legions, provided with an organisation analogous to that of the Roman troops, were attached to the consular armies in such a manner that these were formed, actually, of eight legions, counting, in all, 42,000 infantry.

"The *hastati*, *principes*, and *triarii* of each legion were divided respectively into ten groups; these groups, numerically equal, were called *manipuli*, in memory of the handful of hay which, fixed at the end of a long knotted stick, was used as a banner (enseigne) by the primitive Romans. The tirailleurs were also divided into ten platoons (pélotons). The *manipulus*—the tactical unit—was formed in battle in ten ranks. Thus, on the day of the Trebia, each manipulus of *hastati* or *principes* having an effective of 170 men, presented in battle a front of 17 files; each

The weather was bitterly cold. The passage of the swollen waters of the Trebia had almost frozen the blood of the Roman soldiers. Just as they formed up to take the position Sempronius

manipulus of *triarii* with an effective of only 60 men, presented a front of only 6 files. The platoons of the *levis armatura* employed the manipularian form of a rectangle with a front of 15 men and a depth of 10.

"To form the legion in battle, the *hastati* were ranged in line in a given direction, their ten *manipuli* being separated from one another by a distance equal to the extent of the front of each. Behind this first line and parallel to it, at the distance of a depth of a *manipulus*, the *principes* were ranged in similar order; in a third line, always parallel and at the same distance, the *triarii*. The three arms of the *gravis armatura* thus formed, constituted a system of parallel *lines at intervals* (τριφαλαγγία παράλληλος), but it is essential to observe that these three alignments reciprocally lent themselves to the order *en échiquier* (a formation by alternate squares); that the *spaces* corresponded to the *gaps*. The thickness of the ten ranks of a line occupied 11·10 metres; the total depth of the three lines 55·50 metres, the space between each included. The platoons of the *levis armatura*, were placed, at the time of the formations in line of battle, behind the line of the *triarii*; it is not necessary here to estimate their solidity in figures.

"It is easy, moreover, to calculate the extent of the front of the infantry of a legion in battle array. Each man occupied in the ranks a space of 0·90 metres"—about two feet six inches—"and there was, besides, allowed him a similar space for the freedom of his movements in action. The space occupied by each file was, then, 1·80 metres. It follows from this that, at the Trebia, the front of a *manipulus* of *hastati* occupied 30·60 metres; that the ten *manipuli* of this arm occupied 306 metres; their nine intervals 275·40 metres; the front of this first line of *gravis armatura* 581·40 metres; that this figure accurately expresses the extent of the front of the infantry of a legion, because the *principes* and the *triarii* were formed in parallel lines in rear of the *hastati*, who covered them. If we suppose, then, that the Consuls had left, in numbers, a legion to guard their camps at Statto-Rivolta, and that, consequently, they had only seven legions, allies and Roman, to form in battle array, we see that the troops formed on the right bank of the Trebia presented there a front of more than four kilometres"—two and a half miles—"in length. Ordinarily, the infantry of a legion was covered by a certain force of cavalry specially attached to it, and which might be called divisional, or rather, legionary; but Sempronius departed from this principle, and rendered it independent, in order to throw it by two equal masses on the two wings of the infantry. For this purpose he divided it, according to custom, into *turmæ*, or platoons, of thirty-two horsemen, who formed four ranks of eight files each. As a horseman occupied 1·50 metres in the ranks"—4·14 feet—"the front of a *turma* in battle was necessarily twelve metres. The interval between two *turmæ* equalled the extent of the front of each, or twelve metres. These figures will permit us to calculate with sufficient exactness the extent of the Consular front. The Romans disposed of 4,000 horses, not including those of the Cenomani auxiliaries, whose numbers may be reckoned at 2,000; it was thus a mass of 6,000 horses, divided into 180 *turmæ*, on a line of 4,300 metres"—3,945 yards. "To sum up: the Roman army consisted of from 42,000 to 44,000 infantry, formed in three parallel lines; of from 4,000 to 6,000 horsemen flanking this infantry; or a total of from 46,000 to 50,000 men, showing a front more than eight kilometres"—about five miles—"in extent.

"The Carthaginian infantry, organised in the manner of the Greeks, was formed had chosen, the gusts of a bitter north wind came to intensify their sufferings. They were the less prepared to meet this evil with fortitude inasmuch as, in his haste and excitement, Sempronius had not given them time to partake of their morning meal. They stood there, then, on the right bank of the Trebia, shaking with cold, scarcely able to carry their arms, and suffering from all the pangs of inanition.

In a different manner had the soldiers of Hannibal been employed. That great leader, who had provoked the battle, was well aware that a good meal is a necessity for the soldier on the eve of combat. In the early morning of that day every section of the

uniformly in sixteen compact ranks. It had as its tactical unit the σύνταγμα, or square of sixteen ὁπλίται on each face; sixteen σύνταγματα, united, constituted a φάλαγξ of 256 files; four φάλαγγες a τετραφαλαγγία of 1,024. As one ὁπλίτης occupied a space of 0·90 metre in the rank, the σύνταγμα presented in battle a front of 14·40 metres: the φάλαγξ a front of 230·40; the τετραφαλαγγία a front of 978·80, inclusive of intervening spaces. Now, at the Trebia, Hannibal disposed of 20,000 infantry of the line. If we suppose that he left about 3,600 men to guard his camp at Settima, we may conclude that he had in hand a τετραφαλαγγία of an effective of 16,384 men. Such is the force which he ranged in line of battle, and which occupied an extent equal to a kilometre. The cavalry, equally organised in the Greek fashion, was formed for battle in eight ranks. It had as tactical unit the ἴλη, or rectangular platoon of eight horses on each face; sixteen ἴλαι, united, constituted an ἐφιππαρχία, or brigade, of 1,024 horses, in 128 files, or ranks. As one horse required a space in the ranks of 1·50 metres, the ἐφιππαρχία presented a front of 192 metres. Now, as Hannibal disposed of more than 10,000 horses, perhaps ten ἐφιππαρχία, this cavalry occupied on the wings a space of two kilometres. Consequently, the Carthaginian army, infantry and cavalry, presented a front of three kilometres in a straight line. In front of his τετραφαλαγγία of infantry of the line Hannibal deployed as a curtain quite an ἐπίταγμα of chasseurs armed with lances with long shafts (λογχοφόροι). This light infantry, of an effective of 8,000 men, was divided into four ἐπιξεναγίαι, each comprehending 256 files of eight ranks; each ἐπιξεναγία covered exactly the front of a φάλαγξ of ὁπλίται. The cavalry of the wings was, at the same time, provided with a living counter-guard, designed to break, at need, the charges of the enemy's cavalry. To cover the line, from one end to the other, the General-in-Chief placed in front nine θηραρχίαι, or demi-sections, of elephants. The demi-section consisted of two elephants; and, as the front of a wing of Carthaginian cavalry covered one kilometre, the nine θηραρχίαι formed a line at intervals in advance, each interval extending over a hundred metres. In a word, the Carthaginian army consisted of 28,000 infantry, 10,000 cavalry—in all, 38,000 men, and eighteen demi-sections of elephants; the whole formed in two lines, and occupying only an extent of about three kilometres."—Hennebert, vol. ii. 507–12.

army had been well supplied then with the means of providing for every unit of it an ample repast. The men were partaking of that repast, in front of good warming fires, whilst the famished Romans were wading through the cold waters of the Trebia. No sooner had they finished than they buckled on their armour, and each man stood, either by his saddled horse or close to the tactical unit of his formation, awaiting the order to fall in. It was this moment which Hannibal chose to address their officers. Calling the principal of them around him, he told them that he had need to count upon their devotion and their hearty support. He spoke to them of the gods, of their country, their honour; alluded skilfully to the subjects most likely to excite their passions; and, finally, according to the custom of those times, he guaranteed to all those who should show themselves worthy in the coming action magnificent rewards. On concluding this harangue, the purport of which was instantly made known to the men, he gave orders to fall in. He had under him thirty-eight thousand men, of whom ten thousand were cavalry, and thirty-six elephants. He formed them in two lines occupying a frontage of rather less than two miles.

It was late in the day when the two armies, marching towards each other, met face to face, but yet at a good distance the one from the other. They halted, then, as if by a simultaneous instinct, each assuming an attitude of expectation, whilst the musical instruments on both sides gave out tunes lively and defiant. Then, from the Carthaginian line, there crept to the front the skirmishers bent on engaging in the preliminary combat. These skirmishers—Baliares—naked, on the day of action, to the waist, carried across their shoulder a wallet, or havresack, containing their munitions; each of them was armed with three slings, one rolled round the head like an Arab *brîma*, another round the body, a third in the hand.

Their mode of action was alike skilful and effective. They had slings adapted for long and short distances. At first using that one made to carry far, they took from their wallets one after another the leaden balls which they contained, and discharged them with unerring aim on the ranks of the Roman *manipuli*. It is said that their skill was so remarkable that not only could they count upon striking the face of a particular adversary, but they could hit the very feature they might select.

Advancing, and discharging as they advanced, till they had arrived within a moderate distance of their enemy, the Baliares replaced the first slings, and, taking in their hands the second, suited for a shorter length, discharged from them in a similar manner egg-shaped pebbles which they had collected the previous evening on the banks of the Trebia. When within still closer distance they exchanged the second slings for the third, and poured upon the Romans, at close quarters, stones much larger and heavier, emitted, by dexterous management, with such force that it seemed as though they had been discharged from a catapult.

The Roman light troops, crushed under a storm of missiles of all sizes, disorganised by the crashing of their helmets, of their cuirasses, and of their shields, under the terrific storm, became, after the few first instants, completely demoralised. They had no means of making head against a hurricane, the like of which they had never seen. Their own skirmishers had long since bent under its blast; the Cretan archers, upon whom they, to a great extent, relied, had exhausted their munitions, and besides, their bows had suffered from the immersion in the Trebia. At this conjuncture, Sempronius, to conjure the still advancing storm, brought to the front some machines, then recently invented, which discharged with great force arrows and very large stones. But the rain of the night and the waters of the Trebia had paralysed also their mechanism. It was found impossible to work them!

The preliminary skirmish had thus been all to the advantage of the Carthaginians. But it was not upon his skirmishers that Hannibal counted to win the battle. With them he had caused the first line of the Roman army, the *levis armatura*, to quit its position; and as his own Baliares, their work accomplished, retired round his flanks to the rear at the same time, the front ranks of the Carthaginian τετραφαλαγγία, faced the Roman *hastati*. For a few moments the enemies gazed at each other with fierce glances. Suddenly, as the sound of the instrument which had recalled the Baliares—the σάλπιγξ—died away, the softer music of the flute, the harp and the lyre, playing a march, broke upon the air. Then with a firm and measured step, keeping perfect time with the music, the soldiers of Hannibal, men of Africa, of Spain, and of Gaul, advanced. As they approached the enemy, they briskly closed their intervals to take the compact formation known as συνασπισμός. Halting, foot against foot, shields joined, each man occupying in the rank a space of seventeen inches, they fell into position: the Spaniards ready to thrust; the Gauls waiting the moment to dash to the front, sword in hand; the Africans with lances crossed.* Compact, well supported on its wings, the infantry of Hannibal had the appearance of a monster threatening destruction to all who should oppose it.

The exact movement made by Sempronius to hinder the advance of this formidable array is not recorded. It seems probable, however, that he deployed his columns, bringing the *principes* in line with the *hastati*. It is certain that he marched forward to meet his enemy in good order and with a firm and confident step.

The hostile lines met. But against the serried ranks of the Carthaginian host the Romans could effect nothing. They killed,

* The three races were armed in the following manner:—"Clothed in a red tunic, the African foot-soldier was armed in a solid manner from head to foot; he wore a head-piece of bronze, a steel cuirass, exquisitely chased, a large circular shield (ἀσπίς), glittering with gold and amber, with coral and ivory. As an offensive arm he carried, besides his sword, a pike or lance measuring not less than from twenty to twenty-three feet in length. The Spaniard was proud of his white tunic bordered with purple; the Gaul, content with leg coverings, allowed his body, from the waist upward, to be naked. These men, recruited so far from each other, and whom the chances of war had brought together in juxtaposition, used the same kind of semi-cylindrical shield (θυρεός), nearly four feet in length and thirty-two inches in width; but their weapons were very different. The Spanish sword could be used, at will, either for cutting or thrusting, whilst the large Cisalpine sabre would serve only to cut. Formed thus, of three distinct elements, the Carthaginian τετραφαλαγγία did not the less offer a singularly imposing appearance. The wings indicated a weight, an inert power, a solidity which was proof against everything; the centre, well supported, presented an aspect at once strange and formidable."—Hennebert, vol. ii. pp. 534–5.

indeed, a few men, but the phalanx itself was not to be shaken. Like a strong fortress in a defile it barred the way to further advance, and the Romans could see, as they vainly strove to storm it, the Numidian horsemen gathering in masses on their flanks.

To charge the Roman legions when their men, never very fresh and still fasting, should be exhausted by their efforts to break his phalanx, had, indeed, been the great design of Hannibal. Formed in eight ranks; armed from head to foot; carrying on the left arm a shield strong enough to repel projectiles, and to ward off cut and thrust, in the right hand a lance with a sharp-pointed iron head, the Numidian horsemen, ten thousand strong, charged at a given signal the Roman cavalry. These, less strong in numbers, possessing no cuirasses, only a small leathern shield, and a thin lance with no iron head and which bent with the shock, offered the feeblest resistance. The weight of the Africans overpowered them. The victors, content with breaking their enemy, left to the elephants the task of completing the work they had so well begun, and obeying the guiding signal of their great general, dashed with irresistible fury from the right and from the left on the already shaken *hastati* and *principes*. Uncovered by the defeat of their horsemen, and assailed in front at the same time by the Baliares who, coming up round the flanks, poured upon them at this moment a fire of stones heated in furnaces brought to the spot, the Roman soldiers, fatigued as they were, still displayed all those qualities which enabled them at a later period to become the masters of the world. Suddenly, however, they beheld marching against them the demi-sections of the elephants. These magnificent animals, having crushed out the last resistance of the cavalry, fell now like an avalanche on the already harassed infantry. They moved on, covered with steel armour, trumpeting, with a force which the Romans could not successfully encounter. Still, however, the gallant soldiers of Sempronius offered what resistance was still possible. What though they had the cavalry and elephants on their flanks and the advancing phalanx on their front, they would not yield. Suddenly, however, a sound reached their ears which shewed them that they had yet another enemy in their rear.

It was the sound of the σάλπιγξ—the taratantara. The Romans turned and beheld, crossing the plain towards them, the Africans, the Numidians pure and simple, the fame of whose ferocity had for long been the tale of the Roman camp. They were the men who, under the leadership of Mago, had lain concealed in the ditch of the Trebiola, and who now issued to render the defeat of Sempronius assured and irreparable. They rushed on, gallantly led by the brother of their commander; and the Romans, despite their despairing cries, were soon in the folds of an embrace which left them no avenue for retreat. Then all was over. Ten thousand men, indeed, cut their way to Piacenza; of the others a few dashed into the waters of the Trebia, to be carried away, helpless, to destruction. Of their comrades, comparatively few escaped: the bodies of thirty thousand men, crushed by the elephants or sabred by the cavalry, remained on the plain to testify to the fact that at length it had been possible to conquer even the trained legions of Rome!

Then, from the surface of the plain, there arose shouts of joy. Intoxicated with their triumph, the Africans and the Spaniards, whom we have seen so depressed and cast down during the passage of the Alps, and who entered the plains of Piedmont a demoralised and wearied army, rendered homage by their exulting cries to the genius of their great leader. In that expression of admiration there joined the new allies, the Gauls, freed now for ever, in anticipation, from a yoke which they detested. To all alike, to the first two nationalities especially, the reason for incurring the difficulties they had overcome now appeared clear. It was that Rome should be conquered in Italy; that the horrors of war which the great city had made so many feel should now be brought home to herself. The victory of the Trebia was the first proof of the feasibility, of the greatness, of their leader's plan. Thenceforth everything might be possible!

For Hannibal himself the triumph was complete. "For him" —if I may translate the concluding words of the second volume of the admirable writer * to whom I am so much indebted—"for him, worthy pupil of Hamilcar, a just intuition revealed to him clearly the judgment of ages to come. He felt that no captain would ever attain to the sublimity of his strategic conceptions; that his tactical methods would be for ever admirable; that, especially, his tactics in actual fighting would serve always as a model to adepts in the art. He was pleased one day to proclaim that the victory of the Trebia formed the most brilliant of his titles to glory; an excellent judge in such a matter, he could, with perfect justice, say of that day what Napoleon said of Marengo: 'It is a masterpiece . . . the tooth of envy cannot injure it: it is of granite.'"

Under the walls of Piacenza Sempronius rallied the remnants of his army—alike the ten thousand who had cut their way thither, and the other fugitives who had wandered back by twos and threes to the camp on the left bank of the Trebia. Not for a moment did he admit to the Roman Senate that Hannibal had vanquished him. He had been victorious, he said, but a sudden storm had saved the enemy from destruction. The Senate read his despatch in the spirit in which it was penned. Resolving to show themselves blind to facts, they prepared ostensibly to act as though a victory had been gained. They gave orders, then, for the despatch of armaments to all quarters of their dominions, and, in the midst of the turmoil, they held even the consular election.

But in spite of the boasts of Sempronius the situation of the Roman army was far from enviable. It was shut up in Piacenza, dependent for its supplies on the Po, the navigation of which was only secured to them by the town of Emporium. Hannibal made an attempt to surprise that town, but on his approach its garrison communicated with Piacenza by signals previously arranged, and Scipio, issuing suddenly at the head of his cavalry, well supported by his infantry, checked the ardour of the Carthaginians. A

* Hennebert, vol. ii. pp. 553–4.

slight wound received by Hannibal added to their discouragement, and they fell back. Recognising the difficulty of such an enterprise in the presence of an enemy on the alert, Hannibal then directed his attention to the securing of his communications with Liguria, communications always liable to be threatened by the occupation by the Romans of the town of Vicumviæ. He marched then against that place, and took it by assault.

During his absence on that expedition the plans of the Roman consuls underwent considerable modifications. Their time of office was expiring. They felt all the inconvenience attending an enforced stay in Piacenza. They took advantage, then, of Hannibal's march to separate, Scipio to proceed to Ariminum—the modern Rimini—Sempronius to Arretium, now called Arezzo. They were occupying these positions when they were relieved by the two consuls who had been elected for the year 217. These were Cneius Servilius Geminus and Caius Flaminius.*

* Cneius Servilius Geminus, who replaced Scipio in command at Ariminum, was a leading member of the aristocratic party. He was a man of some ability, a friend of Paulus Emilius, whose opinions he supported when that capable man shared the consulship with Varro.

A still abler man was his colleague, Caius Flaminius. An aristocrat by birth, Flaminius, in his advocacy of popular rights, was the real precursor of the Gracchi. When Tribune in 232, he had succeeded, in spite of the violent opposition of the Senate and of the aristocratic party, in passing a law sanctioning the distribution amongst the people of the Gaulish territory of Picenum (now the province of Ancona), then recently conquered One of the four pretors elected in 227, he received Sicily as his province. The manner in which he administered this province secured for him the gratitude of the people. When, thirty years later, his son attained the dignity of Edile, the Sicilians testified their gratitude for the administration of the father by transmitting to Rome a plentiful supply of corn.

In the year 225 war broke out with Cisalpine Gaul. In the third year of that war Flaminius obtained the consulate for the first time, having as a colleague P. Furius Philus, and the two consuls marched at once towards the north of Italy. Very soon after their departure, the aristocratic party, which always regarded Flaminius as having provoked the insurrection by his agrarian law, caused his election to be annulled, and the issue of an order to the consuls to return. The letter reached the consuls when they were on the point of delivering a great battle to the Insubres on the Adda. Warned, probably by his partisans, of the contents, Flaminius persuaded his colleague to agree not to open the letter till the battle had been fought. The battle was fought, and the Romans gained a great victory. The letter was then opened, and Furius, obeying its mandate, returned to Rome. But Flaminius, who saw that the success of the campaign depended upon the prompt following up of the victory, disregarded the order, and continued the war until he had subdued

The Roman armies, it will be seen, were thus divided from each other by the Apennines, Flaminius being on the south side at Arezzo, Servilius on the north at Rimini. Hannibal, meanwhile, after a vain attempt, frustrated by the severity of the weather, to cross the Apennines, had returned to winter in Cisalpine Gaul. The reader will recognise, then, that the two consuls covered the two roads, one of which it was believed Hannibal must take to reach Rome; the one, the recently made Via Flaminia from Rimini through Umbria; the other, through Etruria.

But the ways of Hannibal were not as the ways of the other generals of that period. Again, with an army of from thirty thousand to forty thousand men, he was between two divided armies each numbering thirty thousand. He had to cross the Apennines, a march which would leave one army on his rear whilst he would be liable to encounter the other on the southern slopes of the range. It was a difficult task; but he was equal to it. His enemies had supposed he would advance by the made roads. He did nothing of the sort. Setting out as soon as the advancing spring permitted him to make his way along paths till then untrodden by armies he marched in the direction of Arezzo. Two routes had presented themselves to him; the one, the easier, along the coast,

the enemy. On his return, then, to Rome he was called to account for his conduct by the Senate; but the people, despite the opposition of the aristocratic faction, decreed him the honours of a triumph.

Subsequently, in 221, he was appointed one of the masters of the horse; but the superstition of the times—from which Flaminius was free—forced him to resign his functions almost as soon as he had taken them up, because the squeaking of a mouse had been heard almost immediately after the election. The following year, however—220—was the most important of his life. Appointed, with L. Æmilius Papus, censor for that year, he caused to be executed two great works which still bear his name—the Circus Flaminius, in Rome, and the Via Flaminia, a road connecting Rimini by way of Umbria and Etruria with Rome. Whilst the construction of these works augmented his popularity with the lower classes, it increased the hatred of the aristocracy. Nevertheless Flaminius was, in spite of this hatred, elected consul for the second time, with Cneius Servilius Geminus. Aware that an attempt would be made to cancel his election on the occasion of his solemn installation at the Capitol, Flaminius did not wait for those ceremonies, but set off at once with reinforcements to Arezzo, where he assumed command from Sempronius, justifying himself on the plea, which was perfectly sound, that the Carthaginians were already traversing Etruria, and that his immediate presence with the army was necessary to check their progress.

a road he could have reached by following the Via Emilia as far as Parma, thence branching off by way of Pontremoli into the Lunigiana. But the shorter way, the way by mountain paths to Fiesole (Fæsulæ) was that which he selected. Progress across these was difficult, almost impossible. He and his men suffered alike incredible hardships. During four days and three nights they marched half way up the leg in water, unable to take a moment's sleep. Hannibal himself suffered more, because he exerted himself more, than anyone. The long deprivation of sleep, and the thick vapours which exhaled from the marshes, cost him one of his eyes. Of his elephants but one had survived the fatigues of the preceding campaign. At length, however, the passage was effected, and he emerged from the mountains into the fertile plains near Fiesole to find the Roman army still at Arezzo.

That place lay in the direct route which Hannibal must take if he would march to Rome. But he had now only one Roman army, and that certainly not superior to his own in numbers, to deal with. The other lay still at Rimini. He pushed on, then, from Fiesole, plundering the country for the support of his army as he advanced, thus making war support war, crossed the Upper Arno somewhere near Florence (Florentia), ostentatiously passed the Roman camp at Arezzo almost within sight of it, and then proceeded, still devastating and plundering, in the direction of Rome.

How, under such circumstances, should Flaminius have acted? He was, as we have seen, a capable man, a strong-willed man, and he had displayed eminent ability on the field of battle. Let us glance at his position. He was at Arezzo with thirty thousand men. His colleague, with a like number, was at Rimini. He was not strong enough single-handed to meet Hannibal who, at the Trebia, had made thirty-eight thousand men vanquish an army exceeding his by at least one-fourth. He acted then in the only possible manner under the circumstances. The moment he heard of the arrival of Hannibal at Fiesole, he despatched messengers to Rimini to urge Servilius to join him. But Servilius considered that his presence at Rimini was necessary to ward off a possible invasion of the Gauls. He contented himself with despatching four thousand horsemen to aid his colleague. With this slight aid, Flaminius considered that he would not be justified in risking the safety of Rome on the issue of a battle in which defeat would be fatal; he allowed, therefore, Hannibal to pass Arezzo; then, finding that the Carthaginian army still pursued its way, plundering, towards Rome, he broke up his camp and followed on its track.

A competent military critic has found fault with this action. "The moment Flaminius learned," writes Colonel Macdougall,* "that Hannibal had entered Etruria, he should have sent off messengers to Ariminum" (Rimini) "to concert measures with Servilius for the concentration of the whole Roman force on the line by which Hannibal was about to operate. For this purpose Flaminius should have fallen back to Cortona, and Servilius should have proceeded by forced marches to join him at that place, which covers the approach to the defiles of the Thrasymene lake, and is about seven miles from them." The criticism continues in the same spirit, and it is unanswerable if one supposition be granted. That supposition is that Servilius would have moved. But Servilius held an independent command. Politically, he was the opponent of Flaminius, and there is no reason to believe that he would have consented to subordinate his own views to those of a colleague for whom he had no liking.

There remained, then, to Flaminius, the choice of two courses. He could precede Hannibal, and, occupying Cortona, cover the defiles leading to Lake Trasimenus; or he could allow the Carthaginian army to move on, and, following it, could seize the earliest opportunity that should offer to take advantage of the first difficulties it might encounter. Certainly, judging by the result, it is to be regretted for his own reputation that Flaminius did not take the first course. But it is not fair to judge by the result. Flaminius was an experienced and capable man; he understood the bearings of the position; probably he thought it would be

* *The Campaigns of Hannibal, arranged and critically considered.* London: Longmans. 1858.

imprudent to expose his army to a battle with narrow defiles in his rear; that by following his enemy he could more advantageously choose the time and the occasion to strike. For, assuredly, the Carthaginians were violating every rule of war by advancing, Rome in front of them and sixty thousand Romans in their rear. In deciding, then, to pursue the second course, Flaminius must be held free from blame. His strategy was the same strategy which Fabius successfully pursued after him.

It is for the manner in which he conducted that strategy that he is to be condemned. Flaminius marched as though his enemy were always in his sight. He detached no light troops to watch his movements, neglected to feel his way, but, keeping his army compact and well in hand, plunged blindly forward, never caring to enquire whether, and certainly never suspecting that, the astute leader of the hostile force might be preparing to take advantage of the peculiarities of the country to deal him a fatal blow. His head, in fact, full of his own designs, had no room for enquiry as to the designs of his enemy.

Of this pre-occupation Hannibal prepared to take advantage. He had advanced by slow marches, still wasting the country, and carrying with him the supplies thus obtained, as far as Cortona, a small town loftily situated above the valley of the Chiana, facing Lake Trasimenus beyond it, and commanding a view of both. Beneath Cortona the road runs to Casa di Piano, traversing for two or three miles the left edge of a level and fertile country. To the left and in front of this level country is a ridge of hills bending down towards Lake Trasimenus, called by Livy "Montes Cortonenses," and now named the Gualandra. These hills are approached at Ossaja. From Ossaja the road begins to rise a little, but does not pass into the roots of the mountains till the sixty-seventh milestone from Florence. The ascent thence is not steep, but perpetual, and continues for twenty minutes. The lake is soon seen below, on the right, with Borghetto, a round tower, close upon the water; and the undulating hills, partially covered with wood, amongst which the road winds, sink by degrees into the marshes near this tower. On a summit to the right is an old circular ruin, which the peasants call "the tower of Hannibal the Carthaginian." Descending, as he advances further, from the highest point of the ridge, the traveller will find himself in a vale enclosed to the left and in front and behind him by the Gualandra hills, bending round in a segment larger than a semi-circle, and running down at each end to the lake, which obliques to the right and forms the chord of this mountainous arc. The position cannot be guessed at from the plains of Cortona, nor does it appear to be so completely enclosed, unless to one who is fairly within the hills. Borghetto is then found to stand in a narrow marshy pass close to the hill and to the lake, whilst there is no other outlet at the opposite turn of the mountains than through the little town of Passignano, which is pushed into the water by the foot of a high rocky acclivity. There is a woody eminence branching down from the mountains into the upper end of the plain nearer to the side of Passignano, and on this eminence stands a white village called Torre. It may be added that there are two rivulets which run from the Gualandra into the lake. The first is crossed about a mile from the entrance into the plain; the second about a quarter of a mile further on. The other part of the plain is covered with thick-set olive-trees in corn grounds, and is nowhere quite level except near the edge of the lake.*

It was, indeed, the very place for a snare. Hannibal recognised the peculiarities of the formation with joy, and, a careful observer of the conduct of his adversary, prepared at once the lure which should entice him to destruction. He accordingly placed his horsemen in the woody hillocks beneath Borghetto, lower than the road, and not visible from either. They thus, unseen, occupied the jaws of the pass, prepared to close it as soon as, after the entry of the last Roman soldier, they should receive the signal. He marched then to the white village of Torre, and, encamping there, despatched his

* The admirable and graphic description of the battle-field given in the text is taken almost *verbatim* from the late Sir John Hobhouse's notes to the fourth canto of *Childe Harold*, stanza lxiii.

Baliares and light troops round through the Gualandra heights to the right, so as to arrive unseen and form an ambush amongst the broken acclivities, and be ready to act upon the left flank of the enemy. The front would be barred to them by the main body; the lake would shut out from them escape to the right; the cavalry would close the pass behind them. Should Flaminius march on as carelessly as he had till then marched he was doomed.

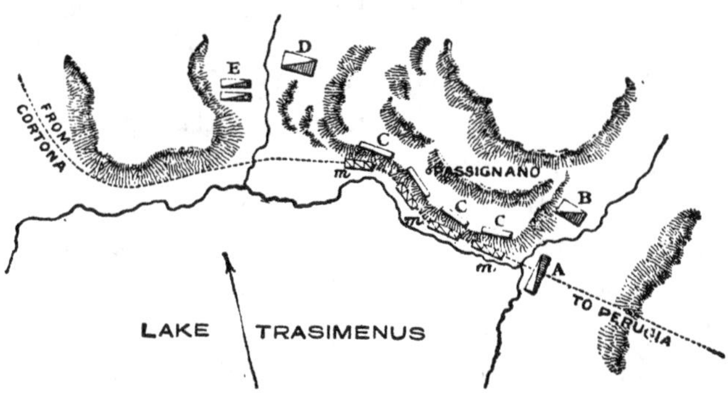

A Spanish and African infantry.
B Gaulish cavalry.
C Hannibal's light troops.
D Numidian cavalry.
E Gaulish infantry.
m Romans in march.

Flaminius, following the track of Hannibal, arrived at Cortona late at night. Doubtless he heard there of the ravages and exactions made by his enemy, and his generous mind became more eager to relieve the peasantry from the scourge of a hostile army which, at their cost, made war support war. Certain it is that he made all his preparations for an advance on the morrow. He was confirmed in his resolution when that morrow arrived; for he beheld, rising from the lake and marshes, a mist thick enough to conceal his advance from his enemy, whilst it did not mount sufficiently high to cover the ranges in front of him. Unsuspicious, then, of the designs of that enemy; neglecting, we must believe, to enquire even as to the precise nature of the pass he would have to traverse; thinking, probably, that under cover of the mist he might recover the plunder which, he knew, encumbered the march of the Carthaginian army, Flaminius pushed forward with the early morn.

As he neared the entrance into the pass—the jaws of it—below which, unseen, the Numidian horsemen were patiently waiting, the sight of the heavy-armed Carthaginians on the hill of Torre beyond—a hill not reached by the mist—greatly encouraged him. "At last I have him!" he must have exclaimed, believing that his own advance was hidden from his enemy. He pushed on, then, the more rapidly. No sooner, however, had his last man passed the level of Borghetto than the Numidians, quietly moving upwards, closed the pass behind them. As he still advanced, Hannibal moved down from Torre to confront him. Then, as the Romans were filing along the narrow road, he gave the signal. At once from the long line of their left flank, on their front, on their rear, they were assailed. Before even they could draw their swords the enemy was upon them. The roll of heavy stones was followed by a charge from men who knew that the victory was already decided. In vain did Flaminius display the courage of which during his life he had given so many bright examples. His men were in a trap. For three hours they were hewn down. It was a carnage—a carnage which the horrors of an earthquake were powerless to intermit.* Then Flaminius fell dead—pierced by a Gaul who had recognised him. From that moment the survivors thought only of how to escape. But escape was all but impossible. Of the thirty thousand men who had entered the pass, fifteen thousand fell to rise no more, nine thousand were taken prisoners, six thousand cut their way through. Of the victors, fifteen hundred only perished, and these were chiefly Gauls!

It was a great victory, but not yet great enough for Hannibal. Of the consular army he had destroyed or taken prisoners four-

* And such the storm of battle on this day,
And such the frenzy, whose convulsion blinds
To all save carnage, that, beneath the fray,
An earthquake reel'd unheededly away!
None felt stern nature rocking at his feet,
And yawning forth a grave for those who lay
Upon their bucklers for a winding sheet;
Such is the absorbing hate when warring nations meet!

Childe Harold, Canto iv.

fifths; but one-fifth had escaped. Resolved to complete his work, he despatched Maherbal to follow up these with all his energy. Maherbal obeyed so well that he caught the six thousand men the next morning at Perugia and forced them to surrender. Nor was this all; the four thousand horsemen whom Servilius had despatched under the Pro-pretor Centenius from Rimini to aid his colleagues underwent a similar fate.

The ambush and surprise attempted by Hannibal at Lake Trasimenus succeeded, then, in every particular. For here not a single man escaped. The entire army was destroyed or captured. Nor was any part of this success due to chance. The result corresponded entirely to the plan; and this happened because the vigour of the execution equalled the subtlety of the design. True it is that the great Carthaginian was so far befriended by Fortune in that he had an opponent who was capable of being deceived, and that the mist which covered the valley greatly aided him in bringing about that deception. Yet Flaminius, it cannot be doubted, was a man of rare ability. Who, under the circumstances, would not have been led astray? He saw the Carthaginian army on the height before him, unable by reason of the mist to watch his movements. What a chance to surprise that army, laden with plunder! Few great soldiers could have resisted such an opportunity. For it can never be too greatly insisted upon, that in war daring is prudence. Not, then, to the want of intellectual capacity on the part of his opponent must the triumph of Hannibal be attributed. It was the combination, I repeat, of subtlety of design with perfect execution which gained him the victory, and which has caused the story of the battle on the shores of Lake Trasimenus to descend to generation after generation as the most perfect illustration of ambushes and surprises of which history gives record.

CHAPTER II.

THE FOREST OF TEUTOBURG.*

ON the walls of the temple of Ancyra—the modern Angora—in the northern part of Asia Minor, may be traced even now, on the tables of brass on which it was engraved, a writing designed by the Emperor Augustus to be a permanent record of his life.† The style of this record is pure; it is concise, free from exaggeration, sublime almost in its simplicity. Translated, it runs as follows:—

At the age of nineteen, raising an army at my own expense, and without any other counsellor but myself, I delivered the State from the yoke of factions. Admitted into the Senate, I took rank among the consular personages with a voice in the deliberations. Soon, clothed with the title of pretor, I shared power with the two consuls, Hirtius and Pansa; then, in my turn, I was nominated consul, and finally triumvir, charged to constitute the Republic. By a just application of the law, I avenged the murder of my father, and condemned the guilty to exile. . . . On sea and on land, at Philippi, at Actium, I fought the enemies of the State. Conqueror, I pardoned; and my clemency reached in foreign countries those who, of their own will, had quitted the soil of their native land. Five hundred thousand Roman citizens have been enrolled under my banners. After their time of service had expired I sent more than three hundred thousand of them into the colonies or to their cities, giving to all lands and money taken from my treasury. Twice has the oration been decreed to me, thrice the triumph; twenty-one times have I been proclaimed IMPERATOR. When, once again, it was wished to decree me the triumph, I did not accept it, but contented myself with deposing my laurels at the capital. . . .

* My principal authorities for this chapter are (1) the Annals of Tacitus; (2) Dion Cassius; (3) Suetonius; (4) Gustav Herzog's *Die Feldzüge der Römer in Deutschland*, 1872; (5) Ledebur's *Das Land and Volk der Brukterer*; (6) *Ueber die Verhältnisse und Wohnsitze der Deutschen Völker zwischen dem Rhein und der Weser zur Zeit der Römer:* Münster, 1835.

† These records were originally discovered in 1554, by Augier de Busbecq; they have since been commented upon by the French traveller, Perrot; by Egger (1844); by Hamilton, *Researches in Asia Minor*, 1842; by Franz and Zumpt, 1845; by Philippe Le Bas, who first gave a literal translation and restored many obliterated parts; by Theodore Mommsen (1865), and by Theodore Bergk (1873)

Thirteen times was I consul; thirty-seven times did I exercise the duties of the tribunial office. Thrice did I receive the powers of censor, and I accepted for five years the title of regulator of the laws and morals. Thrice did I preside at the numbering of the Roman people—a numbering which had not been made for forty-two years—when for the first time I had it made. During my sixth consulate the number of citizens rose to 4,063,000; then to 4,233,000 during the consulate of Censorinus and of Asinius; it was 4,097,000 during the consulate of Sextus Pompeius and Sextus Apuleius. New laws have recalled the Romans to the culture of the virtues of their ancestors. Wishing to surround the magistrates and the priests with a great consideration, I increased their privileges.

After entering into a detail of the manner in which this was carried out, the Emperor thus concluded:

I have pacified and organised the provinces without having made an unjust war on any people. Egypt has been comprised within the empire; Armenia might have been, but I preferred that it should receive a king of my choice. Africa, Sicily, the Spains, Achaia, Asia, Syria, Gaul, Pisidia, have received military colonies. The standards lost by other generals have been recovered by me, and the Parthians, in restoring to us the eagles of Crassus, have implored the clemency of the Roman people. The Kings of India, the Bastarnæ,* the Scythians, the Sarmatians, who dwell beyond the Tanäis [the Don], the Albanians, the Iberians, the Medes, have solicited my alliance. Finally, after having extinguished civil wars, I restored to the Senate and the People the power which, in the interest of the Republic, the unanimous consent had confided to me. To recompense me, the Senate decreed to me the name of Augustus, and suspended above the door of my dwelling a crown of oak. Later, during my thirteenth consulate, patricians, plebeians, knights, combined to call me "the father of the country." That was, in my eyes, the greatest of honours, the noblest of recompenses. Written in the seventy-sixth year of my age.

The year following he died (29th August, 14 A.D.).

Yet simple, truthful, and grand as is this record, there must have been moments during its composition when the mind of its illustrious author dwelt upon an event he did not refer to, but which had caused him intense anguish at the time of its occurrence. Only four years before, the bitter cry, repeated continually, a long time even after the news of the catastrophe had reached him, "Varus, give me back my legions," had told how the defeat in the forest of Teutoburg had rent his soul. It was so strange, so unaccustomed, so unexpected. That his armies, the armies which had recovered the standards taken from other generals and brought back the lost eagles of Crassus, should have succumbed to a horde of barbarians was incredible, unaccountable. Yet incredible, unaccountable as was the fact, it produced a general panic. It was found most difficult

* The tribes inhabiting Podolia and Moldavia.

to raise an army which should proceed to the Rhine to repair the tarnished honour of the country. In the heart of Augustus the blow penetrated deeply: it was keenly felt. How deep, how keenly felt, is displayed most vividly by the omission from the long list of nations in the record I have transcribed of all mention of the German race proper; the race whose homes were nearer to Rome than many of the people enumerated, and which, under the leadership of Arminius, had inflicted this wound on the pride of the first and greatest of the Roman emperors!

Who were these Germani? It is the purpose of this chapter to answer that question, and to describe the ambush and surprise into which their leader, Arminius, enticed the Roman legions under Varus in the ninth year of the Christian era.

The Germani were the people, so called by the Romans, who occupied the lands between the Rhine and the Vistula. A branch of the great Aryan family, these people had migrated, at an unknown period, from their Asiatic homes to the shores of the Baltic. They were certainly domiciled on those shores four centuries before the birth of Christ. By degrees they spread, until, split up into various tribes, each under its own ruler, they came to occupy the entire country between the rivers I have mentioned.

It is difficult to enumerate the names of all their tribes in the pre-Christian era. Tacitus mentions three great divisions: the Ingævones on the shores of the Baltic; the Istævones in the west, the Hermiones in the centre. He distinguishes, further, the Suiones (Scandinavians), and the south-eastern Germans, such as the Bastarnæ, the Rencini, and others. Not to mention more particularly the Bastarnæ, who inhabited Moldavia, it is probable that the tribes who first came in contact with the Romans were the Cimbri and Teutones or Teutoni. Cæsar certainly had to deal principally with the Suevi (Swabians). On the banks of the Rhine dwelt the Sigambri, the Usipii, the Teucteri, the Tubanti, the Batavi; further eastward, the Bructeri and Marsi; then the Chatti or Catti, and behind these, between the Main and the Danube, the Hermunduri. On the northern side, between the Rhine and the Elbe, were the Frisii and the Chauci or Cauci. To the south of these were the Cherusci, who occupied both banks of the Weser and whose territories extended to the Harz and the Elbe. Near to them, on the banks of the Main, lived the Marcomanni. There were also other tribes less powerful, who changed every year the place of their abode. The characteristics of all these tribes were generally the same. They were tall and strong, delighting in war, and possessing an unconquerable love of liberty. Their women were chaste and held in high honour. The people were divided into four classes: nobles, freemen, vassals, and serfs. From the first-named class a king or executive officer was elected; but his authority was very limited, and in case of war breaking out it was often resigned to a warrior considered more competent.

Already, more than a century immediately preceding the Christian era, all-conquering Rome had trembled before the advancing hordes of the Cimbri and Teutoni (113 to 104 B.C.). Rome was delivered by the genius of Caius Marius (102 B.C.), but the name of the tribes which had defeated Roman armies did not the less for a long subsequent period inspire terror. When Julius Cæsar, the political successor of Marius, planned the conquest of Gaul, he was animated by the idea of forming a strong outer barrier against German invasion. Augustus went further still. He would do to Germany what Caius Julius had done to Gaul. With that object he sent his best troops, and, as he believed, his best generals, across the Rhine. At the outset the undertaking seemed to succeed. The border tribes, the Rhœti, the Vindelici, the Norici, the Pannonii, yielded to the arms of Drusus and Tiberius. Further conquests followed, until, in the year 15 B.C., the Roman frontier was pushed on to the Danube. To form a basis for the outposts, and, if necessary for further operations, a line of fortresses—some of which under other names still remain—was built along the valley of the Rhine. During the three years, 12 to 9 B.C., Drusus pushed forward victoriously as far as Aliso on the Lippe, built there a strong fortress, proceeded then to occupy a portion of the lands of the Chatti, and advanced to the Elbe. On his untimely death at the early age of thirty, Tiberius took the supreme command. This able man succeeded, by inciting tribe against tribe, and by forming a Roman party in each, in completing the submission of the whole race. When he had accomplished this he resigned his command, and the Emperor commissioned Quintilius Varus to proceed to administer the newly acquired territories as a part of the Roman Empire.

Quintilius Varus was the son of a man who had espoused the party of Pompeius, afterwards that of Brutus and Cassius, and who had killed himself rather than take a part in a world of which the victors of Philippi would be the masters. The son, however, rallied to the new order of things, became (13 B.C.) consul with Tiberius (afterwards Emperor), and obtained, on the expiration of his consulate, the government of Syria. It has been asserted* that he undertook, a poor man, the administration of a rich province, and that in a few years he retired, a rich man, from the administration of a poor province. Certain it is that he lived there a life of luxury, menaced by no enemies, disturbed by no cares. There could not have been a worse preparation for the difficult task which was to devolve upon him in Germany. The talents which an energetic life would have developed were allowed to decay, and the hard working soldier became a sensualist bent only on self-enjoyment.

Such was the man who was sent by Augustus, A.D. 6, to introduce Roman laws and Roman customs into the newly-acquired territories beyond the Rhine. Varus proceeded thither, taking with him crowds of officials, of place hunters, and of adulators, and fixed his head-quarters amid the Roman fortresses on the banks of the lower Rhine. Hence, whilst he continued the custom of his Syrian administration, whilst he feasted and flattered the heads of the tribe, his officers and followers endeavoured to spread Roman civilization in the country between that river and the Weser, inhabited by the Cherusci. To this end they established tribunals in which the Latin language only was to be used and recognised, imposed the Roman system of taxation, introduced the Roman manner of punish-

* *Velleius Paterculus II.*, 117–20.

ments, even the Roman mode of carrying out the death-sentence. Everywhere the ancient German customs were uprooted to make way for a civilisation which the people did not understand, and which was utterly unsuited to their ideas. The old methods, the cherished traditions, the time-honoured habits of the Germans were rudely displaced in order that the rude men of the forest might aspire to become Roman citizens.

There are no races in the world which cling so much to their own ways as the races which inhabit northern Germany. In another work* I have endeavoured to show how it required more than three hundred years and a vast expenditure of blood to bring the Borussi or Prussi into the fold of Christian civilisation. The pertinacity, the clinging to their own ways, the dislike to dictation from outside which they then displayed, were born in them. The same qualities characterised them a thousand years previously. In this respect the Cherusci and the Catti were the true ancestors of the Borussi. They resented bitterly, deeply, indignantly, the attempt to uproot the customs that gave them a nationality of their own.

Of this bitterness, of this resentment, of this indignation, Varus knew nothing. He had an army of 50,000 men, an army resting on fortresses, engaged in making military roads and forming strong military posts over the country, and the people might well have regarded, as they did regard, their case as hopeless and beyond remedy. It would probably have remained so but for two circumstances; the credulity and carelessness of Varus, and the arrival at his court of one of the master-spirits of the age.

Varus had administered his province for about a year when a contingent of Cheruscan troops which had recently taken part in an expedition against the insurgent Pannonii and Dalmatii, returned to their country. With them came their leader, a young Cheruscan in his twenty-fifth year, who had been educated at Rome, had entered the military service, had thoroughly acquainted himself with the Roman language and manners, and upon whom,

* *The Battle-fields of Germany*, 235–8.

for his services, the Emperor had conferred the right of Roman citizenship and the order of knighthood. The name of this young man was Arminius, the son of Sigimer, a leading chieftain of the Cherusci; he had just succeeded to the headship of the family. This birthright, combined with his great talents, gave him practically the leadership of the clan. To a proved courage he added vast powers of dissimulation, an extraordinary fertility of resources, and the ability, granted only to a few, to dominate over his kind. There is nothing to prove that before his arrival at the court of Varus Arminius had entertained the idea of conspiring against Roman supremacy. Certainly, in the flattering welcome given him by Varus there was nought to inspire hatred. His feelings, then, in the course which he took, were not based upon personal dislike. He came to the court of Varus a Roman citizen, a Roman knight, but possessing a Cheruscan heart. That heart was roused to indignation scarcely controllable at the sight of the attempts made by the satellites of the proconsul to romanise his countrymen. He saw that their indignation, if silent, was deep-seated. It has been said that opportunity makes the man. The sight before him, the sight of an oppressed people and a careless, easily flattered ruler, burst as an opportunity upon Arminius. From that moment he became a conspirator.

His first care was to gain Varus. This, in spite of the opposition of Segestes, the leader of the Roman party amongst the Cherusci, and who had refused consent to the marriage of Arminius with his daughter Thusnelda, he accomplished easily. His Roman manners and apt address made their way with the easy-going Roman consul. His next step was to persuade Varus to move his headquarters to the Weser. He wished to see him cut off from the Rhenish fortresses; forced to distribute his troops to keep open the communications. In this he succeeded: Varus advanced to the Weser. He then passed the word that the system of silent acquiescence in the new system till then displayed by the people was to be exchanged for a more active demonstration in its favour. At the same time he assured Varus that his countrymen were like himself ; that as he, a Cheruscan, had become a Roman citizen, so would they likewise be more than willing to adopt the Roman system, the Roman language, and the Roman manners. There must have been a fascination, very taking, about him ; for, in spite of the warnings to Varus of Segestes, often repeated, that Arminius was deceiving him, Varus did allow himself to be completely deceived.

Then Arminius gave the pre-arranged signal—a signal, not for the defeat, but for the complete annihilation of the Roman legions. It was the summer of the year 9 A.D. Varus, with three legions, some cavalry, and some auxiliaries, in all twenty-seven thousand men, was at Rinteln, ten miles south-east of Minden, on the Weser. Suddenly information reached him that a rising of the tribes had taken place to the south, probably in Hesse. To repress this rising Varus at once took the field. Vainly did Segestes warn him of the events which would happen if he were to set out without taking precautions to secure his base! Unheeded was he when he urged the Roman leader at least to secure the persons of Arminius and the other chiefs of the Cherusci! The advice and the warning were alike attributed to jealousy.

The first destination of the Roman army was Aliso,* a fortress built by Drusus near the mouth of the Lippe. The way to this place lay across uncultivated wastes and primeval forests. The paths were rough and rugged, but so long as the rain held off they presented no insuperable obstacles to soldiers. Varus, however, in the manner of the French generals who, more than seventeen centuries later, marched to encounter Frederick II. of Prussia, encumbered his line of march with a train of waggons and beasts of burden, conveying women, children, servants, and all the paraphernalia attaching to such a convoy. One result of the march of all these

* The exact position of Aliso is still disputed. The historian Ledebur, in his works, "*Das Land und Volk der Brukterer*," "*Blicke auf die Literatur des letzten Jahrzehends zur Kenntniss Germaniens zwischen dem Rhein und der Weser*," places it near the mouth of the Lippe; others believe that it is represented by Else, near Paderborn; others place it at the point where the Ahse and the Lippe unite; and there are still opinions besides these.

impedimenta on a narrow road was to prolong enormously the length of the line; to weaken the military disposition, and render defence, in the event of a brisk attack, almost impossible. To render matters, in such an event, more hopeless still, Varus—on his way, be it remembered, to combat an enemy in his front—was careless enough to allow Arminius and his fellow Cheruscans to march in the rear. To such an extent did his faith in the Cheruscans master him, that when, on the second day, information reached him that they had quitted his camp to bring up the several sections of their clans who had been summoned to the field, he seemed to regard it as a very natural proceeding.

Varus had made the first day's march unmolested. The second day passed without causing him any annoyance, beyond that occasioned by the sudden and mysterious disappearance of his Cheruscan guides. The third day, the 9th September, unaided by men acquainted with the lay of the land, he entered the forest of Teutoburg.

Here his difficulties increased with every step. The unevenness of the ground, the thickness of the forest, the absence of roads, combined to impede his movements. The pioneers were constantly employed to hew a way by which the troops could advance. They were called upon to clear thick underwood, and to fell mighty trees. Often the whole line was stopped by the necessity to cast a bridge over a deep stream or a roaring torrent. In the midst of their labours a terrific storm came to increase enormously their difficulties. Huge branches were blown down; torrents of rain made the earth spongy, and caused the rivers to swell; the order of the march became broken, each man trying to the best of his ability to maintain his footing as he slowly advanced.

Such was the condition of the Roman army when it arrived within sight of a hill eleven hundred and sixty-two feet high, now called, and called since the year 1581, the Grotenburg; known in the middle ages by the name of Teut, and, with its surroundings of tree and brushwood, spoken of by Tacitus as *Teutoburgiensis saltus*. Whilst the Roman soldiers and their long convoy are creeping with difficulty along this slope, let us turn to examine the proceedings of Arminius and his comrades.

Arminius had not quitted the rear of the Roman army until he had ascertained that its front, without guides, was entangled in the rough paths leading to the Teutoburg. Then sped he and his comrades to the point previously fixed as the place of assembly for the national contingents. Thence, at once, war-signals, fire-signals, blasts on the horn, call from every district between the Harz and the Taunus, from the Ems to the Main, every true-hearted German to the chace of the foreigner. They rush in crowds, from far and near, eager for the contest, the confidence of each man increasing as he beholds the constantly augmenting numbers. Boundless enthusiasm prevails. Not even Segestes, the real friend of Varus, can make head against it. He, his brother, and his nephew are forced to join the ranks of the national gathering. One chief alone sternly refuses to be false to his plighted word. This is the Ampsivarian chief, Bojocalus. Him Arminius places promptly in chains. Then with joyous anticipation of victory each section starts for the post allotted to it, some to close up the road by which the enemy had advanced; some to face him; others to hem him in on both flanks. On the ground on which they will attack the Romans there will be no room for manœuvring, no chance for the exercise of discipline. Their numbers aided by the surprise must and will prevail.

That ground, the slopes of the Teutoburg, the Romans had now reached. As they tread the ground, now spongy, now slippery, with hesitating and tired foot, there suddenly appear on all sides the Germans, long and anxiously expected by the still-trusting Varus. For a few moments his hopes mount high—but only for a few moments. The Germans halt at a convenient distance, and then from every thicket, from every brushwood, from behind every tall tree the thick German javelins are hurled into the long Roman line. Not even then is Varus undeceived. At first he deems it a rough horse-play, and orders his men not to return the discharge. Then he thinks it is a band of mutineers, whom an effort can repress. But when, immediately afterwards, heavy columns dash down to break his long line, selecting especially those parts where the soldiers and the camp-followers are mixed in disordered heaps, then the naked truth bursts upon him. He feels that he is betrayed as well as surrounded!

For a long hour his men, slowly advancing, endeavour to repel the ever-increasing foe. At last they reach a broad, open plateau. Here they halt, and here they intrench themselves in the manner familiar to the Roman soldier. Great as have been their losses, their position is now far from desperate. They are in the open, they are formed up, they are covered by intrenchments. In such a position a Marius or a Cæsar, a Drusus or a Germanicus, would have felt confident of ultimate victory. But Varus was still not equal to the occasion. Whatever might have been his capacities in his younger days, a career of idleness and self-indulgence had completely undermined them. Nor among the officers serving under him did any step forward to supply his place. It has been urged that his task was the more difficult, inasmuch as the majority of the men who composed his legions—the XVIIth, XVIIIth, and XIXth—were young soldiers, unused to warfare in Germany. This is true; and, if the entire soldiers had been young, the excuse would have had great force; but they were supported by a backbone of tried and experienced warriors. They required only leading. With a similar material, young soldiers of the Latin race, aligned with and supported by veterans, Napoleon, eighteen hundred and four years later, won, against the descendants of the Cherusci and the Catti, not very far from the same spot, the hardly-contested battle of Lützen. The victory was due to his leading. If, instead of by a Napoleon, those conscripts had been commanded by a Varus, the result would most certainly have been exactly opposite.

During the night of the 9th September the Roman legions, secure behind their intrenchments, were not assailed. They spent that night in burning the carts and the unnecessary articles of baggage. Varus, too, endeavoured by the despatch of men, singly and in small detachments, to discover whether it might not be possible to gain, by a cross-road, the main road leading from the Weser to Aliso. When the day broke, matters looked better. The men were less hampered by *impedimenta,* the order of march in which they formed was of a more practical character. With the early morn they set out, making, as far as their knowledge allowed them, for the great highway of which I have spoken. For a time all seemed to prosper: they were indeed still harassed by the Germans, and their losses were not inconsiderable; but the knowledge that they were prepared for action gave them better heart; the weather, too, had improved, and they reached gradually more open ground, ground upon which they could act more freely. But this gleam of sunshine was merely temporary. Advancing, in the hope of reaching the military road, they became gradually more and more involved in the jungle and brushwood. Then the attacks of their enemies increased in vigour and in intensity. Vainly did the Romans endeavour by repeated charges to drive them back. Whilst the nature of the ground broke their formation, the Germans found an easy shelter in the forest. The absence of leadership made itself painfully felt. Varus, though brave as a lion, acted more like a wolf wearied by a long pursuit.* He gave no orders, but left to each section the task of pursuing the plan which the sectional leader should think best. What wonder that when evening arrived the army had lost confidence! What wonder, that when they reached, with the falling day, a place fit for encampment, the men, wearied out, fainting with hunger, dispirited, and depressed, intrenched themselves in a careless and perfunctory manner! Rest would, under any circumstances, have been impossible; for the weather had changed, and the rain poured down in torrents. They passed a wretched night; the rain not only deprived them of rest, it slackened and rendered useless the bowstrings of the light troops. In the most miserable condition, then, the legions

* In my young days in India I often used, with one or two companions, to ride down wolves. One animal was singled out and chased. When he was completely tired out he would lie down panting. It then became easy to capture him by the mere twisting round his neck of the thong of a hunting-whip.

issued from the intrenchments in the early morning. With difficulty, but in safety, they reached the entrance into the Dören pass. Could they traverse this pass they might still hope, for from its further end Aliso was only distant three or four hours' march; the ground, too, was open, the plain was wide, and, though nature had fought for the Germans, the heavy rain soaking into clayey soil, they might still hope with confidence to reach the fortress whence they could bid defiance to the enemy.

But Arminius had no intention that the victims whom he had enticed so far should ever emerge from that defile. The "Dörenschlucht," as the Germans call it, is thus referred to in the general description of the Teutoburger Wald:*—"The Lippian Forest, or, simply the 'forest,' known in the middle ages as Osneggi, and, as it is likewise now, Osning, by many recognised as the real Teutoburger forest, stretches twenty-nine kilometres wide† through the principality of Lippe from the Völmerstost to the defile of Oerlingshausen, and is, just in its centre, traversed diagonally by the Dörenschlucht, a very deep chasm, through which the road runs to Paderborn. It consists of two almost parallel chains separated from each other by a deep valley." Knowing well the conformation of this pass, and predetermined to make it the grave of the Romans, Arminius had stopped up the further exit from it with abattis covered by palisades, at the same time that he had crowned the heights on both sides, leaving strong divisions handy to act as occasion might require.

Doubtless the danger of the situation struck forcibly Varus and his soldiers, but it was the last danger. On the further side of the pass was comparative safety; once on that side, indeed, they might lose some more men, but the bulk of the army would reach Aliso. They plunged in, then, with all the boldness of despair. Then the real battle began. Surrounded, assailed by groups organised for special purposes, the Roman soldiers fought wildly and desperately, each man for his own hand. The cavalry, led by Vala

* Brockhaus' *Conversations Lexikon*, 12th edition (1879), vol. viii. p. 483.

† About seventeen miles.

Numonius, forced their way through and dashed wildly in the direction of Aliso. Many of them, including their leader, perished in the attempt, but some reached the fortress. Varus, himself wounded, when he saw that all was lost, rather than become a prisoner to the Germans, fell upon his own sword, following the example of his father after Philippi. Many of his superior officers did the same, amongst them Lucius Eggius, one of the prefects of camps (*præfectus castrorum*). The eagle-bearer of the XVIIth Legion, terribly wounded as he was, when he, too, recognised that the day was irrevocably lost, tore the eagle from its staff, concealed it under his dress, and then, throwing himself on the heap of corpses, killed himself. The soldiers, deserted by their leaders, either submitted unresistingly to be slaughtered, or ran upon their own swords. With the termination of resistance all the most savage passions of the Germans reached their fullest height. They thought only of the insults and oppression which had been dealt out to them during the few preceding years. Their one idea was blood. It was not until their hands were weary with slaughter, that, vengeance giving place to avarice, they turned to grapple for the plunder. Those who had survived were then treated as prisoners.

But the horrors of the day were not yet ended. The prisoners made and despatched, and the booty collected, Arminius addressed to the assembled warriors an harangue of congratulation and triumph. The excitement had roused him to fever pitch. Every word that he uttered, every allusion which he made to the contempt which had been showered upon them by the Romans; his repeated references to the valour of the men who had conquered the masters of the world, whetted the barbaric instincts of the crowds whom weariness alone had caused to cease from the slaughter of enemies who no longer defended themselves; and when, at the close of his oration, Arminius pointed to the captured eagles of two of the Roman legions, and other trophies of the fight, and then produced the third eagle, that of the XVIIth Legion, which had been recovered, all bloody, from the body of the man who had hoped to save it from capture by his death; those barbaric instincts passed the bonds of control. Not only did the victors recommence the work of slaughter on the prisoners who still remained in their hands, but they wreaked their vengeance likewise on the dead bodies of those who had fallen before. They hacked to pieces the corpse of the previously decapitated Varus and the corpses of his comrades. Nor was this all. Religious excitement came, not to calm, but to increase the natural ferocity of humanity. On the hedges near the slaughter-ground the Germans improvised altars to their deities, and leading to these the officers of the noblest blood whom they had captured, they sacrificed them as thank-offerings to the unseen Providence who had worked all things for their advantage. Other sections, less piously disposed, erected gallows or crosses, to which, or to trees in the vicinity, they nailed their victims. It is recorded that one man only, a patrician of pure blood and ancient race, escaped their unrelenting fury. This was Caldus Cœlius, who preferred to dash out his brains with the chains by which he was bound. So bloody and so merciless was the slaughter that, six years later, the bones of the Romans still whitened the ground upon which they had been murdered; the trees still bore the heads or the skeletons of the victims.*

A special fate was reserved for the head of Varus. Severed from the trunk, it was despatched by Arminius to Maroboduus, king or chief of the Marcomanni, in the hope that the proof of the vengeance wreaked by Germans on a Roman pro-consul would incite him to join them in the war which the Cherusci certainly would have now to wage against the Imperial legions. But Maroboduus had felt only shortly before the power of the empire, and, believing that the act of Arminius would only draw upon him certain ven-

* "The places in the vicinity of the little town of Detmold," writes M. Stapfer in his article on Arminius in the *Biographie Universelle*, "are still full of souvenirs of that memorable event" (the defeat of Varus). "The field at the foot of the Teutoburg is still called *Winnfeld*, or the field of victory; it is traversed by the *Rodenbach*, or stream of extirpation, and by the *Knochenbach*, or stream of bones; near it is the *Feldrom*, or field of the Romans; a little further, in the environs of Pyrmont, the *Herminsberg*, or hill of Arminius, covered with the ruins of a castle which bears the name of *Herminsburg*; and upon the banks of the Weser, in the same county of Lippe, we come upon *Varusholz*, or the wood of Varus."

geance, he declined the proffered co-operation, and sent the head of Varus to Rome. There it was decently buried.

In this manner did the Cheruscans destroy the army which had held in bondage the country between the Weser and the Rhine. Of that army, which counted with its three legions, its three brigades of cavalry and six auxiliary cohorts, not less than twenty-seven thousand men, but few escaped slaughter; and the fate of the majority of those few, sold into slavery, was less enviable than that of their comrades. But the loss of twenty-seven thousand men was the least part of the injury which the fight of Teutoburg inflicted upon the empire of Augustus. It dealt a death-blow to the prestige of the empire, it announced to the rest of the world that there was a portion of it which would not submit to the supremacy of Rome; and although, in the years that followed, Germanicus splendidly avenged the soldiers of Varus, the triumph achieved on the field of Teutoburg was not the less valid and lasting: it released North Germany for ever from subjugation to Rome. Notwithstanding many defeats which followed, the consequences of Teutoburg were permanent. That battle sent back the frontier of the Roman empire from the Weser to the Rhine!

In Rome the effect was indescribable. More than two hundred years had passed since Hannibal had been in Italy to threaten her with destruction. Then, the courage—the front, bold, almost audacious—displayed by her citizens had been the admiration of the world. At that period she had never wanted a man to send into the field against the invader! But now—was it that increased luxury and dominion had impaired the energy and sapped the self-reliance of her children?—there was seen amongst them no eagerness to avenge the outraged honour of the country. The prevailing fear was lest, at any moment, the conquerors of Varus might appear at the gates of the city—and there was no rush to defend them!

The effect produced upon the Emperor himself has been handed down to us by a historian who was born, indeed, a century and a half after the event, but who enjoyed the very best opportunities of obtaining trustworthy information. "When Augustus," wrote Dion Cassius, "heard of the defeat of Varus, he tore his clothes. Not only was he to the highest degree chagrined for the loss of his army, but he felt the greatest anxiety regarding the next action of the Germans and the Gauls. He dreaded especially lest these should march into Italy and upon Rome. Rome, he knew, could not produce those hardy youths, trained to warfare, who in early days had been her safeguard, whilst the allies who had served for her had been destroyed. He made, then, as well as he could, preparations for the worst. As the citizens of the proper age for service declined to enrol themselves, he forced them to draw lots, and punished every fifth man under the age of thirty-five, and every tenth man above that age, by the confiscation of his property and the cancellation of the rights of citizenship. When this did not suffice, he caused some to be executed. It was only by the use of drastic measures of this nature that he managed, after a long interval, to recruit an army of discharged veterans and emancipated slaves, and these he sent in all haste under Tiberius into Germany."*

This account shows how terribly had the Romans degenerated since the days of Hannibal and the Scipios. In those days the defeat on the Trebia, the slaughter in the defiles of Lake Trasimenus, the crushing overthrow of Cannæ, only served to rouse the spirit of the people, to make them defy with greater boldness the conqueror, to cause the display of the nobler qualities of a free people. All this was changed now. Boldness had given place to superstition, manliness to fear. The executioner had to be called

* Many miracles which were said to have happened at the time tended to the still further discouragement of the Romans. It was declared that the summit of the Alps had fallen in, and that three columns of fire had issued from the cavity; the temple of the God of War on the Campus Martius had been struck by lightning; during the night the sky glowed as though it were on fire; comets traversed the firmament; fiery meteors in the form of spears darted from a northerly direction against the Roman camp; the statues of Victory on the frontier which pointed the way to Germany, had, of their own accord, turned round and now pointed to Italy. These and other supernatural occurrences were regarded by the multitude as connected with the defeat of Varus, and as indicating the wrath of the gods against Rome. Augustus himself was not free from the superstition Months and months after the day on which he received the fatal news he was continually crying: "Varus, give me back my legions."

on to perform the duties of his horrible craft before the youth of Imperial Rome would serve against the enemy who had destroyed one great army!

And yet the prestige of Rome still lived; the discipline enforced in her armies was still stern enough to enable them to defeat the men who had destroyed the legions of Varus. It will be interesting to follow for a brief space the movements of the army formed "of discharged veterans and emancipated slaves," which Tiberius led to restore the fortunes of Rome beyond the Rhine. Their prowess did not, indeed, affect the ultimate result, for the lure of the Teutoburgerwald was decisive; it emancipated Northern Germany for ever from the yoke of Rome; but with respect to the conduct of the Roman legions and to the skill of the Roman generals, the struggle which followed after an interval of five years was not unworthy of the best days of her military renown.

After he had consummated the slaughter of the legions of Varus, Arminius devoted himself to the destruction of the forts which the Romans had built, and to the obtaining of support from the other tribes of Germany for the struggle which he knew to be imminent. In this latter effort he only partially succeeded, but he nevertheless raised an army large enough to threaten Italy from the banks of the Rhine. Meanwhile, Tiberius had, with his newly-raised army of veterans and emancipated slaves, remained during the winter of 9–10 A.D. in Italy. At first he, too, dreaded lest Arminius should cross the Rhine and possibly raise Gaul. But, happily, the German leader had been prevented from executing that project by the skill and conduct of Lucius Apronius. This officer, a nephew of Varus, was, at the time of the defeat of his uncle, with two legions in Lower Germany. He remained there in a strong position, maintaining in subjection the tribes which would otherwise have revolted. His presence there restrained Arminius from aggressive action. With the spring, then, Tiberius marched to the Rhine. But he was on the eve, he well knew, of the succession to the empire. He did not care to compromise either the renown he had acquired or his future prospects in a contest in the marshes and forests which had proved fatal to Varus. During the year 10, then, he was content to maintain the frontier of the Rhine, and to repress a rising feeling of revolt, produced by the defeat of Varus, among the Gauls. The year following he was somewhat bolder; he crossed the Rhine, but, after some skirmishes, in which his troops had always the advantage, he retired to the left bank. At the close of the year, he returned to Rome; leaving one division of the army on the Upper Rhine, under the command of C. Silius, the other on the Lower Rhine, led by Severus Cecina. Germanicus, his nephew and the adopted grandson of Augustus, then Consul for the year, was to succeed him in the command.

One year elapsed, a year of "masterly inactivity." The arrival of Germanicus to assume the command was almost contemporaneous with the news of the death of Augustus (29th August 14 A.D.). This news produced a revolt amongst the legions, and it required all the tact, the judgment, the power over his fellows which Germanicus ever commanded, to repress that revolt. When it was repressed, the soldiers demanded, with loud cries, to be led against the Germans. Germanicus responded to their ardour. He threw a bridge across the Rhine, led across it twelve thousand men—soldiers of the legions—twenty-six allied cohorts, and eight squadrons of cavalry, traversed the Cesian forest by way of the modern Koesfeld,* encamped upon the rampart of Tiberius, and, learning there from his spies that the Marsi—who had been, and were, in league with the Cherusci—were celebrating a solemn festival, he marched, surprised, and fell upon them. To satisfy still further the impatience of his troops, he divided them into four columns, and despatched them to follow up his easy victory. Over an extent of fifty miles they carried fire and the sword. But this barbarous devastation only roused the German tribes in the vicinity. The Bructeri, the Tubanti, and the Usipii (inhabiting the modern provinces of Cleve, Jülich, Berg, and part of Westphalia), hastened to occupy the country between the Romans and the Rhine. But Germanicus, forewarned of their dispositions, led back his troops well

* In Westphalia, twenty miles to the west of Münster.

in hand, and, after a skirmish, in which the XXth Legion, which formed the rear-guard, covered itself with glory, he gained the Rhine without further effort.

Early in the following year Germanicus pushed the war with vigour. A spring of exceeding dryness had made the roads easy to be traversed in April. He had ascertained that divisions were rising amongst the Cherusci, Arminius urging the other tribes to rise against the Romans, Segestes doing his best to prevent them. With a view to take advantage of these divisions, he despatched Cecina with four legions, five thousand auxiliaries, and the German levies, against the Bructeri, the Cherusci, and the Marsi, whilst he himself, with a force not inferior, marched to the Taunus range, restored on that range a fort which had been built by his father Drusus, and then, leaving behind him Lucius Apronius to repair the roads and the dykes, burst into the country of the Catti (Hesse). The Catti, susprised, submitted or dispersed, and allowed the conqueror to burn their capital, Mathuin (Maden, on the Eder). Cecina, meanwhile, had, whilst holding in check other hostile tribes, gained a victory over the Marsi. Germanicus and Cecina then returned to the Rhine; but the former had scarcely reached his head-quarters when a deputation arrived from the Cherusci, headed by Segimund, son of Segestes, demanding help against Arminius, who was besieging him in his stronghold. Germanicus at once returned, attacked and defeated the besiegers, made a great number of prisoners, and delivered Segestes. This chieftain proceeded at once to Rome, where he was received with great distinction. Amongst the prisoners taken on this occasion was Thusnelda, whom Arminius had married despite the opposition of her father, Segestes. The annals of Tacitus give no more touching picture than that of the attitude of this noble woman. Expecting soon to suffer the pangs of maternity, she displayed a loyalty to her husband and her fatherland which has caused her name to descend with honour to posterity.* The capture of

* Germanicus sent her to Rome, where she bore a son, whom she named Thumelicus. Mother and son graced the triumph of the conqueror on his return.

his wife, and the opposition of his father-in-law, greatly embittered Arminius. Declaring that Segestes, if he liked, might till a servile soil, but that the German people would never forget that they had seen the rods, the axes, and the toga between the Elbe and the Rhine, he redoubled his exertions. In a short time he induced almost all the tribes round about him to join the Cherusci. Germanicus felt the danger of the situation, but he was equal to it. To divide the enemy he despatched Cecina, with forty Roman cohorts, through the country of the Bructeri towards the Ems; directed the Prefect, Pedo, to march with his cavalry along the borders of East Friesland; whilst he himself embarked with four legions on the Zuyderzee. Soon he had his army re-united and well in hand on the Ems. The people of East Friesland, the Chauci or Cauci, described by Tacitus as the best and noblest of the German tribes, who were then at war with the Bructeri, offered their alliance, which was accepted. Germanicus attacked the Bructeri, who were actually devastating East Friesland, defeated them with great loss, and recovered one of the eagles, that of the XIXth, lost at the massacre of Teutoburg. Following up the Bructeri, Germanicus ravaged their country between the Ems and the Lippe. Learning that the bones of the soldiers who had fought under Varus remained, unburied, on the spot where those soldiers had fallen the Roman leader determined to render to them the last honours due to mortality. Preceded by Cecina, charged to cover the advance, and guided by some of the men who had escaped the carnage, he arrived, the sorrow of his soul depicted in the sadness of his countenance, on a plain white with human bones, except where, ever and anon, bent and broken armour disturbed the painful uniformity. To dismount, to have the bones collected, to cause to be dug a vast grave, to deposit there the bones, and to cover the mound with turf, of which with his own hand he laid the first sod, was the first action of Germanicus; the second to rush in pursuit of the man who, though born a German, had been trained in Rome, had accepted service under and honours from Rome, had eaten Roman salt, and whose treachery had caused the destruction of which the sight he had just witnessed reminded him. Arminius, meanwhile, retreating by ways nearly impenetrable, had reached a plain flanked by thick forest on the one side, on the other by a spongy morass. Here, placing his chosen troops in the forest, he resolved to meet the Romans. Germanicus hurried into this plain, the troops led by Cecina forming the van-guard. Arminius fell back till the Roman army was well compromised. He then gave the signal, and from the side of the forest the Germans rushed on the flank of their enemies, and drove them towards the morass. The day had been lost but that Germanicus, his reserves well in hand, brought up his legions, checked the too-confident charge of the Germans, and forced them in their turn to retire. But this was all he could accomplish. Neither side could claim the victory, and, as the bad season was approaching, Germanicus hesitated to risk his army in an unknown and difficult country. Ordering, therefore, Cecina to fall back on the Rhine, he led his own corps to the Ems, and employed the means by which he had advanced to return.

After his departure Cecina had to make head against Arminius, who was inclined to press him hard. A very capable soldier, possessing a cool head, Cecina fell back, constantly fighting, before his enemy. He suffered much, however, from the attacks repeatedly directed upon him, and he was forced at last to abandon his baggage. Scarcely hoping to be able to prolong his retreat, and determined, if he were to die, to die with his face to the foe, he halted in a plain which the enemy must traverse, prepared to give them battle. In reality, his fate lay in the hands of the German leader; for Cecina had exhausted his supplies, and had no means, under the constant and harassing assaults of his enemy, to procure any. In war, there is a time to attack and a time to refrain from attacking. This was a time when a great leader would have exercised a wise restraint over himself, and have preferred to condemn his enemy to die by inanition rather than to risk all the fruits of his labours by assailing him. It is due to Arminius to state that he was very sensible of this truth, and that he used all his efforts to persuade his countrymen to remain quiet. But the fiery spirits about him preferred the more violent counsels of his uncle, Inguiomer. Under the influences of those counsels they rushed upon the all but exhausted Romans. The battle once joined, Roman discipline prevailed. The German assailants were routed with great slaughter, and Cecina was able to fall back on the Rhine without further molestation.

The loss at sea, in a storm, of two legions, commanded by Vitellius, more than counterbalanced this success; but that loss, again, was so far indirectly favourable to the cause of Rome in that it evoked from Gaul, from Spain, and from Italy, offers to Germanicus to supply his army with arms, with horses, and with money. Germanicus accepted only the arms and the horses. His own funds were sufficient to reward his soldiers. He endeared himself to them further by the manner in which—a precursor in that respect of the illustrious Turenne—he interested himself in their individual wants and sufferings, and sought at all costs to lessen their privations.

The campaign, though it had been barren of real success, had been very fruitful of experience; and of that experience Germanicus disposed himself, the following year, to profit. He had realised that whilst the Roman soldiers were more than a match for the Germans when they once fairly met them on the battle-field, yet that the knowledge of the country possessed by the latter, their superior numbers, their greater power of movement, unencumbered with heavy armour, in a country woody and marshy, every turn of which they knew, enabled them to force upon his soldiers long and often painful marches, and to harass them in a manner which prevented him from reaping the fruits of many a well-concerted plan. He found, moreover, that the last campaign had exhausted the resources of Gaul for the supply of baggage animals. In this extremity Germanicus conceived the bold plan of transporting himself and his army to the seat of war by sea. Adopting such a plan he would be able to begin the campaign earlier; the march of his legions would not be retarded by their baggage; ascending the course of the rivers they would meet the enemy fresh and eager for the fray.

He set himself at once to construct vessels which should be specially adapted for this enterprise, and directed that they should be ready at a fixed date, off the island of the Batavi* (Insula Batavorum.) Before his vessels, which numbered a thousand, had assembled there, and whilst C. Silius, who had replaced Cecina, was marching to repress an outbreak of the Catti, information reached Germanicus that a large body of Germans were besieging a Roman fortress on the Lippe, presumably the fortress of Aliso, Germanicus marched thither with six legions, compelled the enemy to raise the siege—they retired in fact on his approach—and united the fortress more firmly than ever by a string of impregnable posts with the Rhine. He then returned to the island of Batavia, and, having seen that his transports were thoroughly well furnished with provisions and other requirements, he embarked his troops, and entered the canal which bore the name of Drusus, invoking, as he did so, the spirit of his father to guide his enterprise. Pushing on with all speed he arrived safely in the Ems. But instead of ascending that river, as he had probably originally designed, he disembarked at Ancisia, on the left bank, and proceeded to throw bridges to enable his army to cross to the right bank. Thence he marched to the banks of the Weser. Scarcely had he reached that river when he learned that the Angrivarii, the near neighbours of the Cherusci, had revolted in his rear. The repression of their outbreak by Stertinius, detached for that purpose with the cavalry and light troops, occupied some time, but it was thorough. Meanwhile Arminius, warned of the danger, had approached the opposite bank of the Weser. Knowing that his brother, Flavius, held a high command in the Roman army, Arminius demanded permission to have an interview with him. The permission having been accorded, Arminius, surrounded by his chiefs, advanced to the brink of the river. Noting that Flavius had lost

* The island inhabited by the Batavi, formed by the Rhine, the Waal, and the Maas. The chief town was Lugdunum, the modern Leyden, between the Maas and the Waal. According to Tacitus, who is loud in his praises of their courage, the Batavi were originally a branch of the Catti, by whom they were expelled from their home, and then took refuge on the island which, under the Romans, gave its name to the country now known as Holland.

an eye, and being told that he had lost it in the service of the Romans, Arminius asked in what manner he had been rewarded. Flavius replied that his pay had been increased, that he had received a chain of gold, a crown, and other military honours. Arminius mocked his brother for the cheap rate at which he had sold his services, and asked him what compensation he found in such rewards for the betrayal of his country. Flavius replied with equal heat, denouncing the treason of Arminius. So bitter did the wordy contest become, that, but for the intervening river, the two brothers would have come to blows.

The next day Germanicus crossed the Weser. He had but just effected the passage when he learned from a deserter that Arminius had massed his army in a neighbouring wood, and that he intended to surprise the Roman camp that night. This information was shortly afterwards confirmed by scouts. Germanicus formed his army, then, in a manner to resist such an attack, covering the camp with earthworks and palisades. Having done this, he went amongst his men, as nearly eighteen centuries later Napoleon went, before Austerlitz, to assure himself of their dispositions. To disguise himself he threw a bear's skin over his shoulders, and was accompanied by but one man, similarly wrapped up. The result, in the main, but confirmed the parallel. Everywhere he heard nothing but praises of, and confidence in, himself: one man praised his high birth, another his noble carriage, a third his patience, a fourth the affability which knew no distinction of persons; all declared that in the coming fight they would prove their gratitude and their appreciation. To praise of Germanicus, the soldiers added expressions of contempt for the enemy; they derided their want of discipline, the roughness of their armour, and the lightness of their weapons of offence. Whilst they were thus engaged in conversation, the voice of a German was heard in front of the palisades, pouring abuse on the Romans. The latter answered by shouts of defiance. Then about 3 o'clock in the morning the German masses rushed on. After a brief conflict they were repulsed, and for the rest of the night the camp was not disturbed.

On the morrow, both armies, alike eager for battle, marched into the plain of Idistaviso—the modern Hastenbeck, in Hanover. It is an irregular plain, bounded on one side by the Weser, on the other by a chain of hills. One end of it is shut in by forest-clad hills, in the form of a semicircle, the other is comparatively open. The Germans occupied alike the forest-clad hills and the chain; the Romans, who were the attacking party, the open.

Germanicus ranged his army in this manner. The Gaulish and German auxiliaries formed the vanguard; behind these came the slingers and archers; then the legions, the cavalry, and the light troops; in the rear of all, the other allied contingents.

As the Roman army advanced, the Cherusci, who occupied the chain of hills opposite the river, descended with loud cries to attack them. Germanicus saw his advantage. He promptly ordered Stertinius to interpose, with one half of the cavalry, between them and the rest of the German army on one flank, whilst he commanded the other half to gallop round their right, and take them in rear. Wheeling, at the same time, his infantry to the right, he advanced in firm order against the forest-clad hills, in which Arminius was posted with his main army. The double attack succeeded. Whilst Stertinius crushed the Cherusci and drove them back to their original position, and beyond it, Germanicus, on his side, carried all before him. In vain did Arminius, who was wounded in the head early in the day, strive to restore the combat. Whilst the energies of his troops were already lessening, a fatal occurrence took the remainder of the fight out of them. The Cherusci, driven from their position, dashed, panic-stricken and breathless, pursued by Stertinius, at the critical moment, into their already wavering ranks. Then all was over. The Germans, forced to flee, were cut down by thousands. For ten miles from the field of battle the plains were covered with their slain. Many were forced into the Weser. The fugitives cast away arms, and lost all their baggage. Amongst the latter were the chains they had brought for their Roman conquerors.

On the field of battle the enthusiastic Roman soldiers hailed

Germanicus as "Imperator." On it, likewise, they erected a trophy, bearing the names of the tribes they had defeated. For a moment they may have thought that the victory had been decisive.

But it was far otherwise. It was a great victory gained on ground chosen by the Germans, but, in the matter of decisiveness, it was not a sufficient answer to Teutoburg. On the contrary, the news of the defeat, and especially of the erection of the trophy, roused the tribes of Germany to further and stronger efforts. From the east and from the west, from the north and from the south, they rushed eagerly to the war. Almost on the morrow of the battle, when Germanicus had hoped to see the country cleared as far as the Elbe, a new army sprang, as it were, from the earth, to dispute with him the very ground on which he had fought!

Germanicus recrossed the Weser; the new German host followed him. At length, by a flank march, they gained a plain, a causeway across which formed the boundary line between the lands of the Cherusci and the Angrivarii, and offered battle. Germanicus could not refuse it, for the Germans were between him and his base. They had chosen a splendid position. The plain was narrow, shut in between the river and thick forests, and the approach to it covered by marshes. The forests, likewise, were covered on all sides by marshes—one space alone excepted, the ground upon which the demarcation causeway had been thrown up. Arminius took the fullest advantage of the position. He formed his infantry on the causeway to face the Romans, whilst he placed his cavalry in ambush in the wood, to dash upon them as they should advance. This time he could not but feel certain of victory.

But in Germanicus he had to deal with the most skilled leader of the day. That great soldier had detected in an instant the plans of Arminius, and had decided how to baffle them. Charging his lieutenants to protect his flank, he led his choicest infantry against the causeway. Long and bitter was the contest. At length the disciplined and solidly-armed Romans penetrated into the wood. Once in that wood it became for both parties a battle for dear life; for there was no retreat for either. But here again superior discipline, aided by unsurpassed leadership, asserted itself. In the close conflict the long spears of the Germans had no chance against the short swords of the soldiers of Rome. In vain did Inguiomer, uncle of Arminius, whose wounds had kept him from taking an active part in the battle, rush from rank to rank to encourage his warriors. Germanicus, who had taken off his helmet and fought bareheaded, to be the better recognised by his men, urged his men to the slaughter, declaring that the only mode of finishing the war is to exterminate. Not till evening fell did he relax his pursuit.* He then drew back his troops, "satiated with gore," to a palisaded camp beyond the causeway. The next day, after having thanked his soldiers for their conduct, he erected with the arms captured from the vanquished, a trophy bearing the inscription: "Victorious over the people between the Rhine and the Elbe, the army of Tiberius Cæsar has consecrated this monument to Mars, to Jupiter, and to Augustus."

The monument having been erected, Germanicus, though the summer was still in its glory, fell back, the presence of his army being sufficient to repress a threatened rising of the Angrivarii. On reaching the Ems he re-embarked his troops on the vessels which had conveyed them thither, and set sail for the island of Batavia. But hardly had his fleet gained the ocean, when a terrible storm set in, which dispersed his ships, and drove many of them on rocks and islands. The vessel which bore Germanicus was driven on the coast of the territory inhabited by the Cauci. Finding himself alone with his immediate crew, and fearing lest all the others had been lost, the illustrious Roman leader was in despair, and loudly accused himself of being the cause of the mis-

* It is difficult to give, with perfect exactitude, the details of this battle, even to reconcile the accounts of the fighting with the description of the lay of the ground. There is reason to think that though Germanicus gained the victory, that victory was not so decisive as Tacitus would have us believe. For, in the first place, during the battle there was a cavalry fight, which, by the admission of the Romans themselves, was indecisive; in the next, although it was the height of summer, and the season was favourable for manœuvring, Germanicus, in spite of his victory, continued to fall back towards the Ems.

fortune. At length the weather moderated, and gradually Germanicus recovered the greater number of his soldiers. But the mischief had been done. The news of his misfortune had spread, in exaggerated form, over the northern coasts of Germany, and the terrors caused by the recent victory had given place to renewed hope. To nip this feeling in the bud, Germanicus seized the earliest opportunity to despatch Silius with thirty thousand infantry and three thousand horse against the Catti, whilst he led the rest of his army against the Marsi. In a preliminary skirmish against these, he not only had the advantage, but recovered another —the second—of the lost eagles of Varus. He then ravaged the country, the Marsi carefully avoiding a pitched battle. The campaign then terminated.

It was the last campaign of Germanicus in Germany. Tiberius Cæsar, long jealous of his renown, took the opportunity of the conclusion of a campaign which had been glorious for the Roman arms, to recall him. In vain did Germanicus demand one year more to finish the work so well begun, to replace Rome in the position she had occupied before the defeat of Varus. "You have fought great battles," replied Tiberius, "but your losses have caused a greater drain on the resources of the empire. I, when on nine occasions I was sent by Augustus to Germany, obtained greater results by prudence than by force. As the Roman honour has now been avenged the Cherusci and the other refractory tribes may well be left to stew in their own juice." Such, at least, was the main purport of his reply. He added, that if Germanicus thought it absolutely necessary that the war should be continued, he should leave the glory of concluding it to his brother by adoption, Drusus, who, in default of any other enemy, could hope, only in Germany, to merit the title of *imperator* and to gather noble laurels. Germanicus did not insist further; he returned to Rome to meet there the reception due to a hero.

The campaigns of Germanicus were the last efforts of Rome to recover the position lost by Varus in the forest of Teutoburg. These efforts, despite many victories, had failed. Rome had avenged the prestige of her armies, but she had not recovered the position she had lost. The battle of Teutoburg had been decisive of the question which lay nearest to the hearts of the Germans; it had prevented Germany from sharing the fate of Gaul; it had baffled the policy which Varus had been sent to carry out—that of romanising the Germans. After Teutoburg that policy had become impossible. Not even the extra year so earnestly demanded by Germanicus would have sufficed. There is a vast difference between the retaining under a foreign yoke peoples who have always recognised the superior prestige of the conquering power, and the re-subduing of those peoples after, by their patriotic efforts, that prestige has been destroyed. Under all these circumstances, then, and considering the nature of the battles fought by Germanicus during the last year (17 A.D.) of his command, how a victory such as that gained at Idistaviso (Hastenbeck) led, almost immediately, to another battle, more bloody and more desperately contested, a contest in which defeat would have been ruin, I dare not presume to question the wisdom of the conclusion arrived at by Tiberius when he wrote: "The Roman honour has been avenged; let the Cherusci and the other refractory tribes stew in their own juice." He was right. Rome had still a noble frontier. The moment she ceased to attempt to subdue the refractory tribes they began to stew in their own juice with a vengeance. The uncle, Inguiomar, jealous of his nephew's glory, leagued with Maroboduus, chief of the Marcomanui, against him. To baffle these, there sided with Arminius the Longobardi and the Semnones, and with their aid the conqueror of Teutoburg gained a great victory. This victory so elated him that he aspired to supreme power in Northern Germany. His plans were, however, discovered and he was assassinated (20 B.C.). He was then only thirty-eight years old. "Arminius," wrote Tacitus, just about eighteen centuries ago, "was uncontestably the deliverer of Germany; he had not fought the Roman people in the rise of their power, as had other kings and generals, but when Rome was at the summit of her glory, when the empire had reached her greatest splendour. He was not always fortunate, but he did not cease for a moment to impose on the enemy by his attitude and by his army. During twelve years arbiter, with the consent of his fellow-citizens, of the affairs of Germany, he was the object of their veneration after his death."

The history of the ambush and surprise of Teutoburg would have been incomplete unless it had been followed by a narrative of the events which prove the immense significance of that battle. The slaughter of the Roman legions treacherously led into an ambush would have been an affair of comparatively little moment if that slaughter had been promptly avenged and if the lost position had been as promptly regained. But the slaughter was not promptly avenged. In the presence of it Imperial Rome trembled; it shortened the life of Augustus, and it gave a living proof that the Rome of the Scipios was no more. Eight years of murderous warfare were required to wring from Tiberius the admission that the honour of Rome was satisfied. As for the position—that was lost for ever. Those eight years confirmed the aspirations of the German tribes, fortified in them the conviction that the free wild life of the forest was preferable to gilded slavery. The struggle which followed Teutoburg proved the value of the battle waged in that forest. The history of that struggle forms, then, a necessary continuation of the history of the battle.

CHAPTER III.

RONCESVALLES.*

ALTHOUGH he died only Mayor of the Palace, Charles Martel was not the less the founder of the Carlovingian dynasty. The illegitimate son of Pépin, properly Pippin or Pipin, of Heristal, Mayor of the Palace to the puppet sovereigns of the Merovingian race, Charles, hated by his father, and left by him at his death a prisoner in the fortress of Cologne, possessed all the qualities which enable a man to hew his way to greatness. The Neustrians, or inhabitants of what is now Western France, resenting, after a short experience, the dispositions of the deceased Pipin, drove the new mayor and his *entourage* from their midst and pursued them into the very heart of Austrasia—the kingdom of which Metz, the Divodunum of the Romans, was the capital. Assailed simultaneously by the Neustrians and their allies, the Frisians, the Austrasians bethought them of the bastard boy shut up in Cologne, the boy whom they recollected as bearing himself manfully when he used to fight by the side of his father, and, in the hope that he might

* The principal authorities referred to in this chapter are: (1) *Annales*, d'E'ginhard; (2) *Vie de Charlemagne*, par E'ginhard; (3) *Des Fait et Gestes de Charles-Le-Grand, Roi des Francs et Empereur, par un Moine de Saint-Gall*; (4) *Histoire de France*, par Michelet; (5) *Histoire de France*, Henri Martin; (6) *Histoire du règne de Charlemagne*, Leclerc de la Bruère; (7) *Histoire de Charlemagne*, Gaillard; (8) *La Chanson de Roland*, par Léon Gautier (ouvrage couronné par l'Académie Française et par l'Académie des Inscriptions et Belles-lettres), 11th edition, 1881.

lead them to victory, they released him from his prison and proclaimed him to be their duke. Charles was then in his twenty-sixth year. Strong, active, full of energy, he was just the man for the occasion. Rallying his troops, he attacked the Neustrians, and although he was repulsed on the first assault, he profited by the check to infuse new ardour into his men, and then redeemed it by a victory at Amblef (Amblève) near Cologne (716). He gained a second the year following at Vincy, not far from Cambray, and pursued the Neustrians to Paris. Having thus completely freed Austrasia, Charles crossed the Rhine, turned his arms against the Frisians, defeated them with great slaughter and carried the war into Saxony. Whilst Charles was fighting the Saxons on the Weser, the Neustrians, recovering from their defeat, re-armed, and calling to their assistance Eudo, Duke of Aquitaine (South-Western France), marched towards Austrasia. But Charles, warned of their preparations, had quitted the Weser the instant he had realised the danger, and, detaching Eudo from the alliance by recognising him as king, hurled back the Neustrians in disorder from Soissons as far as Orléans (719). He then forced Raganfred, or Rainfroy, the Neustrian Mayor of the Palace, to resign his office, and on the death the same year of the puppet Merovingian King, Clotaire IV., whom he himself had nominated on the flight of his predecessor, Chilpéric II., after the battle of Soissons, restored Chilpéric, and was appointed by him to the vacated office which his father, Pipin, had held. On the death, the year following, of Chilpéric, Charles elevated to the gilded throne Thierri IV., son of Dagobert III., a child of seven years! Charles thoroughly understood how to deal with such a master. Surrounding him with all the glitter of royalty, he allowed him every kind of recreation. He drew the line only at politics. But he arranged so well that the young king did not feel the exclusion. To his last hour—and he reigned twenty-four years—Thierri believed that he ruled as well as governed, so great was the number of his palaces, so magnificent were his purple robes, so numerous were his horses, so luxurious was his table, so exciting were the pleasures of the chase, and so beautiful and so compliant were the ladies of his Court!

For the few following years Charles was engaged in contests with the Alemanni and the Bavarians who had revolted, and with the Saxons who had never been subdued. From this task he turned for a moment (731) to repress the growing power of Eudo, titular king of Aquitaine. In two successive invasions he made his presence felt in the province of Berry. He was about to proceed further, when suddenly a piercing cry for help from the very prince whose territories he was assailing reached his ears. To understand the cause of this cry, I shall endeavour to lay before the reader a very brief outline of the previous history of the ruler of Aquitaine.

Eudo had succeeded at an early age to the inheritance of the Duchy of Aquitaine. The Kings of Neustria were weak. Eudo was young, ambitious, powerful, and uncontrolled. By the exercise of his great qualities he succeeded, sometimes by arms and sometimes by treaties, in greatly increasing his dominions. Long before the death of Pipin, Eudo had added Gascony and the Duchy of Toulouse to his ancestral dominions. He had acquired, besides, the territory of which Bourges was the chief town; the Auvergne, the Velay,* the Limousin†; in a word, the whole of eastern Aquitaine as far as the Loire. Beyond that river he had annexed that portion of Neustria which was afterwards called the province of the Nivernais, whilst upon the left bank of the Lower Rhone he had taken the eastern portion of Provence, and upon the right bank the territory subsequently known as Vivarais.‡ He was sovereign duke of these territories when he came in contact, in the manner described, with the bastard son of Pipin. Recognised as King by Charles, Eudo withdrew from the contest and set to work to consolidate his dominions. The position he occupied was one entailing great responsibility. South-western France had become, by the progress

* A division of France of which Le Puy was the capital. It is now comprised in the Department Haute-Loire.

† Now forming the department Corrèze and part of Haute-Vienne.

‡ A district in Languedoc of which Viviers was the capital. It is now comprised in the departments Ardèche and Haute-Loire.

of the Arabs, the border-land of Western Europe, and there devolved upon its ruler the duty of protecting Europe against the invasion of a people alien in race and in creed. The invasion, long threatened, came in 723. A Saracen chief, El-Samah-ben-Abdul-Malik, crossed the Pyrenees with the largest Muhammadan army till then seen in Europe, and laid siege to Toulouse. Eudo rushed to the rescue of the threatened city, assailed the invaders on the road between Toulouse and Carcasonne, and inflicted upon them a complete defeat. Two years later, 725, he repelled a second invasion, defeating in Provence, after a hotly contested battle, an Arab chief named Aubessa. Then followed four years of respite. Towards the close of these four years, Eudo received overtures of alliance from Othman-ben-Abou-Neza, the Saracen governor of the Spanish frontier nearest to his own territories, and who proposed to revolt against the Divan of Cordova. Eudo accepted these overtures; gave to the governor the hand of his daughter, Lampagie,* and, when, as agreed, the rebellion broke out, and the Saracens were engaged against each other, prepared to increase his dominions by the annexation of Languedoc. But his ally, Othman, was too promptly and too easily subdued to suit his ambitious plans. At a moment when Charles was threatening him with an invasion (730) the news reached him that his son-in-law had been taken prisoner and beheaded, and that his daughter, famous for beauty, had been sent to Damascus to adorn the seraglio of the Commander of the Faithful. The following year, 731, Charles made two successive invasions of Berry, burning and destroying all before him. Eudo was too much occupied by the projected invasion of the Saracens to oppose him. At last, 732, that invasion burst upon him. In the spring of that year, Abdurrahman, Governor-General of Spain, entered France at the head of the largest and best accoutred Saracen army which had ever crossed the Pyrenees. His progress was a triumphal march. Crossing the Garonne, Abdurrahman defeated an army which

* Some have it that Lampagie, called also Numeránce and Menine, was not given, but stolen.

Eudo had raised to bar his progress, then pillaged and ravaged the country as far as Bourdeaux, again defeated Eudo, took possession of that city, and burned its churches. In despair Eudo fled to implore the aid of the rival who, but the year previously, had tried to despoil him.

There was but little time for Charles to reflect. The danger was pressing. Every day brought news of the further progress of the Saracen army. Charles then, whilst exacting very onerous conditions from the supplicant, responded favourably, incorporated in his own army a large number of the fighting men from beyond the Rhine who had but just submitted to him, and hurried to meet the Saracen invader. He found him, on the 10th October, in the plain between Poitiers and Tours, approaching the latter place. The battle instantly joined. It was fierce, desperately contested, and the first day victory hung in the balance. But, on the second, the energy of Charles—known to the world from that day forth, for the hammer-like force of his blows, as Charles Martel—and the active courage of Eudo, prevailed; the Frankish host conquered. But the carnage had been terrible; the Saracens left, it was said, 375,000 corpses on the field of battle.

By this victory, known as the victory of Tours, Charles Martel saved Europe. It was the culminating point in his wonderful career. Not that he ceased to conquer. As one fruit of his victory he asserted his overlordship over Aquitaine and her dependent territories, and forced Eudo to pay him homage. He subsequently fought many battles, not always successfully, with the Saracens for the possession of Languedoc and Provence. In the midst of these he was recalled to the Rhine by an invasion of the Frisians. With them, with the Alemanni, the Bavarians, and the Saxons, he waged war till the year 739. In that year the rapidity and force of his blows had exhausted his enemies, and he enjoyed a breathing time which lasted till his death, three years later. In the interval Charles had reconquered Burgundy, and, Eudo having died in 735, and the sons of that prince, Hunold and Atto, having asserted their independence, had engaged in war with Aquitaine, a war which terminated the year following by the recognition by the two revolted princes of the overlordship of Charles.

Charles Martel died peaceably in his bed at the age of fifty-three (741). He left three sons, Carloman, Pipin, and Griffon. To the first he bequeathed Austrasia, Thuringia, and Swabia; to the second, Neustria, Burgundy, and Aquitaine; to Griffon, son of his second wife, a princess of Bavaria, a very small part of the province of Champagne.

I do not propose to enter in any detail into the history of two of the brothers. It will suffice to state that, in 747, Carloman, disgusted with the world, entered a monastery; that Pipin, who had continued the system of ruling, as Mayor of the Palace, under the shadow of a puppet Merovingian king, at once seized his elder brother's inheritance; and that Griffon, after many struggles for the enlargement of his borders, during which he, for a time, settled down peacefully as ruler of Le Mans and other districts on the Loire, was killed, to the profit of his brother, in 753. In a word, we have to deal only, or mainly, with Pipin.

Pipin's first difficulties were with Aquitaine, Hunold having at once, on the death of Charles, asserted his independence. The struggle upon which Pipin entered lasted two years. Then Hunold, weary of the world and its trials, retired into a monastery, leaving his son, Waifer, to continue the battle for independence. Of this war it will suffice to state that it lasted nearly throughout Pipin's reign; that though, in 766, Pipin had thoroughly conquered the country, yet a rebellion almost immediately afterwards broke out, which was only subdued in 768. Then, however, Waifer, still unsubdued, was assassinated, and Aquitaine was formally united to the kingdom of the Franks. This consummation was reached only a few months before the death of Pipin. Waifer left one son, Lupus, of whom we shall hear further when we come to the period of the successor of Pipin.

Whilst Pipin, on the one side, was contending for the complete submission of Aquitaine, he was engaged, with his brother Carloman, in securing the Frankish possessions beyond the Rhine. With this view they made joint expeditions into Germany in 742–3–4 and 5. In the last of these years Carloman, as already stated, retired into a monastery. Pipin, left alone, asserted his authority throughout his brother's dominions in 747 and 748. In the following year, however, his attention was fully engrossed by a matter which affected the future government of the Frankish kingdom.

Up to that time Pipin had borne only the titles which he had inherited from his father and grandfather—those of Mayor of the Palace and Duke and Prince of the Franks. When, however, he had united under the nominal rule of the Merovingian prince, Childeric III., the whole of the Frankish dominions, he despatched (751) ambassadors to Rome to consult Pope Zachariah on the question as to whether it was proper that the kings of the Franks should reign without ruling.* The Pope replied that it were better that he who wielded the power should be also the King, and, in virtue of his Apostolic power, he authorised Pipin to assume that rank. Upon this reply Pipin acted. He convened an assembly of nobles and commons at Soissons in 752, and was there crowned King by the hands of Boniface, Bishop of Mainz. The change was considered so necessary for the peace of the kingdom that not a single voice was raised against it. As for the puppet King, Childeric, he was shaven and shorn and transferred to the monastery of Saint Bertain, near St. Omar, where he died, the last of the Merovingian dynasty, in 755.

The service thus rendered by the Pope to the Carlovingian required reciprocity on the part of the latter, and Pipin showed no disposition to evade the obligation. Pope Zachariah died a few months after he had given his famous answer to the ambassadors of Pipin (14th March 752). His successor, Stephen III., finding that the Papal territories were menaced by the Lombards, crossed the Alps to invoke the aid of the Frankish king.† Pipin, who had

* "Afin de consulter le pontife touchant les rois qui alors étaient en France et qui rien possédaient que le nom, sans en avoir en ancune façon la puissance."—*Annales d'Éginhard.*

† He presented himself, according to the old chronicles, covered with ashes, wearing a hair-shirt, accompanied by a number of priests in deep mourning.

recently returned from a victorious expedition against the Saxons, responded according to the wishes of his visitor, and summoned a national assembly to confirm his promise. It gives us a striking picture of the free institutions then prevailing amongst the Franks, when we find that their national assembly, convened for the purpose of sanctioning the wishes of the King, had the boldness to reject the plan altogether. An invasion of Italy, said its members, promises no national advantages, and we will not give it our sanction. Pipin, whilst recognising the power of the assembly, did not altogether abandon hope. He appeared for a second time at its bar to urge his plan. On this occasion his pleadings, urged with force and eloquence, were successful, and the assembly yielded. Pipin then crossed the Alps, defeated Astolphus, King of the Lombards, at Susar, the point where the routes by Mount Cénis and Mount Genèvre join; besieged him in his capital, and forced him to restore to the Pope the dominions he had taken from the Holy See. No sooner, however, had the Frankish king recrossed the Alps on his return, than Astolphus renewed his preparations. The Pope, therefore, despatched a letter to his protector, urging him in the most pathetic terms to return to defend him. Pipin responded, returned with extraordinary celerity, disconcerted the plans of Astolphus, and once again besieged him in Pavia (756). This time he was resolved there should be no mistake. He therefore formed of Ravenna, Pentapolis, the Emilia, and the Duchy of Rome, the domain of St. Peter, to be thenceforth united to the Papal Chair. Astolphus, whilst preparing for a new war, died the same year. He was succeeded by his constable, Didier, Duke of Istria, a man equally ambitious and equally unscrupulous. The Pope confirmed Didier in his sovereignty, on condition that he should carry out the conditions upon which Pipin had insisted. Didier promised much, but the history of his reign is the history of a continued struggle to evade his obligations.

The remainder of the reign of Pipin was occupied in struggles with the Saxons, with the Saracen invaders of Languedoc, and with Waifer. Over the first he signally triumphed in 758. In the south he made some progress, taking Narbonne after a siege of six months and a blockade of three years. His most serious struggle was for Aquitaine, and this terminated, in the manner already described, in 768. Then, when his success was at its height, when he had acquired all that he had tried to acquire, Pipin died (24th September 768). He had reigned sixteen years.

He left two sons; Charles, afterwards called Carolus Magnus and Charlemagne, and whom we shall refer to by the name by which he is best known in history, and Carloman. In a general assembly convened by the late King just before his death, Pipin had divided his vast empire between the two brothers, bequeathing to Charlemagne Neustria, Burgundy, and Aquitaine, to Carloman Austrasia and all Frankish Germany, and had bound them to act in concert. Between two such natures, however, concert was impossible. Whilst both were ambitious, Carloman was jealous of the superior address, influence, and mental qualities of his brother. An opportunity soon offered for the display of this defect. As on the accession of Pipin, so on the accession of Charlemagne did Aquitaine give the signal for civil discord. This time it was Hunold, father of the Waifer with whom Pipin had waged his long and stubborn contest, who threw down the gauntlet. Emerging from the monastery to which he had voluntarily retired some twenty-three years before, Hunold entered Aquitaine, assumed the title of Duke, and proclaimed a war of independence. All the discontented flocked round him; he was joined even by many who, uncertain of the character of the new Carlovingian ruler, then in his twenty-seventh year, thought that the chances of their old sovereign were by no means despicable. For a moment, then, Hunold was formidable. Charlemagne, new to the situation, was comparatively weak. He sent, then, to invoke the aid of his brother. Carloman responded, and reached, with his army, the camp of Charlemagne at Duasdives, supposed to have been a town in the province of Poitou. Then there took place an interview between the royal brothers. What passed at that interview cannot with certainty be recorded; but it can scarcely be doubted that Carloman insisted, as the condition of his support, upon concessions to which Charlemagne would not consent. This, at least, is certain, that on the conclusion of the interview Carloman returned with his army to his own dominions.

Thus left to himself, Charlemagne marched against Hunold, forced him, by a skilful manœuvre, to evacuate the position he had taken between the Dordogne and the Garonne, and to flee into Gascony. Charlemagne summoned Lupus,* the reigning duke of that province, to surrender his guest, and Lupus, fearful of drawing upon himself the anger of the Frankish king, complied. Thenceforward Hunold ceased to be formidable. After some adventures, not necessary to the story, he perished at the siege of Pavia (774).

The death of Carloman in 771 (4th December), and the flight to Italy of his widow and two children, removed a great difficulty from the path of the ruler of Neustria. Thenceforth he could claim to be the sole ruler of the Frankish empire. Being sole master, he proceeded without delay to give evidence of those great qualities which placed him in the very front rank of the world's great men.

A contemporary scholar, born the very year which made Charlemagne sole sovereign over the Franks, 771 A.D., known at the time variously as Heinhardus, Agenardus, and Eginhardus, but who has descended to posterity as Éginhard, wrote not only a life of the great monarch, but the annals of his reign and of the reign of his successor. These works, rich in detail, afford incontestable evidence of the events which marked the illustrious career of the new Cæsar. Without further preface I shall proceed to introduce him as one, though not the only nor even the principal, figure in this chapter, to the reader.

* It seems certain that this Lupus was no other than the son of Waifer; if so, he was the grandson of Hunold. The old chronicles nowhere state that the son of Waifer became Duke of Gascony, but they do affirm that Lupus, son of Waifer, took a prominent part in the surprise of Roncesvalles, and was afterwards taken and hanged. Now, it is a well-authenticated fact that Lupus, Duke of Gascony, took a prominent part in that surprise, and was afterwards taken and hanged. It seems just to infer, then, that there was but one Lupus, and that he was the son of Waifer, and Duke of Gascony.

"Charles," writes Éginhard, "was strong of body and made on a large scale. He was tall, but his height was not out of proportion to the rest of his body, for it is certain that it did not exceed seven times the length of his foot. The crown of his head was round, his eyes were large and bright, his nose was somewhat long, his hair, white in colour, was thick and luxuriant, his expression cheerful and pleasant. In all his attitudes, whether he were standing or sitting, there reigned an air of nobleness and dignity; and, notwithstanding that his neck was short and thick, and that he had a prominent stomach, he was, in other respects, so well proportioned that these defects were not noticed. His gait was firm and all his movements gave evidence of a manly nature; his voice, however, was too thin for a man of his size and height."

When he succeeded to the complete inheritance of his father, Charlemagne was in his thirtieth year. He had married, during the lifetime of his father, Hemiltrude, or Hilmetrude, a lady whose parentage is unknown. By her he had two children, Pipin the deformed, who died in 792, and a daughter. Partly, probably, on account of the deformity of the son, Charlemagne, in spite of the protestations of Pope Stephen IV., had acceded, 771, to the prayers, based on political grounds, of his mother, Bertrade, or Bertha, a daughter of Caribert, Count of Laon, and put away Hilmetrude,* to marry Désirée, daughter of Didier, King of the Lombards. He had, in spite of the refusal of his brother to co-operate with him, made prisoner of Hunold, and now, sovereign of all the dominions of his father, he found himself face to face with the Saxons—the Pagan people who had been the most redoubtable enemies of King Pipin.

The annalist, Éginhard, is careful to inform us that the great council of the Franks, called the General Assembly, met every year to consider the business of the nation. In the year 772, the

* Hilmetrude survived her disgrace many years. She was buried at St. Dénis, where the epitaph may still be traced: "*Hic jacet Hilmetr. reg. uxor Caroli Magni.*"

General Assembly met at Worms, a city which had formed a part of the appanage of Carloman. It would appear that, on this occasion, its members discussed the question put before them by their new sovereign—the question of war with the Saxons. It is certain that it was decided to undertake that war.

Charlemagne entered Germany, and devastated the country he traversed with fire and sword. It is clear that he marched towards the Weser, and that, proceeding thither, he came upon the track of the ill-fated Varus. For the annalist records that "he took the castle of Ehresburg and destroyed the idol called *Irminsul* by the Saxons." This *Irminsul* was a national pillar erected by the Germans in honour of Arminius, whom they called Hermann, and the correct name of which was Hermann-Säule. Charlemagne remained for three days on the spot, marched then to the Weser, and received there twelve Saxon hostages as a guarantee of the submission of the entire people. He then returned to France.

Meanwhile Pope Stephen IV. had died, 1st February 772. His successor, Adrian, the first of that name, was a remarkable man. It was said of him after his death, that he had known how to reconcile the strong character of the ancient Roman with the suppleness and adroitness of the new society. Adrian had scarcely entered upon the duties of his sacred office when Didier, King of the Lombards, breaking the solemn engagements he had made with Stephen, marched into the Papal territories and took possession of the towns of Ferrara, Faenza, and Comacchio. Against this tyranny Adrian appealed to the King of the Franks. The ambassador sent by Adrian reached Charlemagne at Thionville—the German Diedenhofen—where he had arranged to pass the winter. The question was fully discussed between the two. There seemed indeed no argument whatever against intervention. Charlemagne was bound by the engagements of his father to protect the Papacy against Lombard aggression. Didier was, indeed, his father-in-law, but not only did that position confer no rights cancelling the solemn obligations of a King of the Franks, but Charlemagne had already tired of Désirée, who had borne him no children, and he was upon the point of severing his marriage with her in order to take to wife Hildegarde, a princess of the Suevi. Charles resolved, then, on war. He proceeded to Geneva with his army, and, after much deliberation, resolved to follow the example of Hannibal, and cross the Alps. With this object he divided his army, and, committing one-half of it to the command of his maternal uncle, Bernard, with instructions to march by the pass of La Cluse, led the other half across Mount Cenis, into the plains of Turin. Details of this march are unfortunately wanting. "I might describe here," writes Éginhard in his *Life of Charlemagne*, "the immense difficulties which the Franks experienced in traversing the Alps to enter Italy, the hardships they had to endure to cross those all but inaccessible summits, those rocks which rise to the very Heaven, and those rough blocks of stone; but my aim in this work is to transmit to posterity rather an account of the manner of life of Charles than the details of his wars." It is to be regretted. This, however, is certain, that, in spite of the difficulties and dangers, Charles led his army safely into the plains of Italy.

Didier, taken by surprise, acted as, a thousand years before, Scipio had acted in the presence of Hannibal. Recognising, as Scipio had recognised, in the line of the Po the true line of defence for Italy, and the Pavia–Stradella–Piacenza point as the true point of defence of the Po, he concentrated his army at Pavia, prepared there to offer a strenuous resistance. Charlemagne proceeded to besiege him (October 773), but, despite of all his efforts, the town resisted. The winter passed, but still Pavia held out. Charlemagne then left a sufficient body of troops before the city, and proceeded to reduce Verona, and the other towns north of the Po. He then returned to Pavia, but, as that place still resisted, he made a pilgrimage to Rome to implore the prayers of the Pope. Meanwhile a strict blockade was maintained, and, on his return, Didier and his garrison, reduced by hunger, surrendered (May 774). Charlemagne deprived his captive father-in-law of his sovereignty, carried him and his wife, Ansa, prisoners to Liège, whence, shortly afterwards, he transferred Didier to the monastery of Corbie, where he died. Lombardy was annexed to the kingdom of the Franks.

The Saxons had taken advantage of the absence of Charlemagne to ravage his borders. On his return he despatched three several corps against them. But, although these slew many of them, Charlemagne felt that he would never be secure so long as his enemies in Germany should remain unconverted. It became, then, one great object of his life to force Christianity upon them. With this view he summoned, the year following, 775, a General Assembly at Düren—about twenty miles from Aix-la-Chapelle—and, having explained to its members his views, crossed the Rhine at the head of his army, took Siegburg—at the mouth of the Agger—by assault, and rebuilt and re-garrisoned the fort of Hohensyburg. Thence he marched to the Weser, attacked and defeated a large body of Saxons who attempted—near the existing watchtower on the Brunsberg—at Hoxter, the castle of Bruno, brother of the Saxon Duke, Witikind, to bar to him the passage of that river, and, leaving there a part of his army, pressed on with the remainder as far as the Ocker.* There he was joined by the chief of the East Saxons, Hassius, with offers of submission of himself and his tribe. Charlemagne accepted these offers gladly, and turned his steps back to the Weser. On his way he received similar tenders of fidelity from the descendants of the Angrivarii—the people who, the reader will recollect, dwelt in the time of Germanicus on the banks of the Weser.

Meanwhile the West Saxons, inspired by their chief, Witikind, had taken advantage of the absence of Charlemagne to perpetrate their crafty designs on the division he had left behind him on the Weser. As the Frankish foragers were returning to camp with their supplies, the Saxons joined them as friendly companions, but, on entering the intrenchments, they threw off their disguise and massacred many before they were expelled. Charlemagne, on

* The Ocker rises in the Harz mountains, and, flowing in a northerly direction past Brunswick, joins the Aller, after a course of sixty miles, ten miles to the west of Gifhorn.

learning this, went after them and cut up a great number. But their chief escaped.

The war, however, was not finished. The Saxons remained unconverted. Nor was it possible for the King, in 776, to complete his work of the preceding year. A rising in Italy, fomented by a Lombardian nobleman, Rotgaud, whom Charlemagne had made Duke of Friuli, called him once more across the Alps. It would appear that the campaign was very short and very decisive. The conspirators, surprised, surrendered, and Charlemagne, having filled their places by men upon whom he could depend, returned hastily to the Rhine. His return was opportune; for the Saxons had taken advantage of his absence to recover Hohensyburg, and were besieging Siegburg. As he advanced, however, the happy news reached him that the garrison of Siegburg had made a sally, and had not only completely defeated the Saxons, but had pursued them as far as the Lippe.

Charlemagne resolved to take advantage of this event to finish with them at once and for ever. He convoked, then, a national assembly at Worms, placed before it his plans, and, these having met with universal approval, set out at the head of his army for the Lippe. As he approached the source of that river, a tributary of the Rhine, the Saxon battalions, instead of showing a hostile attitude, rushed forward to implore pardon for their faults and to offer to become Christians. Charlemagne complied with their desires, had many of them baptised on the spot, and forgave them.*

* A distinguished orator and statesman used very skilfully this incident to describe the defection of a considerable section of a great political party at the bidding of their leader. "That was a scene," said Mr. Disraeli in 1846, "I believe, unprecedented in the House of Commons. Indeed, I recollect nothing equal to it, unless it be the conversion of the Saxons by Charlemagne, which is the only historical incident that bears any parallel to that illustrious occasion. Ranged on the banks of the Rhine, the Saxons determined to resist any further movement on the part of the great Cæsar; but when the Emperor appeared, instead of conquering, he converted them. How were they converted? In battalions—the old chronicle informs us they were converted in battalions and baptised in platoons. It was utterly impossible to bring these individuals from a state of reprobation to a state of grace with a celerity sufficiently quick. When I saw the hundred and twelve fall into rank and file I was irresistibly reminded of that memorable incident on the banks of the Rhine."—*Selected Speeches of the late Lord Beaconsfield*, vol. i. p. 169.

He then again repaired and re-occupied the fort of Hohensyburg, built another on the Lippe, and, crossing the Rhine, passed the winter at Heristal.

To make quite sure of the Saxons, Charlemagne, after holding the national assembly at Paderborn early in the spring of 777, again entered Saxony. Once more was there a repetition of the scene of the previous year. All the chiefs of the Saxons, except Witikind, the most persistent of the resisting lords, who had fled for refuge to the Court of Siegfried, King of the Danes, presented themselves, with their principal nobles, and demanded to be baptised. Charlemagne caused their requests to be complied with, and restored them to their honours and dignities, on the condition that if they should rebel again they should lose both, and their liberty in addition.

This year brought the great Sovereign of the Franks in communication with the people whose inroads into Europe had been decisively checked by his grandfather, Charles Martel, at Tours. This was Ibn-al-Arabi, a Saracen chief, who, discontented with his Sovereign, Abdul Rahman, Khalif of Cordova, had come to implore the assistance of Charlemagne, and to offer, in exchange for that assistance, to admit his troops into the towns of which he and his friends disposed. It would appear that the idea commended itself to the mind of the Frankish king, and he gave to the consideration of the best means of carrying it into effect the winter months of 777–8. As soon as the passes of the Pyrenees were reported practicable—in the spring of 778—he traversed Gascony, entered Navarre, attacked and took Pampeluna, then pushed on as far as Saragossa. But that place was strong, and repulsed him. He therefore returned to Pampeluna, caused the walls of that town to be rased, and recrossed the Pyrenees. It was during this return to France that there occurred, at Roncesvalles or Roncesvaux, that ambush and surprise which forms the main subject of this chapter.

The opposition he encountered at Saragossa had profoundly chagrined Charlemagne, and this chagrin was augmented by the absence of any very definite information as to the conduct of the

Saxons. He marched, therefore, with all haste to the foot of the Pyrenees. There he divided his army into three divisions, traced out, after the manner of Hannibal, a separate route for each, and began the ascent of the passes. His own march was quickened by information, which reached him very soon after he had set out, that the Saxons had taken advantage of his absence to rise in revolt. The second division followed him almost as expeditiously; the third, commanded by a hero whose name has descended to posterity in all the poems and songs of that stirring period, by the illustrious and immortal Roland, more leisurely. It is the march of the last which concerns us in these pages!

Who was Roland? To answer this question we can only examine the far from explicit testimony of Éginhard and sift the legends and the traditions of the age in which his exploits were celebrated. According to those traditions, Roland was a nephew of Charlemagne, the son of his sister Bertha* and the Seneschal Milo of Aglant. Not having been born in wedlock, his parents endeavoured for a long time to conceal his existence, and his early days were the reverse of bright. Nature, however, had endowed him with a joyous and cheerful disposition, and it is related that whilst Charlemagne, on his return from his first visit to Rome, was occupying the Palace of Sutri, the youth presented himself there, was received, and delighted everyone by his wit and gaiety. The chief counsellor of the King, Duke Naimes, at once suspected that the boy must be of noble birth. He had him traced to his home, and discovered the intimacy of Bertha with Milo. Charlemagne, in his first moments of anger, wished with his own hands to strike his sister; but Roland rushed to the defence of his

* Thus in the *Charlemagne* of Venice (twelfth and thirteenth centuries). But she is called "Bacquehert" in *Acquin* (twelfth century), and "Gille" and "Gilain" in other poems. According to another legend (*Karlamagnus Saga*, thirteenth century), he was a son of Charlemagne himself. It has been imagined that the cause of his birth may have been the great sin which the Emperor designedly omitted in his confession to Saint Gilles, and to which other writers have mysteriously alluded. But, regard being had to the amorous character of Charles the Great, to the fact that he had nine wives, and that, despite of that fact, his morals scandalised even that dissolute age, this legend may be dismissed.

8

mother, and, fastening furiously on his uncle's hand, made the blood spurt from his nails. "He will be the falcon of Christianity," exclaimed Charles, already proud of his nephew's prowess. He then pardoned his sister, sanctioned her marriage with Milo, and took them both into favour.

Such is the legend, or rather, one of the legends. More diverse still are the stories of the life which followed, most, if not all of them, the outcome of the poetic license of the age. But this, at least, is certain: the name and exploits of Roland spread into all lands. In France, in Spain, in Germany, in the Netherlands, in the countries of Scandinavia, in Italy, his achievements were sung. He was the hero, the paladin, of a heroic age. On the eve of the battle of Hastings, Taillefer chanted his achievements before William of Normandy and his nobles. There was not a warrior of renown, from the seventh to the twelfth and thirteenth centuries, who had not revelled in the inspiring song. The cause of that song must have been more than a myth. The hero of many nations and many ages must have lived to inspire by his example. Roland was for those ages a reality, a bright particular star, the embodiment of all that was grand, and chivalrous, and noble, and heroic. We may, then, follow his fortunes on the field of Roncesvalles with the absolute conviction that the groundwork of the story is true; that the surprise of the Frankish army did take place; that Roland commanded that army; and that he was killed whilst fighting nobly and gloriously. These main facts granted, I shall cull the details from the only authentic records which have reached posterity.

Roland had held high offices under the King, and, at the time of the expedition into Spain, was Prefect of the Marshes of Brittany. On the return of the army from Spain he had been, as I have stated, commissioned to lead across the Pyrenees the third or rear division of the Frankish army. Serving by his side were Eggiard, master of the King's household, Anselme, Count of the Palace, and many other nobles of distinction. The rear division had quitted Pampeluna the 22nd August, and, there being no fear of an attack upon it by the Saracens, was marching carelessly and in loose order. It had never occurred to the minds of Roland and his companions to anticipate opposition or hostility from the people whose country they were traversing, and who, on their march to Spain, had displayed no feelings which were not friendly. At length, the 15th August, the Frankish rear division reached the little valley of Roncesvalles, or Roncesvaux, a valley about twenty-three miles to the north-east of Pampeluna, on the road to the place now known as St. Jean Pied de Port. It constitutes a long defile between two lofty ranges extending in undulating plateaus on both sides of it, and covered with trees and forest. It was just the place for an ambuscade.* But the idea of an ambuscade had never occurred to Roland and his comrades. Their enemies were the Saracens, not the Gascons.

But the Gascons, or Basques, as they are indifferently called, had watched the return of the Frankish army with curious and longing eyes. There is no reason to suppose that the rank and file of the nation had any other object than plunder; but with their duke, Duke Lupus, it was far otherwise. The son of Waifer had many personal wrongs to avenge, and to avenge these he knew well that he might appeal to a race of wild mountaineers, to adventurers who owed allegiance neither to Saracen or Frank, who lived mainly by robbery, to enrich themselves by despoiling an army laden with booty. Lupus and his Gascons had allowed the first and second divisions of the army to pass unheeded. These, lightly equipped, were hurrying on to other wars; but the third division—the division commanded by Roland, was comparatively isolated. To intercept this division, then, he and they devoted all their care and all their energy. They occupied in great force the wooded ranges on both sides of the valley. But of their troops they made no display. On the contrary, the greatest care was displayed in arranging that

* The student of the campaigns of the Peninsular War will recollect that it was by this pass that the French army under King Joseph and Marshal Jourdan retreated after the battle of Vittoria, and that subsequently Soult attacked the English there.

not even the rustle of a leaf should give evidence of the existence of a single man. Covered by the forest screen, they awaited with ardent expectation the moment when the Frankish host should be well entangled in the valley below them.

At the narrowest part of the defile there stands at the present day a little chapel called Ibaneta. It is just at the point where

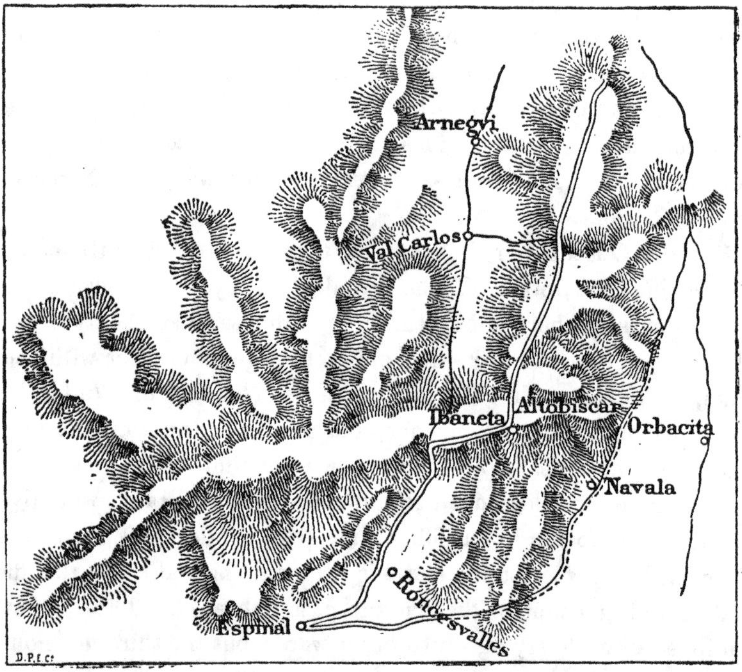

the road or path, which till then had a broad space of open ground on its right, enters the narrow gorge leading to Roncesvalles almost at a right angle.

Through this gorge the third division of the Frankish army entered, joyous, careless, dreaming least of all of an enemy! Suddenly, from the flanks, from the front, from the rear, the mountaineers rush upon them. According to the legend, Olivier, the comrade of Roland, who had ascended a height, sees them coming. He descends and thus makes his report:

Olivier dit: "J'ai vu tant de paiens,*
Que nul homme n'en vit jamais plus sur la terre.
Il y en a bien cent mille devant nous, avec leurs écus,
Leurs heaumes lacés, leurs blancs hauberts,
Leurs lances droites, leurs bruns épieux luisants.
Vous aurez bataille, bataille comme il n'y en eut jamais.
Seigneurs Français que Dieu vous donne sa force;
Et tenez ferme pour n'être point vaincus."
Et les Français: "Maudit qui s'enfuiera," disent-ils,
"Pas un ne vous fera défaut pour cette mort!"

He then urges Roland to sound his horn; his famous horn, Olifant, the only horn of the kind in the Frankish army; a horn, which, made of ivory, possessed the virtue of sending its sound to enormous distances.† He tells him that, on hearing its sound, Charlemagne will return with his army. But Roland is confident of success, and he wishes to reap all its glory. He replies:

"Je serais bien fou," répond Roland;
"Dans la douce France, j'en perdrais ma gloire.
Non, mais je frapperai grands coups de Durendal.‡
Le fer en sera sanglant jusqu'à l'or de la garde.
Nos Francaise y frapperont aussi, et avec quel élan!
Félons païens furent mal inspirés de venir aux défilés;
Je vous jure que, tous, ils sont jugés à mort."

But Olivier has seen the enemy. How can the twenty thousand men under Roland, surprised in a defile, resist five times their number about to rush upon them? He therefore renews his entreaties:

'Ami Roland, sonnez votre olifant;
Charles l'entendra et fera retourner la grande armée
Le Roi et ses barons viendront à nostre secours."

But Roland will not. Still obstinate and defiant, he answers:

"A Dieu ne plaise," répond Roland,
"Que mes parents jamais soient blâmés à cause de moi,
Ni que France la douce tombe jamais dans le déshonneur!
Non, mais je frapperai grands coups de Durendal,
Ma bonne épée, que je ceinte à mon côte.
Vous en verrez tout le fer ensanglanté.
Félons païens sont assemblés ici pour leur malheur;
Je vous jure qu'ils sont tous condamnés à mort."

* In the legend, the enemies were the Saracens.

† Il y a soixante mille *cors* dans l'armée de Charles, mais il n'y a qu'un *olifant*. Note to Gautier's *Le Chanson de Roland*, quoted in the text.

‡ The name of the sword of Roland.

But Olivier is not convinced ; he still insists :

"Ami, Roland, sonnez votre olifant ;
Le son en ira jusqu'à Charles, qui passe aux défilés,
Et les Francais, je vous le jure, retourneront sur leurs pas."

Then Roland :

"A Dieu de plaise," répond Roland,
"Qu'il soit jamais dit par ancun homme vivant
Que j'ai sonné mon cor à cause des païens !
Je ne ferai pas aux miens ce déshonneur.
Mais quand je serai dans la grande bataille,
J'y frapperai mille et sept cents coups ;
De Durendal vous verrez le fer tout sanglant.
Français sont bons ; ils frapperont en braves ;
Les Sarrasins ne peuvent échapper à la mort."

Clearly seeing destruction inevitable, unless the two first divisions can receive information of the pressing danger, Olivier thus retorts :

"Je ne vois pas où serait le déshonneur," dit Olivier,
"J'ai vu, j'ai vu les Sarrasins d'Espagne ;
Les vallées, les montaignes, en sont couvertes ;
Et les landes aussi, et tout les plaines.
Qu'elle est puissante, l'armée de la gent étrangère,
Et que petite est notre compagnie ! "

Roland, with the dogged resolution of the man who would risk destruction rather than share his glory—ever a striking feature in the character of French commanders—answers :

"Tant mieux," répond Roland, "mon ardeur s'en accrôit,
Ne plaise à Dieu, ni à ses très saints anges,
Que France, à cause de moi, perde de sa valeur !
Plutôt la mort que le déshonneur.
Plus nous frappons, plus l'Empereur nous aime ! "

Olivier makes one more effort, but it is in vain. The enemy is upon them, and the battle begins.

Such is the story. Out of it we can admit this one fact, that the Franks were surprised in the manner I have already described, and that they perished to a man. "The army," writes Éginhard, "was defiling in a long and thin line, shaped according to the nature of the country. The Gascons placed themselves in ambush on the crest of the mountains, which, by the number and thickness of its woods, favoured their designs. Thence, precipitating themselves on the baggage, and on the rear-guard protecting it, they thrust them down into the valley; killed, after an obstinate combat, every man even to the very last; pillaged the baggage, and, protected by the shades of night which were already deepening, scattered themselves with an extreme celerity in many places. In this engagement the Gascons were served by the lightness of their arms and the advantage of position. On the other hand, the weight of their arms and the difficulty of the ground rendered the Franks in all respects inferior to their enemies."

It is thus evident, according to the chronicler of the period, that the defeat was overwhelming. The same authority adds that the enemy dispersed so completely after the battle, that it was impossible for Charlemagne to discover and punish them. More recent inquiries prove, however, that the conspiracy was arranged between Inigo-Garcia, who commanded the Gascons of Navarre; Fruela, who ruled over those of the Asturias; and Lupus, who was lord over those of Gaul; that, on learning that they had destroyed a third of his army—the entire force under Roland—Charlemagne returned, defeated the Gascons, caused Lupus to be hanged, and divided the territories of the three allies among his own sons. It is certain that the victors of Roncesvalles did feel the weight of the vengeance of the great King; and it is a fact beyond a doubt that, including this first invasion of Spain, Charlemagne, during his reign, made seven campaigns against the Saracens, and although in more than one these were the invaders, yet they were defeated and pursued into their own territory. It is, indeed, incredible that a powerful monarch, such as Charlemagne was, should have allowed a disaster like that of Roncesvalles—a disaster so great that the story of the heroism of those who perished became the great epic poem of the age—to remain unavenged and unpunished! This view is borne out by the fact that since the year 602 the Gascons had been subject to the Dukes of Aquitaine! In 770, when, on the death of Pipin, Charlemagne succeeded to a moiety of his father's dominions, Hunold emerged from his monastery to head a revolt, and, beaten, had been forced to flee for refuge to Lupus, Duke of the Gascons. Charlemagne had informed that chieftain that he must deliver his guest or be prepared to feel the weight of his arms. Lupus had then delivered up Hunold. In connection with this circumstance we have the fact that immediately after the defeat of Roncesvalles, Aquitaine, in sympathy with Lupus, prepared once more to rise in revolt. The province did not rise, because of the prompt measures immediately taken by Charlemagne. He defeated the Gascons, disposed of their lands* in the manner related, and then, entering Aquitaine, declared it to be a kingdom of which his third son by Hildegarde, of whose birth he had but just received information, was to be the King! Three years later the young Prince, Louis, accompanied by some of the wisest councillors and best administrators at the disposal of his father, took formal possession of the kingdom.

I do not propose to follow the great Sovereign into all his wars. The summary of his conquests given by Éginhard will suffice to prove that the surprise of Roncesvalles was an incident which in no way affected his glory or his prosperity. "His father Pipin," summarises that writer, "had bequeathed to him the kingdom of the Franks, already considerable and powerful. Charlemagne almost doubled it. Before his time the territory of the Frankish nation comprised only that part of Gaul which stretched from the Rhine to the Loire, and from the ocean to the Balearic Sea, and that part of Germany which, comprised between Saxony and the Danube, the Rhine and the Saale, was inhabited by the East Franks. In addi-

* Gascony, or Vasconia, was, literally, the land of the Basques. These people, driven during the sixth century by the West Goths from their homes on the southern slopes of the Pyrenees, took refuge in the district called by the Romans *Novempopulania*, between the Atlantic ocean, the Garonne, and the Western Pyrenees, now known as the departments of Landes, the Upper Pyrenees and Gers, and the southern portions of the Upper Garonne, Tarn Garonne, and Lot-Garonne. In 602 the Gascons submitted, after a hard resistance, to the Franks. They were placed under the control of the Dukes of Aquitaine; who, however, shook themselves free from the Frankish rule, until King Pipin, and afterwards Charlemagne, subdued them. The second of these, Charlemagne, gave them dukes dependent on the Carlovingian kingdom of Aquitaine; but these, using the love of freedom innate in the blood of the Basques, made repeated but vain efforts to shake off the Frankish yoke.—Vide *Brockhaus' Conversations-Lexikon*, Art. *Gascogne.*

tion, the Germans and Bavarians had submitted to the Franks. By the wars which we have recounted Charles conquered Aquitaine, Vasconia, and the chain of the Pyrenees as far as the Ebro, a river which takes its origin in Navarre, traverses the most fertile fields of Spain, and empties itself into the Balearic Sea near Tortosa. He conquered, besides, all Italy, from Aosta to Lower Calabria—the frontiers of the Greeks and the Beneventines. He conquered, also, a considerable part of Germany, Saxony, that Saxony whose territory, it is said, equalled that of the Franks in length and doubled it in breadth; then the two Pannonias, Dacia, beyond the Danube; Istria, Libumia, and Dalmatia; and, if he did not annex the maritime towns it was because, by reasons of friendship, he abandoned them to the Emperor of Constantinople; finally, all those barbarous nations who dwell between the Rhine and the Vistula, the Ocean and the Danube, and who, scarcely allied by language, are entire strangers to each other by their character and habits. The principal of these are the Valatabes, the Sorbes, the Abodrites, the Bohemians. These last he reduced by arms, the others offered their submission."

But Charlemagne has higher titles to the respect of posterity than those which are the boast of the mere conqueror. He introduced into war, which prior to his advent to power had been conducted on a brutal system, influences of a civilising character. The softening power of Christianity, and with Christianity a purifying civilisation, accompanied his arms. He was the creator of a new order of things, the founder of a new empire, in which intellect took its place as the manipulator of brute force. He gave a signal instance of this in the manner which he adopted to assume the Imperial Crown. He might, like the Roman emperors before him, have accepted that Crown from the hands of his soldiers on the field of victory. But he foresaw that such action on his part would only consecrate the claims of force. He recognised that the Papacy was then actually, what it claimed to be afterwards, the supreme arbiter in secular as well as in spiritual matters. From the hands of its representative, then, he preferred to take the Crown which his position and his character entitled him to wear. For that purpose he proceeded to Rome, and there he was crowned by Pope Leo III. on Christmas Day of the year 800.

Of the reforms which this illustrious sovereign introduced into the administration of justice, in the ecclesiastical customs, in the civil law, how he pushed the spread of letters and of education, this is not the place to write. If I have dwelt at some length on his character and on the events of his reign, it is because an account of the one great misfortune suffered by his arms, the surprise of Roncesvalles, would otherwise have been incomplete. That surprise was but an untoward incident in a singularly successful career. Far from producing any permanent evil influence on the fortunes of the sovereign whose army was destroyed, the blow rebounded on those who had delivered it. The surprise of Roncesvalles, in fact, was the direct cause of the more complete conquest of Aquitaine and of the lands of the Basques by the great ruler whose arms had been insulted. The incident, however, is not the less immortal. The story of its hero Roland roused the hearts of the Normans before Hastings; that it still excites the sympathy of Europe is sufficiently proved by the fact that the edition of the *Song of Roland* to which I have referred repeatedly in the course of this chapter bears upon its title-page the date 1881!

CHAPTER IV.

KERKOPORTA.

When Chengíz Khán died in the year 1227, he died lord of one of the greatest empires the world had ever seen. All the cities of Central Asia, of China, and of Persia, had submitted to his will. The ancient kings and rulers had been, on all sides, slain or driven into exile. During the latter years of his life there had not been a man in the entire Asiatic world who would have ventured to raise his finger against him.

To escape the rule of such a master, Sulaimán, Khán of Persian Khorásán, had endeavoured, about the year 1224, to flee, with the Turkish clan which remained true to him, into Armenia. These Turks were the brethren of the forefathers of the men who now, under the name of Turkomans, or Turkmans, occupy Turkistán, recently conquered by Russia. Sulaimán and his Turks so far succeeded in their venture that they reached Armenia in safety, and were allowed to remain there unmolested. But the love of home never died out in their bosoms, and when, three years after the death of Chengíz Khán, the news reached Sulaimán that his mortal enemy was no more, and that his policy had died with him, he prepared to return to his native land. Crossing the Euphrates on

* I have consulted, amongst other works, for this chapter, Leonicus Chalkondylas's (*Corpus scriptorum historiæ Byzantinæ*); Ducas's *Historia Byzantina*, edition E. Bekker; Von Hammer's *Geschichte des Osmanischen Reiches*; Engel's *Geschichte des Ungrischen Reichs*; Mailath's *Geschichte der Magyaren*.

that return journey, Sulaimán was drowned, and his followers, terrified at the catastrophe, renounced the further prosecution of the idea. Some of them resumed the habitations they had occupied during the seven years of their exile, but a large portion of them, following the fortunes of Ertogrul, son of Sulaimán, migrated further westward, and pitched their permanent camp only when they had reached the lands between Angora and Broussa, in the dominion ruled over by Kaikobád (Alla-u-din), the Seljukian Sultan of Iconium. For permission to occupy and hold as vassals these lands the wanderers engaged to render feudal service to the Sultan. Matters prospered with them. When Ertogrul died, at the age of ninety (1288), though still a humble Khán, unempowered to coin money, or to pronounce the Friday prayer,* his tenure was thoroughly recognised, and he expired dreaming that his successor would become the founder and independent ruler of a great nation. The dream was realised. His son, Osmán, or Othmán, surnamed the Ghâzi, the Conqueror, became the founder of the dynasty which reigns still at Constantinople, the first chief of the warriors who, in memory of, and through him, are still known as Osmánli and Othmáns, or Ottomans!

It happened in this manner. Kaikobád, Sultan of Iconium, had died in the year 1235, just the year after Ertogrul had become his vassal. His son and successor, Kai Khosrou II., reigned only nine years, and was followed by his son, Kai Kous II. Under the reign of this prince the Sultanate of Iconium was exposed to many dangers, and Kai-Kous himself was forced to share his dominions with his two brothers, Khilij Arsan and Alla-u-din Kai-Kobád, Kai-Kous retaining Iconium. Alla-u-din died whilst on a mission to Tartary, and the two surviving brothers, each jealous of the other, took to arms to contest the possession of the entire heritage. Victory decided in favour of Kai-Kous, who at once consigned Khilij Arsan to prison. But Houlagon, the first Mongol King of Persia, defeated Kai-Kous, delivered Khilij Arsan, and, after an interval, divided the Sultanate between the two brothers. Then followed a

* That is, possessing neither supreme civil nor military authority.

series of intrigues, of depositions, of restorations, too long to detail. It must suffice to say that in the year 1278 a band of Tartar invaders carried Kai-Kous and his son prisoners to the Court of their Khân. Then the Mongols strangled Khilij Arsan, and recognised his son, a boy four years old, as Sultan of Iconium, with the title of Kai-Khosrou III. This boy, however, was soon displaced in favour of Masoud, the then liberated son of Kai-Kous.

Masoud was the last Seljukian Sultan of Iconium. He died in the year 1294, the sixth year after Othmân had succeeded his father Ertogrul in his peaceful inheritance.

In all that related to Othmân and his Turks those five years, though uneventful, had been passed in preparation for the result which Othmân himself saw to be inevitable. When Masoud died, he was one of the most considerable of the chiefs who had paid homage to that ruler. The breaking-up of the Seljukian Sultanate found him, then, strong enough to claim a considerable share of the dominions of which it had been constituted. Othmân claimed, and was allotted, the province of Bithynia. After four years spent in settling the portion of the province which had submitted, he prepared to march against the larger part, then occupied by the Greeks. With this view, he summoned his Turks to arms, forced the passes of Mount Olympus (July 1299), invaded the territory of which Nicea was the capital, and conquered it, the city of Nicea excepted (1300 to 1304). Spending three years more in firmly establishing himself there, Othmân then subdued (1307) the province of Marmara, and pursued his conquering career until he had mastered the whole of Bithynia, Nicea and Broussa excepted. By this time he had become a Sovereign Prince, "having the right to coin money and pronounce the Friday prayers"; but he did not use either privilege, and never assumed the title of Sultan.

Othmân had made his capital at Kara-Hissar — at the foot of a hill crowned by a ruin, supposed to be the site of the ancient Cybistra—but he had long coveted Broussa (Prusa ad Olympum), the real capital of his territories, a city beautifully situated at the foot of Mount Olympus, and had built castles in its vicinity in order to hem it in and force its surrender. Broussa, however, resisted his arms so long that, unable, from severe and continued attacks of gout, to keep the field, Othmán retired to Kara-Hissar, leaving the prosecution of the siege to his son Orkhán. A few days before his death (1326), he enjoyed the satisfaction of hearing from Orkhán that Broussa had fallen. The city had surrendered without bloodshed, and on favourable terms. Thither Othmán caused himself to be immediately transported, and there he died, in the seventieth year of his life, and the twenty-seventh of his reign, an independent prince. He left to his son a title greater, in the actual power it bestowed, than that of Sultan, the title of "Ruler of the kingdom of the Osmánli."

Great as was Osmán, his son and successor, Orkhán, was still greater. It is related of him that, on his father's death, he proposed to divide his dominions with his younger brother, Alla-u-din, whose great abilities he had long recognised. Alla-u-din refused, alleging, as a reason, his father's will that there should be but one ruler over the territory which had become Turkish, and that he had designated Orkhán for the post. He refused even to take a share of the private property of the family, and demanded only a village in which he could live in peace. Then Orkhán said: "As you will have neither the horses, nor the cattle, nor the sheep, I beg you will at least become the shepherd of my people —that is, that you will take the office of Vizier." Alla-u-din consented. It was a fortunate circumstance for his countrymen, for, amongst the able men who have filled that office for the ruler of the Turks, not one displayed greater ability than the second son of Othmán!

To the internal administration of this illustrious man the Osmánlis were indebted for many of the institutions which made them great. I can refer here but to one of these. Having designed a coinage, bearing the stamp of his brother, he set to work to build up, on foundations which should endure, an army which should prove irresistible. His colleague in this work was the commander of the forces, Kasa Khalil Schandar Ali. Basing their idea alike on the words of the Prophet, "Every new-born child brings with it into the world the germ of the Muhammadan faith," and on the right of the conqueror to dispose of the property and person of the vanquished, these organisers insisted upon the forcible conversion to Muhammadanism, and the enlistment in the ranks of the army, of the fifth child of all the Christian children taken captive. In this manner, every year witnessed the increase of a corps of new troops, trained and disciplined almost from their cradle, who were to become the terror of the enemies of Islam, the mainstay of her power, until, after the lapse of many centuries, a new policy was to cause their dissolution!

When this corps was yet in its infancy, Orkhân, accompanied by some of its members, proceeded to his religious adviser, the Derwésh Hadji Bégtasch, and asked for them a blessing, a standard, and a name. The Shekh replied: "Let their name be 'the new warriors' (Jenitscheri); let their head-covering be white, their arms victorious, their sabres cutting, their spears piercing, and may they ever return with victory and prosperity." The number of this chosen band was originally fixed at twelve thousand, but it increased in subsequent years. The remaining infantry of the Osmánli army consisted of men who held their lands on the terms of feudal service, and who were liable only in case of war; and of irregular masses who generally acted as skirmishers. The cavalry was formed likewise of three bodies: the irregular, those rendering feudal service, and those regularly organised and paid. The last came to be well known in Europe under the designation of "Sipáhis."

With the germs of such an army the two brothers began their conquest. Their natural enemy was the Greek Emperor, Andronicus III.* They defeated the armies of that prince and took Nicea (1330) and Nicomedia (Ismid) (1338). The conquerers then, enticed by the divisions of the two brothers who ruled it, moved

* 'Ανδρόνικος Παλαιόλογος, grandson of the Emperor of the same name, with whom he had been associated as joint-Emperor in 1325, and whom he succeeded in 1332.

against Karasi, one of the ten independent provinces which had been formed on the dissolution of the Seljukian empire in Asia Minor. They subdued this province. Then they held their hands, and devoted the twenty years which followed to the consolidation of their Asiatic dominions. Not, indeed, that during this period they did not make many attempts to effect an entrance into Europe. Many times baffled by the Byzantines, they at length succeeded. In 1357 Sulaimán, the eldest-born son of Orkhán, took possession of the castle of Zimpe,* not far from Gallipoli. The Greek Emperor, John IV.,† who had assumed that title in 1341, offered Sulaimán ten thousand ducats if he would restore the place. Sulaimán accepted the offer, and the money had already reached Gallipoli, when a terrible earthquake occurred which inflicted incredible damage on the coast of Thrace. The walls of Gallipoli, then as now a most important town at the entrance to the sea of Marmara, were thrown down; the affrighted inhabitants fled in dismay, and the Osmánli, who saw their opportunity, entered unopposed. Vainly did John Cantacuzène demand the restoration of the town. Under various pretexts Orkhán refused to comply. The main argument upon which he rested was this: that the place had come into his hands, not by violence or the action of man, but by an operation proceeding from God. With this argument the Greek Emperor, too weak to appeal to arms, was forced to be content. Gallipoli remained in the hands of the Turks.

The year after its capture (1358) its governor, Sulaimán, was killed by a fall from his horse whilst hawking. His father, Orkhán, followed him not long after, 1359, to the tomb, having lived seventy-five and reigned thirty-five years. He had done much for his clan. Though he could neither read nor write, he had been a wise and capable ruler, a lover of justice, and a warrior of the

* Now called Dschemenlik or Tschini.

† Andronicus III. died on the 15th June 1341, and was succeeded by his son, John Paleologus, a minor; but John Paleologus was expelled the same year by John Cantacuzène, who was crowned on the 21st October

first rank. His ways were very simple; he was in the habit, it is recorded, of, with his own hand, serving the soup and lighting the lamps in the charity kitchens of Nicea. His brother, Alla-u-din, had died before him; the sovereignty, therefore, devolved upon his younger son, Murád or Amurath.*

Amurath was thirty-four years old when he ascended the throne. His first military act was to take possession of Angora—the ancient Ancyra—a place of great renown in the time of Augustus, and in the three centuries which followed his death.† It was not for its earlier renown, of which probably he had never heard, that Amurath marched against Angora. In his eyes its importance lay in its position. It was the point where all the commercial roads, the roads from Syria and Armenia to the coasts of Thrace and Cilicia, met and crossed. It was on that side alike the emporium of the industry of Syria and Armenia and the key of his dominions. He took it without much difficulty.

Having secured his territories on the Asiatic side, Amurath turned his attention to Europe. Here, too, he had a long career of conquest. Town after town fell unresistingly into his hands. He penetrated into Thrace, took Adrianople, after gaining a battle before its walls (1361), and made that city the capital of his dominions (1362). But his rapid conquests roused against him enemies alike in the East and in the West. Whilst in Asia Minor the Prince of Kermian—the territory nearest to the Osmánli, of the ten territories into which the Seljukian empire had been divided—set on foot intrigues against him; in Europe the Servians, the Bosnians, the Albanians, the Wallachians, and the Hungarians combined (1363) to oppose the new conqueror. But Amurath was equal to the occasion. Learning that Louis I., King of Hungary, known in history as Louis the Great, was marching with twenty thousand men on Adrianople, the

* Correctly, the name is Murád; but long European custom has sanctioned the corrupt form of spelling.

† *Vide* pp. 66–7. Councils were likewise held at Ancyra in the years 314, 358, 375 A.D.

Turkish ruler, who was preparing to march at the moment against the Prince of Kermian, sent his Vizier, Hadji Ilbeki, with ten thousand men to deal with him. The Hadji made careful inquiry, and ascertained that the enemy was encamped on the Maritza, two days' journey from Adrianople; that, confident in his numbers, he kept no watch or guard at night. He made, then, his preparations, fell upon the Christians without warning, and surprised them so completely that, to use the words of the Turkish historian, Saiad-ud-din, "like wild animals, seized by sudden terror in their lair, they took to the most rapid flight, dashed heedlessly into the Maritza, as quickly as the wind before the flame, and sank beneath its waters." The victory was complete. King Louis saved himself with difficulty. He was wont to attribute his escape to the fact that he carried on his person a portrait of the Virgin Mary.

The victory—gained in the first battle which brought in hostile contact the Osmânli and the Magyar—was decisive. It was followed by five years of conquest in Roumania, Servia, Wallachia, and Asia Minor. Nor did Amurath neglect other means of adding to his dominions. The marriage of his son, Bajazid, surnamed Ilderim (The Flash of Lightning), with the daughter of the ruling chief of Kermian, secured to him the greater part of the possessions of the bride's father. He purchased also from the Prince of Hamid the territories which he possessed as the inheritor of one of the divisions which had once formed part of the Seljukian empire.

Then he had peace, six years of profound and uninterrupted peace. He employed this period in effecting various internal reforms in the administration and in the army. He commenced with the latter. He caused the sipâhis to be formed into squadrons, each squadron being placed under the command of a squadron-leader. Under the commandant-in-chief of the corps of sipâhis were placed four general officers. He perfected the organisation of a special corps, called the Woïnaks, whose duties correspond to a great extent with that of the horse-keepers in India. Upon them, that is to say, devolved the care of the horses and the stables. This corps was composed solely of Christians, who, as a compensation for the drudgery of their duties, were exempted from tribute. To reward the services of the sipâhis, Amurath created feudal tenures in the greater part of the provinces of his empire. These estates were cultivated by peasants, whether Muhammadan or Christian, called *raïas,* to whom was assigned the ownership of the soil on the conditions (1) that they should be subject to the seigneurial jurisdiction of the sipâh ; (2) that the profits of the cultivation should accrue to the latter. The son of the raïa was to inherit the property of his father; but in the case of failure of direct issue, a collateral could only succeed with the consent of the sipâh, and on the payment of certain dues: were there no living relative, the property passed, under similar conditions, to one of the neighbours of the deceased—the sipâh, however, retaining the right to dispose of it otherwise, should he think fit so to do. The sipâhis were bound to reside in their fiefs in time of peace, and to furnish, in time of war, one cavalry soldier for every seventy shillings of annual income (three thousand *aspres*).* The fiefs were hereditary in the male line, but in default of male issue they reverted to the domain of the Crown. The pasha of the province was bound, then, to confer the lapsed fief upon another sipâh, or upon an old soldier.

Other decrees which Amurath issued during his reign tended to the settlement of matters, till then disputed, upon a fixed and intelligible basis. It was natural, he being a conqueror and the leader of a young and progressive people making their way to supreme power upon the ruins of an empire worn-out and effete, that these decrees should affect mainly the spoils of victory. He prescribed, for instance, the principle upon which booty taken in war should be apportioned. One-fifth was assigned to the priests, one-fifth to the poor, the remainder to the Crown to be disposed of as the Sultan should order. In consequence of the remonstrances made to him by the chief priest, or mufti, upon the little public reverence paid by him to religion, Amurath began the erection of the famous

* An *aspre* is a Turkish coin, the equivalent of rather less than three centimes. Three thousand aspres would amount, therefore, to something under ninety francs.

mosque of Adrianople. He proved that the East is, in very truth, "the land of the bath," by the erection of numerous public baths in that city at a time when such incitements to cleanliness were little thought of in Western Europe.

At this period of his reign Amurath had reason to congratulate himself upon his increasing influence. In every quarter his arms had been victorious. He crushed, by the power of his voice,* a rebellion hatched by one of his sons, in alliance with Andronicus, son of the Greek Emperor, against himself and that sovereign, and inflicted condign punishment upon the authors of the outbreak. Another son of John Paleologos, the second, Manuel by name, Governor of Thessalonica, undeterred by this example, made a sudden attack upon the fortified town of Seres.† Amurath baffled his design, and forced the young conspirator to flee for refuge to Constantinople. John Paleologos, dreading the wrath of the Turk, refused to receive him. Manuel then fled to Lesbos. Refused shelter there, he threw himself, in despair, on the generosity of Amurath. The result justified his action. Amurath received him kindly, and sent him back to Constantinople, with a request, equivalent to a command, that a kind reception might be awardéd to him.

But the days of peace passed by only too soon. In 1386, the ruler of Karamania or Iconium, the most powerful chieftain, next to the lord of the Osmánli, who had carved out a kingdom from the ruins of the Seljukian empire—jealous of the increasing power of the Osmánli, raised an army to attack him. With a promptitude which he had ever recognised to be, on such occasions, the secret of success, Amurath marched in person to quell this rising. The two armies met near Iconium, then a great city, well known to the readers of the Bible as the city in which Paul and Barnabas preached, now known as Koniyeh, or Konieh, and familiar to modern politicians

* He rode, alone, during the night, to the camp of the insurgents, and with a loud voice promised them pardon should they at once return to their duty The rebellious sons were instantly deserted by almost all their followers.

† Forty-seven miles north-east of Salonica.

as the spot near which the Pasha of Egypt defeated the Turkish army in 1832. Amurath ranged his army in a manner which was to serve as a precedent and as a pattern for his successors in command of the Osmánli for ages to follow. His Asiatic infantry formed the front line, a little to the right, the European the second, in echelon to the left; the Janissaries were in the centre, behind these; and behind the Janissaries again were the regular cavalry. The battle joined, and the Prince of Karamania was defeated. Amurath behaved with great wisdom and generosity, forbade all plundering on the part of his troops, and restored to the now penitent prince a large portion of the territory he had forfeited. Some of his councillors, flushed with the heat of conquest, advised the Sultan to take the opportunity of annexing the small territory of Teke. Amurath, however, refused, saying: "The Prince of Teke possesses only two cities, Istanos and Attalia:* it would be a disgrace to me to make war upon so small a prince: the lion cares not to chase flies." When this speech was reported to the Prince of Teke, he at once surrendered to Amurath all his territories except the two cities the Sultan had specially mentioned.

Amurath then disbanded his army. But, as the troops composing it, 20,000 in number, were returning carelessly to their homes, the Servians, in alliance with the Bulgarians, instigated by Sismán, chief of the last-named and father-in-law of Amurath, fell suddenly upon them and destroyed three-fourths of them (1387).

The chief—called in those days the Kral—of the Servians, was Lazarus, a man of considerable influence. Well aware of the storm he had provoked, this chief hastened to strengthen himself by alliances against his powerful enemy. He induced not only the Bosnians, the Albanians, and the Poles, but, likewise, the Hungarians to make common cause with him. In the month of

* Now called Adalia, and inhabited by the most fanatical of all the followers of the Prophet. The city, which has always flourished, was founded by Attalus II., King of Pergamos, and, in the Middle Ages, was called Attalea, Sattalea, and Sattalje.

June 1389, an army composed of these peoples, the like of which, in point of numbers, the Turks had never before encountered, stood ready for action on the broad plain of Kosowa.

Amongst these people the Bulgarians were not. Amurath, furious at the behaviour of his father-in-law, had, as soon as he could raise new forces, despatched an army under his Vizier, Ali Pasha, to lay siege to Nicopolis, in which city Sismán had taken refuge. Seeing the hopelessness of his position Sismán demanded pardon. To obtain it he was obliged to surrender Silistria. These terms had been barely settled when aggressions on both sides caused a renewal of the war. This time Amurath resolved there should be no mistake. He incorporated the whole of Bulgaria in his dominions, and gave Sismán a pension.

And now he had to encounter five nations on the plain of Kosowa. That plain, called by the Germans the *Amselfeld*, the blackbird-plain, has a width of nearly fourteen miles, a length of upwards of forty. It is traversed by the little river Sitnitza, and bears upon its surface many villages. Round it rise pleasant and well-wooded hills, which become to the east and to the south gradually higher, till in the former direction they culminate in the Ljubotini, in the latter in the Schargebirge (Tschardagh). As Amurath marched on to that plain and beheld before him the countless numbers of the Christian hosts, exceeding his own by at least three to one, he might have doubted the expediency of committing the future of the rising nation whose fortunes he guided to the issue of one single battle. At a council of war he at once summoned of his principal officers, many expressed great fears regarding the result. Some counselled to form a line of camels in front of the army that these might perplex the enemy unaccustomed to look upon them. Others opposed this advice on the ground that the camels, more terrified than the enemy, might turn and break through the Turkish ranks behind them. Bajazid Ilderim, the favourite son and destined successor of the Sultan, strongly opposed the proposition on the ground that such a device argued mistrust in the God of battles, and that it was proper, confiding in Him, to oppose sword to sword.

To end the controversy the Grand Vizier produced a Koráu, and, after the usual invocation, sought counsel from its leaves. At the first attempt his eye lighted on the following verse : "Fight with unbelievers and hypocrites." The second was equally decided; the verse ran: "Often, certainly, will a great multitude be beaten by a lesser number."

The enthusiasm aroused by the result of this appeal to the sacred book was prodigious. Amurath alone appeared still to doubt. He realised more than anyone in his camp the importance of the issue at stake; that for the Osmánli the battle meant supremacy over Eastern Europe or destruction. Without pronouncing a decision, he broke up the council, and, retiring to his tent, passed the night in imploring God for assistance, and in beseeching that if a martyr were required to secure victory for the true belief, his life might be the sacrifice.

In the camp of the confederate Christians a different tone prevailed. There all was joy and confidence. To such an extent did the latter feeling prevail, that when it was proposed to fall upon the Ottoman camp during the night, Ivan Castriota,* the leader of the Albanians, caused this proposition to be rejected on the ground that the darkness of the night would prevent the complete destruction of the enemy!

A strong wind which, during the night, had blown the dust of the plain in the direction of the Turkish camp, was succeeded, as the day broke, by a soft rain. Under its influence the battle joined. Long and fiercely was it contested. At length the left wing of the Ottomans seemed to give way. There commanded Bajazid Ilderim—the Flash of Lightning. Acting with the celerity which gave him his surname, Bajazid flew from rank to rank, rallied the soft-hearted, and led them again to the charge. This

* Mailath (vol. ii. p 131) says "Georg Castriotta." But this is impossible. George Castriota, better known as Skander Bég or Iskander, and who is the hero of Lord Beaconsfield's story, *The Rise of Iskander*, youngest son of Ivan Castriota, was only born in 1404, fifteen years after the battle of Kosowa. I shall have occasion further on to refer to him and to the remarkable accuracy of the historical facts related by Lord Beaconsfield in his charming novelette.

time successfully. The Christian right wing gave way in its turn. Its leader, the author of the war, Lazarus, Kral of Servia, was taken prisoner. A forward movement along the line completed the well-begun work, and, after five hours of fierce fighting, Amurath stood the victor on the hard-fought plain!

The battle over, Amurath was crossing the plain covered with corpses, in pursuit of the enemy, when suddenly there sprang from the ground a man armed to the teeth, who, rushing at him, plunged his sword into his breast and then attempted to escape. Three times did the assassin force his way through his foes, but he was at last surrounded and taken. He proved to be a Servian noble, Miloch Kibilowitch by name, a son-in-law of the Kral, Lazarus. Falsely accused of having an understanding with the Turks, he had resolved to give this striking proof of his fidelity to the national cause. Amurath lived long enough to see his murderer slain, and his enemy, Lazarus, beheaded: he died, then, with the conviction that God had heard his prayer, and had accepted him as the offering for the success of the Ottoman arms!

Bajazid Ilderim was proclaimed Sultan on the field of victory. He was forty-two years of age, strong of body, ambitious, resolute, merciless. The death, already noticed, at an earlier period, of his elder brother, Sulaimân, had left him the true heir to the throne.

But, to make assurance doubly sure, his first act was to remove from his path, by the convenient form of strangling, his younger brother and possible competitor, Jakub. "The insurrection of many is worse than the execution of one," he exclaimed, as he gave the order, destined to form a precedent in his family. Bajazid then made peace with Servia on hard conditions for that principality. The new Kral, Stephen, was forced to agree to pay him a yearly tribute, to supply him with troops in time of war, and to give him his sister in marriage. The lady constituted his fourth wife, Bajazid having already allied himself with the Kermian family, with a Byzantine princess, and with a European princess of unknown family, whom, destined for the Greek Emperor, his admiral had captured at the entrance of the Dardanelles!

As soon as he felt himself firm in his father's seat Bajazid began that career of conquest which astonished and alarmed the Christian world. First he sent his army against Allah-Shahr, the city of Allah, the Philadelphia of Holy Writ, the only city in Asia Minor which still belonged to the Greeks. It is a striking proof of the state of decadence and humiliation to which the Eastern Empire had been reduced, that when the commandant of Allah-Shahr refused to surrender to a Turkish host, Bajazid compelled the Emperor John VI. (Paleologus) and his son Manuel to undertake the siege, and to storm the place for the Turks with their own troops.*

The capture of Allah-Shahr was the prelude to other conquests. The reader has been told how, on the ruins of the old Seljuk Empire several petty states, ruled over by independent sovereigns, had risen. Three of these, Aidin—of which Allah-Shahr formed a geographical part—Sarakhan, and Mentesche, formed the coast lands of Anatolia. These were quickly subdued (1391). The Prince of Karamania was the next victim to the ever-victorious Osmânli. As the price of peace he was forced to surrender the larger portion of the territories still remaining to him. Meanwhile John VI. had returned to Constantinople, leaving his son Manuel in the Turkish city of Broussa. Arrived at the capital he began, in dread of the designs of Bajazid, to restore the fortifications. To obtain easily materials for this purpose he caused three of the finest churches in the city to be pulled down. As soon as the news of this action reached Bajazid he despatched a peremptory order to John to discontinue the works, threatening, in case of refusal, that he would cause his son Manuel to be blinded. John obeyed, but the mortification killed him: he died a few days later. Manuel managed to escape betimes from Broussa and to seize the reins of government at Constantinople,

* Bajazid had espoused the cause of Andronicus, the rebellious son of John, and whom John, on the command of Amurath, had partially blinded. Having possessed himself of the persons of John and Manuel, Bajazid had entered with them into an agreement whereby, whilst Constantinople and its environs should be secured to them, Andronicus should hold, as fiefs of the crown, the towns and districts of Selimnia, Heraclea, Rodosto, Damias, and Panidas; of the island of Marmara, and of the town of Thessalonica. Manuel was to remain with him as hostage for the due execution of this convention.

now almost the last refuge in Europe of the Byzantine Empire. He was speedily followed by the army of Bajazid, who, furious at the action of Manuel, vowed that he would not retire till he had occupied the city and killed the usurper.

In his extremity Manuel made a despairing appeal to the Christian powers, especially to the King of Hungary. The prince who then ruled over the border-land of Christian Europe was Sigismund, afterwards Emperor of Germany. Son of the Emperor Charles IV. and of Anne of Silesia, he had married Mary of Hungary, and in virtue of that marriage had become sovereign of the country. Touched by the appeal made to him by Manuel, Sigismund now proclaimed a new crusade, enlisted under his banner generous and enthusiastic volunteers from France, from Germany, and from Poland, and, at the head of an army of 100,000 men, entered Bulgaria (1396) and laid siege to Nicopolis. Bajazid, who, in the interval, whilst pressing the siege of Constantinople, had possessed himself of the shores of the Black Sea, subdued Wallachia, and traversed Greece to the southern range of the Peloponesas, flew to the rescue of the beleaguered city. On the 28th September, 1396, he attacked and totally defeated the Christian host. The overthrow was so decisive that instead of pursuing the all but annihilated enemy, Bajazid returned to Greece to complete the conquest of that country and of the Morea.

From his triumphs there Bajazid was called by the invasion of a new and more dangerous foe than any he had till then encountered. Taimúr Lang, sometimes strangely called Tamerlane, whose conquests throughout Asia had proclaimed him to be a warrior of the first rank, was earnestly solicited by Manuel, and by the princes of Asia Minor, whom Bajazid had dispossessed, to save them from the destroyer. Taimúr listened favourably to their entreaties, and, writing a menacing letter to the lord of the Osmânli,* prepared to execute his threats.

* "Know," he wrote, "that my armies cover the earth from one sea to the other; that princes are my servants, and, formed in ranks before my tent, humbly await my orders; that the fate of the world is in my hands, and that Fortune is my

Bajazid had overrun Greece, had taken Athens, and whilst his army was still pressing Constantinople, was enjoying the sweets of repose at Broussa when he received the letter of Taimúr. He answered it in a strain more likely to invite than to repel attack, and, not content with that, entered Armenia, and, taking by storm the cities of Erzeroum and Sivas Sebastea, placed in them strong garrisons. Believing that this aggressive action would deter Taimúr he returned to Adrianople. But the ejected chief of Erzeroum had fled to the camp of Taimúr, and had incited him to vengeance. The conqueror marched into Armenia, took Sivas by storm, recovered Erzeroum, and was about to seek out Bajazid, when the hostile action of the Mameluks forced him to divert his course to Syria. Arrived there, he twice defeated the army of the Sultan of Egypt—before Aleppo and before Damascus—sacked and destroyed Bagdad; then, finding his hands free, sent a message to Bajazid, in which he left it to that prince to decide whether he would have peace or war. For a moment it appeared as though a collision between the two great Mussulman rulers would be avoided, for Bajazid, alarmed at the enormous preponderance of his enemy, demanded peace. But, a little later, Taimúr, having discovered that Bajazid was encouraging his enemies, marched against him. The two armies met in the plains of Angora (28th July 1402). The disproportion between their numbers was great, the host of Taimúr counting nearly a million, whilst Bajazid disposed of only a hundred and twenty thousand. Of these, moreover, eighteen thousand were Tartars, who had been tampered with by Taimúr, and ten thousand were Servians. The remainder were discontented, in consequence of the long arrears of pay due to them.

inseparable companion. Who art thou to brave me? Poor Turkoman ant! do you dare to attack the elephant? If in the forests of Anatolia thou hast gained some insignificant victories; if timid Europeans have taken flight before thee, thou oughtest therefore to thank Muhammad, and not thine own valour. . . . Listen, now, to the counsels of reason! Confine thyself to the narrow limits of thy patrimony: step not beyond, or thou art lost!" The reply of Bajazid was couched in terms not less haughty and vain-glorious. "For long," he concluded, "have I burned with desire to measure swords with thee." He signed his letter in a manner which of itself conveyed an insult to the great conqueror of Asia.

The battle was ranged in the following manner. On the side of Taimúr, the right wing was commanded by his eldest son then living, Prince Miran Sháh, having under him his son, Prince Abubekr; the left by Taimúr's grandsons Sháhrokh and Khálil. The centre—the post of honour and danger—was led by Taimúr's grandson, son of his deceased eldest son, Muhammad Sultan, having equally on his right and left forty regiments, and immediately in front of him the sacred standard. Taimúr himself commanded the reserve.

The Osmánli were thus ordered. The front or right wing was commanded by Bajazid's eldest son, Sulaimán, governor of Aidin, Sarakhan, and Karasi; the left, or second line, was occupied by the Servian troops. Bajazid, with his three sons, Isa, Musa, and Mustapha, led the centre. The reserves were led by the most capable son of Bajazid, Muhammad.

At 6 o'clock in the morning, Taimúr, prompted by a dervésh, dismounted, and said his prayers: on their conclusion he gave the order to attack. Mirza Abubekr, who commanded the advanced division of the right wing, fell with great fury on the Servians, who, it will be recollected, formed the left of the Turkish army. The Servians, though combating for a master alien in religion and race, fought, however, like lions, and inflicted very severe loss on the assailants. Then it was that Mirza Muhammad Sultan, the grandson of Taimúr, who commanded the centre of his army, begged permission to go to the aid of his hardly-pressed cousin. But before Taimúr answered him the battle had practically been decided in another quarter. The right wing of the Turkish army, commanded by Prince Sulaimán, was composed mainly of troops from the recently subdued provinces of Aidin, Sarakhan, Mentesche, and Kermian. Unlike the Servians, these had no heart in the cause; long arrears of pay were due to them; and they had no inclination to sacrifice their lives to help Bajazid to gain a victory. First, then, the men of Aidin, then those of the other three places I have mentioned, went over to Taimúr. They were followed by the Tartars with whom he had tampered during the night. This desertion decided the battle. The leader of the Servians, Prince Stephen, noting that the day was lost, cut his way through the masses of the foe to the place where Bajazid stood, and implored him to save himself by flight. But Bajazid was a man of extraordinary obstinacy. He would not care, he said, to sit upon a degraded throne, and that throne would be degraded were he to flee before an enemy whom he had insulted and defied. He had still under him ten thousand Janissaries, probably at that time the best troops in the world, and with these he had occupied a height which assured to its defenders a strong position. Stephen, finding his entreaties fruitless, covered the retreat of the few troops who had remained faithful to Sulaimán on the right, and fell back in a westerly direction, whilst Prince Muhammad, covered by the cavalry of the Amírs of Amasia, retreated to the mountains towards the east. Bajazid, thus deserted, held the hill he had occupied till nightfall. "Hot had been day," writes Von Hammer, "like that of Honain, when the Prophet held out so bravely against the superior forces of the unbelievers; hot as the day on the plain of Kerbela, when Husein and his followers, overcome by thirst, fell into the hands of the enemy. Bajazid was as determined as the Prophet and as his grandson Husein. His ten thousand faithful Janissaries fell around him, either parched with thirst or consumed by the sword of the Tartars. It was only when night had fallen that Bajazid, yielding to the entreaties of Minnet Beg, consented to mount his horse. But his horse fell, and the Sultan of the Osmánli was taken prisoner by Mahmud Khan, a descendant of Chengíz Khán, the titular chief of the Schagatai."

Taimúr had returned to his tent the moment victory had pronounced itself decisively, to rest after the fatigues of the day, when suddenly the tent door was opened, and the Sultan of the Osmánli was brought into his presence, bound hand and foot. Seeing his illustrious captive—the man who but the day before had been the lord of thousands—so manacled, Taimúr could not restrain his tears. He caused him to be unbound, and, though he kept him a prisoner during the period of life—less than two years

—which yet remained to him, he treated him with unvarying kindness.

Had there been any vitality left in the Byzantine empire, it now had a chance of recovering its position. The conquering Turk had been struck down; four sons of Bajazid were competitors for empire; the Emperor, Manuel, released from a long siege, stood between them. It is due to Manuel to admit that he was not wanting on the occasion, but a far greater man than he would have failed to reanimate a corpse.

Of the sons of Bajazid, engaged in the battle of Angora, Sulaimán had, after the battle, retreated by way of Constantinople to the European provinces held by the Osmáuli, and, to secure the support of the Emperor Manuel, had confided to him, as hostages, his young brother Kasim and his sister Fatima. Muhammad, his father's favourite and the most capable of the family, had taken refuge in and assumed rule over Amasia and Tokat. Isa had, as we have seen, fled to Broussa, where, about the time of Bajazid's death, he asserted his independence, and cemented an alliance with the Greek Emperor by marrying into an illustrious Greek family. Musa had been taken prisoner. After his father's death, Taimúr despatched him with his father's corpse, to the care of the restored independent ruler of Kermian. Mustapha had disappeared. It was supposed he had been killed, though his body had not been found.

It was but natural that each of the surviving princes should cherish the desire to gain for himself the undivided empire of his father. The news of the death of Bajazid (8th March 1403) and of the return of Taimúr to Central Asia, preluded the outbreak of the storm. Muhammad was the first to break the peace. His troops, led by the son of the man who had defended Angora against Taimúr, attacked and defeated the army of Isa in the defiles of Ermeni. This victory caused a considerable portion of Isa's forces to desert to Muhammad. Thus encouraged, Muhammad, on the advice of the chief who had come over, proposed to Isa a partition of their territories. Isa refused; whereupon Muhammad continued his victorious march, beat the army of his brother at Ulubad, thirty-two miles west of Broussa, and took Broussa and Nicea. Isa fled to Constantinople, and proceeded thence to Adrianople to implore the aid of his brother Sulaimân. Supported by Sulaimân, he made friendly overtures to Muhammad; then, suddenly marching on Broussa, endeavoured to capture that place. The citizens, however, shut the gates in his face. Muhammad then inflicted upon him two severe defeats near Broussa, a third at Angora, then a fourth, very decisive. After this battle Isa disappeared for ever from the scene; no one knew, and no one particularly inquired, how.

In the interim Muhammad had sent to the ruler of Kermian to demand the body of his father and the person of his brother Musa. The ruler of Kermian had complied, and whilst Muhammad had caused the body of the renowned "Ilderim" to be interred in a picturesque spot on the banks of the foaming mountain stream, the Akschaglau, he had spared his brother. But he was resolved to be sole ruler. Isa having been subdued, and having disappeared, he turned his attention to Sulaimân, formidable less from his character, which was pleasure-loving, than from the strong position he held as ruler of the provinces of which Adrianople was the capital, and the firm alliance he had made with the Emperor Manuel, to whom he had restored Thessalonica and his father's conquests in Macedonia as far as the River Struma or Kara-su, as well as the coast-line between Panis and Varna. Rather alarmed at Muhammad's undisguised intentions, Sulaimân had levied an army. This army he had been forced to lead against a rebel vassal, the lord of Smyrna. Triumphing over that vassal, he sent his Vizier, Ali Pasha, against Muhammad. That general captured, through the treachery of one of the adherents of the latter, the city of Angora. Muhammad, who arrived in the vicinity of the city a few hours after it had surrendered, learning that his brother's army was commanded by Ali Pasha, whilst his brother was indulging in the pleasures of wine and women at Adrianople, left Angora to its fate, and, with the instinct of a true general, hastened to the point where the contest would have to be decided. His plan was foiled by the treachery of one of his generals, who went over to Sulaimán, and he was forced to return to the vicinity of Tokat. From here he concluded an alliance against Sulaimán with the ruler of Iconium, to drive him out of Asia Minor, whilst he stirred up his brother Musa to attack his European possessions.

Musa fulfilled his commission with alacrity and with a certain amount of success. Sulaimán, pressed by his Vizier, had joined his army in Asia Minor. Musa's movements recalled him to Europe. The two brothers first came in contact in the vicinity of Constantinople. The superior generalship of Sulaimán prevailed on this occasion, and Musa fled into Wallachia. The victory seemed to confirm the power of the conqueror, who then invaded Carniola and concluded a treaty with the republic of Venice. A little attention on his part to affairs would have secured for him complete sovereignty over the dominions of his father.

But Sulaimán was too inveterate a lover of pleasure to succeed. In spite of the warnings of the Greek Emperor to be on his guard against his brother's machinations, he allowed sensuality to be his master. It thus happened that one day, when Sulaimán was, as usual, absorbed in his pleasures, an army led by Musa appeared before the gates of Adrianople. The character of Sulaimán appeared from the manner in which he received the news. Informed, whilst indulging in his pleasures, by Michael Ogli, of the unwelcome news, he replied with a Persian verse. Again, roused by an old and confidential minister, he answered: "Art thou mad, old man, to disturb my pleasure with such vagaries? Who is Musa, with his rabble, that he should dare to contest the throne with me?" The old minister, rebuffed, implored Aga Hassan, chief of the Janissaries, to impress upon their master the danger of the situation. Aga Hassan made the attempt, when Sulaimán, infuriated at his boldness, gave the order, the most insulting to a Muhammadan, to cut off his beard. Reeling under this insult Aga Hassan mounted his horse, and, telling his tale to the chiefs and generals, urged them to join Musa. All but three consented. Adrianople surrendered to the new chief, whilst Sulaimán, roused at last, fled, accompanied by the three faithful chiefs, towards Constantinople. On his way thither the magnificence of his attire attracted the attention of the inhabitants of the village of Dugundschi, who had suffered much from the exactions of his people. Five brothers, renowned for their prowess and their skill as archers, rode close to him to assure themselves who it was. Sulaimán, alarmed by their attitude, shot first one, and then a second; whereupon the other three shot him dead with their arrows and decapitated him.

Musa then became Sultan. One of his first acts was to cause the three brothers who had killed Sulaimán, and all the inhabitants of their village, with their wives and children, to be barred within their houses and huts and burned to death. He perpetrated this act of cruelty to prove that the blood of his royal house would always be avenged. But, in point of fact, he was by nature hard, unfeeling, merciless. It would be too long to follow his course of cruelty. Suffice it to state that it gradually roused against him the great Osmánli nobles and the vassal princes of the empire. One of these, Ibráhim the son of the late Grand Vizier, Ali Pasha, whom he had sent to Constantinople to demand from the Emperor Manuel the payment of overdue tribute, bearing with him the secret adhesion of many others, advised Manuel to refuse, and, instead of returning, proceeded direct to the court of Muhammad at Broussa, to ask for aid and support. Upon this, Musa having by stratagem secured the person of Orchán, son of his brother Sulaimán, laid siege to Constantinople. Meanwhile the Ottoman fleet had been beaten by that of the Greeks, commanded by Emanuel Paleologus, a natural son of the Emperor John IV., and Muhammad, listening to the counsels of Ibrahim, whom he had made his Vizier, had agreed to make war against Musa. The Greek Emperor sent him ships to convey his army to Constantinople. The first three days after his arrival at that city were spent in festivities and rejoicings. On the fourth Muhammad sallied fourth at the head of his own troops, and a detachment of Greeks, and attacked the besieging army. He was, however, repulsed. The

10

same fate marred a second attempt. Before he was ready to try a third attack intelligence reached him that his own dominions had been invaded by the Governor of Ochrida.* He returned thither; then, allying himself with the Prince of Sulkadr, the territory bordering the eastern possessions of the Osmanli, with the Kral of Servia, their nearest neighbour in the west, and bound, moreover, to the Greek Emperor occupying the centre, he marched into Europe and encamped at Visa, south of Adrianople. Musa, meanwhile, forced to raise the siege of Constantinople, had marched into Servia and taken from the Kral all the border fortresses which his father, Bajazid, had lost after the battle of Angora. He now occupied a position between the Servian army and that of Muhammad. Seeing this, and to entice Musa to move towards him, Muhammad marched on Adrianople, defeated a detachment of his brother's army in front of the city, and summoned it to surrender. The reply he received to this summons was an invitation to beat Musa in the field: that accomplished, the city would willingly open to him its gates. Acting on this hint Muhammad marched against his brother, forced the Balkan passes in spite of the efforts of Musa's troops to prevent him, and reached the plain near Sophia. Here he received assurances of support from many pashas. He marched then to Nissa, and thence to the banks of the Morava. Joined here, first, by many influential chiefs, and, finally, by Stephen, Kral of Servia, he set out, confident of success, to seek Musa, who, forsaken though he was by one chief after another, still retained the formidable Janissaries, till finally he encamped on the plain of Schamurli.

Muhammad halted in that plain two days. On the third he beheld the army of Musa, now reduced to seven thousand Janissaries, defiling from the mountains. He then ordered the battle, placing the Kral of Servia on his left, his adherents from Adrianople on his right, and standing himself in the centre. The dramatic scene which followed is thus related by Von

* A town of Albania, the then capital of the Turkish dependencies in that province.

Hammer: "Whilst the armies stood opposite to each other, Hassan, the Aga of the Janissaries, who had left Musa to join Muhammad, called with a loud voice to his former comrades, 'Why do you delay, my sons, to come over to the most just and the most virtuous of the Osmánli princes? Why remain, oh miserable men, injured and abused by a man who is not in a position to protect himself, much less others?' Musa, when he heard the summons, could not restrain himself or his troops. He rode furiously at Hassan, who turned to flee. Musa pursued him, caught him, and, striking his head with the full force of his sword, cut him in two. He was preparing to repeat the blow, when a man who had accompanied Hassan met it by striking at Musa with his sabre and cutting off his right hand. Then Musa rode back into his camp, but the Janissaries, seeing his condition, his hand severed, deserted him, upon which he fled towards Wallachia. He was pursued by horsemen, at the head of whom was Sarudsche Pasha, and was found dead in a marsh, possibly strangled by his escort."

The death of Musa left Muhammad master of the situation. Thoroughly qualified was he to fulfil the duties which devolved upon the supreme lord of the Osmánli. Well educated, fond of literature, more especially of poetry, and of the arts, Muhammad I. loved, above all things, justice—that justice which knows no distinction of race, of sex, or of creed. He was merciful by nature, and preferred peace to war. The rigid followers of the Prophet reproached him, indeed, with laxity in his mode of life, but, weighed against his many virtues, this was but a small fault. By his people he was called Tschélébi—the urbane. Throughout his reign he deserved their affection.

Muhammad was twenty-six years old when he became sole master of the Ottoman empire. His first care was to fulfil his obligations to those who had befriended him in the years of his comparative obscurity. Foremost amongst these was the Greek Emperor, Manuel. To him Muhammad restored the strong places which the Osmánli had seized on the borders of the Black Sea, in

Thessaly, and on the Sea of Marmara. He sealed the alliance with fresh protestations of friendship, and dismissed the ambassadors of Manuel laden with presents, and conveying to their master the following words: "Tell my father, the Greek Emperor, that by his assistance I have recovered my father's empire; that I am ever gratefully thinking of that act of kindness; that I am devoted to him as a son is to his father; and that I shall ever place myself with joy at his disposal." He received, at the same time, ambassadors conveying the congratulations of the rulers of Servia, of Wallachia, of Bulgaria, of the Duke of Janina, of the Despot of Lacedemonia, of the Prince of Achaia. These he invited to his table, drank to their health and prosperity, and dismissed them with these words: "Tell your masters that I give peace to all, and that I accept peace from all. May the God of Peace punish those who shall disturb it!" He made, likewise, a treaty with the Venetians, securing the safety of their colonies. The representative of Venice was Francis Foscari.

But the tranquillity which was possible in Europe was still absent from Asia. The Prince of Karamania had seized the opportunity of the contest between Muhammad and Musa to march upon and besiege Broussa. Unable to take the city by assault, the besiegers had endeavoured to divert the water of the famous springs to the north-west, which constituted the main supply for the inhabitants, and to undermine the castle. The fertility of resource and the gallantry of the commandant, Aiwaf Pasha, baffled these attempts; but the Prince of Karamania still remained before the city, devastating the neighbourhood. To avenge his father's death, who had been executed by order of Sultan Bajazid, he dug up the grave of that Sultan, and caused his remains to be burned. Whilst he was engaged in such unworthy warfare, burning and destroying all around him, it happened that the convoy escorting the corpse of Musa to its final resting-place arrived in the neighbourhood. Then occurred a curious circumstance. Whether it was that the arrival of that convoy gave to the invader the first intimation of the defeat of Musa, or that he felt sure that the victorious Muhammad would follow closely on the body of his brother, or that the appearance of the corpse of the son on the morrow of the burning of the bones of the father aroused within him superstitious terrors, this is certain, that he at once fled in dismay. A confidant who, surprised at his action, ventured on the remark, "If thou fleest thus before the dead Osmânli, how wilt thou stand before the living one?" scarcely lived long enough to repent the rashness of his words!

It will suffice for the purposes of this work if I state simply that the Sultan experienced no difficulty in repressing the risings, not only of the Prince of Karamania, but of all the chieftains in Asia, great and small, who had acknowledged his father's supremacy. But it deserves to be recorded that his treatment of the conquered differed widely from that of his predecessors and of most of his successors. He never struck a fallen foe. There was a richness of generosity in his nature which made the bestowal of pardon one of his greatest pleasures. After establishing the supremacy of his arms in the greater part of Asia Minor, Muhammad was besieging the last of his enemies in Smyrna. In his camp before that place he entertained the Genoese governors of Phocæa, of Mitylene, and of Chios, the Grandmaster of the Knights of St. John, the Princes of Kerman, of Tekke, and Mentesche. On the tenth day of the siege the mother, the wife, and the children of the rebel chief who held the city against him, came to implore his mercy. He granted all that they asked on the sole condition of the surrender of the place. Smyrna surrendered. Muhammad at once caused its towers and walls to be razed, and the tower, which the Grandmaster had built at the entrance of the harbour, to be destroyed. The Grandmaster, greatly annoyed at this, presented himself to the Sultan the following morning, and informed him that the destruction of the tower which his Order had built, would cause war with the Pope and the sending of a hostile fleet and army to the coast. Muhammad calmly answered him: "I should like," he said, "to be the father of the Christian race throughout the world, to distribute amongst them rewards and honours, for it is the great duty of a ruler to reward the good and to punish the evil; but, then, at the same time I must have consideration for my own people, such as the Mussulmans require of me. Although, they tell me, Taimúr devastated all Asia, yet his destruction of the castle of Smyrna is a monument to his fame; for, in that castle all our slaves found a safe refuge, whilst freemen, travelling by land or sea, were constantly seized, taken there, and treated as slaves; all this time the brethren of your Order and the Turks were waging in the country constant war against each other, To put a stop to that, Taimúr, in other respects a godless man, deserved praise. And wouldst thou, now, that I should be more godless than that tyrant? But, in order to satisfy you, and at the same time not to hurt the interests of the Mussulmans, I will give you a place in the principality of Menteschе where you may build a castle after your own heart." Such was Muhammad I., noble and generous in all his sentiments, careful for the interests of alien races as he was watchful over those of his own!

Sending a peaceful message to the Sultan of Egypt, Muhammad marched against and defeated the Prince of Karamania (Iconium), and forced him to accept terms of peace. He directed his course thence to the Black Sea to retake possession of the town of Schanik on its coast. He was still there when he heard that the Karamanian chief had again rebelled. He had reached Angora, on his way to attack him, when he was struck down by a disease which baffled all the science of his physicians. In their perplexity these called to their aid a renowned doctor attached to the court of the Prince of Kermian, Sinan by name, but who was still more widely renowned under his *nom de plume*, as a writer of poetry, of Scheichi.* This learned physician very soon discovered that the Sultan's illness was simply depression of spirits, and he declared

* Scheichi is best known as the author of the earliest and most romantic poem of the Osmanli, called "Chosrú and Schirin," in which, treading in the footsteps of the great poet of the Persians, Nizámi, he sings the happy love of Schirin with King Chosru, and her unhappy love with Ferhad. A German translation of the poem was published at Stuttgart in 1809. The original Persian poem bearing the same title, by Nizámi, appeared in a German translation in Vienna in 1812.

that the news of a victory over the Prince of Karamania would effect a more perfect cure than any prescription. The Vizier, Bajazid Pasha, a man who had been true to the Sultan from his early boyhood, undertook to furnish this remedy. He sent a confidential messenger to the Karamanian Prince, with whom he had cultivated friendly relations, to invite him to quit Iconium with his army, in order to be at hand in case the illness of the Sultan should terminate fatally. The rebel prince fell into the trap, quitted Iconium, was suddenly attacked by Bajazid, and defeated. He fled to Taschel in Cilicia, but his son, Mustapha, was taken prisoner. The news of this important victory produced all the anticipated effect upon Muhammad. He shook off his melancholy; then, with his usual generosity, he promised to Mustapha, who swore to him never again "to cast a look on the possessions and lands of the Osmánli," the restoration of the cities he had taken from his father, and sent him back to his capital laden with presents. The young prince proved himself to be unworthy of so many favours. Scarcely had he quitted the vicinity of the Osmánli camp, than, declaring that "between Karamán and Osmán there must be war from the cradle to the grave," he made a swoop upon the horses of the Sultan and carried them off. Muhammad, to punish his perjury, followed on his track, besieged him in Iconium, took the city, and again pardoned him. The Sultan then returned to Europe.

Important events, indeed, called him thither. A misunderstanding between his Admiral, Schali Beg, and the Venetian Governor of the Cyclades, had caused a war with the Venetians. That governor, Pietro Zeno, considered that the islands of the Grecian Archipelago had not been included in the last treaty between the two nations, and he had not hesitated to capture Turkish vessels wherever he found them. This action had caused reprisals. A Turkish fleet of forty-two ships sailed at once from the Black Sea, and laid siege to Chalcis. Upon the appearance, however, of fifteen Venetian galleys, led by Pietro Loredano, with whom were two commissioners sent to treat, the Turks took shelter in the harbour of Gallipoli. It was said that the Turks answered all the overtures of the Venetians by the discharge of poisoned arrows. It is certain, however, that an understanding had been arrived at between Loredano and the Turkish admiral, and that, but for an accident, a peaceable solution would have been secured. It happened, however, (29th May 1416,) that a disturbance took place between a Venetian galley and a Genoese vessel which was on the spot. The Turks, regarding the Genoese vessel as part and parcel of their own fleet, sailed to its rescue. The battle then joined. It resulted in the complete triumph of the Venetians, who took twenty-seven of their enemy's vessels. Shortly afterwards Loredano entered the Dardanelles, and, finding the places on the coast well defended, sailed to Constantinople. By this time the Sultan had returned to Europe, and his peace-loving nature at once manifested itself in acts. Peace was concluded on the condition that prisoners should be restored; that all hostilities should cease; and that the right should be allowed to the Venetians to treat as enemies all Turkish corsairs who should disturb the public traffic of the Archipelago or of the Dardanelles.

Muhammad was now at peace with all the world. He exchanged letters of amity with all the rulers between the Tigris and the Oxus, and with those even beyond the Oxus. He received assurances of friendship not only from Shahrokh, the grandson of Taimúr, but likewise from the chiefs of the two great Turkman clans of Central Asia.

I pass over the desultory warfare from 1416–19, between the Osmánli and Sigismund, King of Hungary, signalled as it was by some bloody conflicts, and terminated by a victory gained by the Hungarian King over a Turkish army between Nissa and Nicopolis (4th October 1419), because it remained entirely without result. Almost on the very morrow of the victory referred to, the hostility of the Venetians and the domestic troubles in Bohemia caused by Ziska and his followers forced Sigismund to return to his own territories.

Far more dangerous to the empire of the Osmánli, threatening, indeed, the very existence of Islam in the lands conquered by the descendants of Othmán, was the rebellion of Bidr'udín. A short history of this religious revolt is necessary to a right comprehension of the events which led to the terrible wars with the Christians which filled the reigns of the two immediate successors of Muhammad I.

Bidr'udín of Simaul, a town seventy-five miles to the south of Broussa, had been one of the faithful adherents of Musa. On the death of that prince, the generous Muhammad had granted him not only his life, but the continuance of his salary as Army Judge Advocate, and had fixed Nicea as his residence. But Nature had made Bidr'udín a mystic, a lover of the occult. For him the then existing systems of religion and of culture had ceased to have a charm, and he at once devoted all his brain-power, which was considerable, and all his energies, which were enormous, to the thinking out of a system to supersede the old one. For this end he left Nicea, and fixed his abode on Mount Stylarios.* Thence he despatched a fanatical Turk, called Böreklüdische Mustapha, to proclaim his doctrines in one direction, whilst to promulgate them in another he selected an apostate Jew, called Torlak Kemal.

But it was less to the qualities of his two disciples than to the doctrines he had taught them to preach that Bidr'udín trusted for success. In all ages and in all races those doctrines have appealed, never unsuccessfully, to the needy and the indigent. They are, in the present day, undermining Northern Germany, and they are cherished by the great masses in the northern parts of this island. Bidr'udín preached, indeed, pure communism. Declaring that the God of the Christians and the Jews was likewise the God of the Turks, he invited the adherents of those religions to join him upon the principle of the absolute community of all goods. "I may use thy house as mine; thou mayest use my clothes, my arms, my horses as thine; everything except the wife." The better to win the Christians, he instructed his disciples to proclaim that whoever

* The foreland of the gulf of Smyrna, better known as Kara Bournon, opposite the island of Chios.

should say that the Christians did not honour God was himself a godless man.

He succeeded. The doctrines of communism preached far and wide over the land brought him crowds of followers. The Anchorite communities of Chios despatched to him an ascetic from their body to declare their sympathy and union.* From all sides adherents flocked to him, full of zeal for the new doctrines, anxious to put them in immediate practice for their own benefit, and actually practising them in many cases at the expense of their richer co-nationalists who had preferred to adhere to the old order of things.

The movement attained to so great a strength, and was producing so disastrous an effect upon the empire in general, that at last even the peace-loving Muhammad was forced to take measures to crush it. He despatched orders to Sismân, governor of Sarukhan, son of the King of Servia, and himself a convert to the Muhammadan faith, to crush the insurrection. Sismân set out with a body of six thousand men towards Mount Stylarios, the head-quarters of Mustapha, but found the passes to that mountain so strongly occupied that he hesitated to attack. Whilst he was still hesitating, the partisans of the new prophet rushed upon him and his men and destroyed them.

This victory encouraged Mustapha to enlarge alike his doctrines and his pretensions. He abolished the turban, directed that only one cloth should be used for the covering of the body, and proclaimed that it was preferable to belong to the Christian rather than to the Muhammadan community. Meanwhile, Sultan Muhammad had commissioned Ali Bey, the new Governor of Sarukhan and Aïdin, to attack the innovators. Ali Bey obeyed, but was so decisively beaten that he had great difficulty in fleeing, with a thin remnant of his followers, to Manissa. This defeat roused the Sultan to a supreme effort. His despatched his son Amurath—who was to succeed him—then only twelve years old, guided by his

* The message brought was: "I am, as thou art, an ascetic; I worship the same God; and, in the night, I come to you, striding with dry feet through the sea."

trusty vizier, Bajazid Pasha, with the strongest and best-equipped army he could raise, against the rebels. Bajazid resolved to make short work of it. Carrying all before him, and sparing neither age nor sex amongst his opponents, he reached the passes which had frightened Sisman. He stormed these, then attacked the hill itself—the nest of the sectarians—carried it after a most obstinate resistance. He gave no quarter. Mustapha was amongst the prisoners. Though tortured in the horrible manner common to those barbarous times, Mustapha steadfastly refused to renounce his faith. His enemies, at length, renouncing the hope of converting, nailed him to a board, his hands and feet outstretched so as to form a cross in the shape of the letter X, paraded him through the city on a camel, and then hewed him to pieces before the eyes of his disciples. These were as steadfast as their master. With the words: "Father Sultan, may we come into thy kingdom," they rushed on the swords of their persecutors and died to a man!

The defeat and death of Mustapha and his immediate adherents was followed by the pursuit, capture, and execution at Manissa of his disciple, the Apostate Jew, Torlak Kemal. Amurath and Bajazid then crossed into Europe to pursue the original author of the new religion, Bidr'udín of Simaul. They caught him and his followers at Seres, nearly fifty miles north-east of Salonica, defeated him and hanged him.

Thus ended the religious revolt—the only outbreak, as Von Hammer truly remarks, in the history of the Osmánli, until the rise of the Wahábis in our own time, which rested its foundations on the slippery ground of religious reform, the only one absolutely of which the leading disciples and promulgators belonged to religious orders. Sultan Muhammad had, indeed, crushed it, but the spirit which had been roused was soon to display itself in a different shape and under a form almost as dangerous.

The reader will recollect that after the battle of Angora, Mustapha, the fourth son of Sultan Bajazid Ilderim, had disappeared. It was supposed that he had perished. At all events, his body had not been found, and it is at least curious that if he had been the real

Mustapha—a point in dispute between the Greek historians, who affirm, and the Turkish, who deny, his claim—his existence should have been unknown until the time when the religious outbreak of Bidr'udín and his followers had proved that the empire of Muhammad was assailable. However this may be, it is certain that scarcely had the religious outbreak been crushed, than the information reached Sultan Muhammad that a chief, styling himself Mustapha, and supported by the Prince of Wallachia, and the twice-pardoned Governor of Nicopolis—formerly Lord of Ephesus and Smyrna—had entered Thessaly with an army, claiming the empire on the ground of elder brotherhood. Muhammad, hastening to meet the new claimant, encountered and defeated him, near Thessalonica (1419). The defeated leader and his chief adherents took refuge within the walls of that city, and demanded the protection of the Greek Emperor, to whom it belonged. Manuel accorded that protection. The real greatness of Muhammad never showed itself more prominently than on this occasion. Not only did he respect the orders of Manuel, but, to prove to him that he bore him no ill-will for thus protecting a man who had been, and who might again become, a formidable rival, he returned to his capital by way of Constantinople, trusting implicitly, in an age when treachery was a recognised weapon of warfare, to the good faith of the Greek Emperor. There were not wanting men at the court of Manuel, who urged that prince to seize the chance which the confidence of the Sultan had placed in his hands; but the Greek Emperor had had too much experience of the generosity of his guest, and too lively a sense of honour, to play him false. Amongst the arrangements which had been previously concluded between the two sovereigns was one whereby, in consideration of an annual payment by the Sultan, Manuel took upon himself the safe-guardianship of Mustapha and his few adherents. The visit to Constantinople followed the conclusion of this agreement.

In the spring of the following year (1421) Muhammad returned to Adrianople. Three days later he was struck, whilst riding, by apoplexy. Feeling his end approching he sent for his faithful

Vizier, Bajazid Pasha, and conjured him to display to his son Amurath the same attachment and devotion of which he had given so many proofs to himself. Fearing lest the ordinary fate of younger sons of an Ottoman sovereign should befall his two younger boys, he begged Bajazid to send them at once to Constantinople, to be there under the tutelage and protection of the Greek Emperor. The next day he showed himself at his window to his troops, to whom the news of his illness had caused the deepest sorrow. The day following he died. The two Viziers, Bajazid and Ibrahim, dreading the effect which the news of his death, before Amurath, who was then at Amasia, should have reached Broussa, might produce on the soldiers, deemed it prudent to conceal the occurrence. When, as usual, the troops came before the palace demanding to see their Sultan, they returned satisfied. The Vizier had dressed the corpse in the royal robes, placed it on a chair near the window, and caused a page to move its hand as if in salutation. This deception was maintained for forty-one days. It became useless only when news arrived that Amurath had reached Broussa and had proclaimed his succession.

Few men have possessed greater qualities than Muhammad I. Few sovereigns have been more truly king-like. He had reigned only eight years. Coming to the throne after an interregnum he had consolidated the Ottoman Empire. A patron of the arts, he had finished the great mosques of Adrianople and Broussa, erected a second, which contains his remains, at the latter place, and encouraged poetry and the cultivation of the science of medicine. He avoided, as much as possible, external wars. To wage these became the first business of his son and of his grandson; and it is to an occurrence in one of their wars that this brief account of the rise of the Turkish Empire owes alike the reason of its existence and its title.

Amurath II., the eldest surviving son of Sultan Muhammad, had been born in 1403. He was in the freshness of vigorous youth, then, when he received in 1422, at Amasia, the news of his father's death. He proceeded with all haste to Broussa, and was there proclaimed and acknowledged as Sultan. He remained at that place until the arrival of the corpse of his father. Having caused him to be ceremoniously interred in the Green Mosque, and having fulfilled the necessary days of mourning, he despatched ambassadors to the Courts of the Emperor Manuel, of King Sigismund of Hungary, of the Princes of Karamania and Mentesche, to announce his accession and to offer to renew the treaties which had been made with his father. It is curious to note the results of these several negotiations. The replies from Karamania and Mentesche were satisfactory; with Hungary a truce for five years was concluded; but from Constantinople there proceeded difficulties. These it is necessary to explain. The Greek Emperor Manuel had been allied by the closest bonds of friendship, personal and political, with Sultan Muhammad I. The latter had given a proof of the great trust he felt in the loyalty of Manuel, when, feeling his death approaching, and dreading the instinct which would prefer the taking of two lives to the repression of two revolts, he had conjured his Vizier, Bajazid Pasha, to commit his two youngest sons to the care of the Greek Emperor. Bajazid had promised, but had delayed compliance. Sensible of the delay, the Emperor, before he received from Murad the notification of his accession, had despatched two ambassadors to Adrianople to require from Bajazid the execution of the promise he had given to the deceased Sultan, and threatening, in case of refusal, to call into the field, Mustapha, the pretended son of Sultan Bajazid. Bajazid Pasha took it upon himself to reply to the ambassadors in haughty terms: "It is not becoming," he replied, "and is contrary to the precepts of the Prophet, that the sons of Musulmans should be educated by infidels." He added that it was necessary that the Emperor should renounce the proposed guardianship if he would care to maintain peace and friendship with the new Sultan.

The action that Manuel then pursued is conceivable only on the belief that he entertained a very poor opinion of the new Sultan and his minister. Otherwise it is difficult to understand why a sovereign so generally sagacious, seated on an undermined and decaying throne, should have deliberately provoked the hostility of a nation of warriors; a people who had made their way in Europe, step by step, at the expense of the empire over which he ruled. No sooner, however, did he receive the reply of Bajazid Pasha, than he despatched a fleet of ten galleys, well armed and well manned, under the orders of Demetrius Laskaris, to Lemnos, with instructions to Demetrius to release the pretender Mustapha, and his principal adherent, Schunéid, then under surveillance, and, on obtaining from the former certain stipulated guarantees, to support him in his endeavours to obtain for himself the succession to the dominions of Muhammad I. The chief of the guarantees demanded in case of success were the restoration to the Greek Emperor (1) of Gallipoli, (2) of the entire northern coastland from Constantinople to Wallachia, (3) of the Thessalian cities as far as Erysos and Mount Athos. To these terms Mustapha freely assented, and swore to fulfil them.

It seemed at first as if Manuel had made a prudent forecast. Mustapha appeared with a fleet and army before Gallipoli. Though the city shut her gates, the people of the surrounding country received him with acclamations. Leaving the fleet to finish with the city, Mustapha marched towards the isthmus of Athos, taking possession of every town which lay in his way, and increasing his army with every step. Meanwhile, Amurath had commissioned Bajazid Pasha, who had joined him at Broussa, to march against the pretender. Bajazid took ship with a small force, increased it, after landing, to thirty thousand men, and, taking the road to Adrianople, finally encamped on a marshy plain in sight of that city. He was in this position when Mustapha, strengthened by the accession of many important chiefs, and commanding a force greatly superior in number, moved to attack him. Before trying the ordeal of battle, Mustapha, who, evidently, possessed great readiness of speech combined with personal daring, rode forward to the very front line of the army of Bajazid, and, addressing the soldiers in a loud voice, directed them to obey the order now given to them by the true heir to the throne, and to lay down their arms! It was a scene not dissimilar to that which was enacted four hundred years later when Napoleon landed in France from Elba. With one consent, though there had been no previous consultation, as if inspired simultaneously by an irresistible and dominating persuasion, the soldiers of Bajazid came over in a body, and acknowledged Mustapha as their master. Bajazid and his brother Hamsa Beg, who alone refused to follow their men, were taken prisoners. The former was beheaded on the spot; the life of the latter was spared, to the cost, as it happened, of him who spared it.

The news of the success of Mustapha produced events which seemed to complete the revolution. Gallipoli at once surrendered, and it seemed as though all the hopes of Manuel would be fulfilled; that he would descend to posterity as the restorer of the empire. But there is a weak point in almost every treaty. That made between Manuel and Mustapha was not exempt from the general condition. That weak point, moreover, had now come to be the main point. It was whether a promise made by a pretender when he was poor and helpless, and without consideration, would be considered by that pretender to be binding when, in virtue of its existence, he had attained power, and wealth, and position. Mustapha came rapidly to the conclusion that such a promise was not binding. When Gallipoli had fallen, and when Demetrius proposed, in virtue of the sworn agreement, to place a Greek garrison in the city, Mustapha objected, saying, that he had not conquered Gallipoli for the Greeks, and that Demetrius was free to return to Constantinople!

Manuel at once recognised that he had played a false card, and, withdrawing all support from Mustapha, endeavoured to come to terms with Amurath. As, however, he always persisted in the fulfilment of the late Sultan's wishes regarding the guardianship of his two younger sons, and as Amurath as persistently refused, the negotiations came to nothing. Manuel remained thenceforth a simple spectator of the contest, which Amurath, still supreme in Asia, and having as allies only the Genoese of Phocœa, was now to carry on with the man whom he had incited to rebellion.

Amurath was young, but he was pure-minded, honourable, and religious. The desertion of his army and the death of his father's friend, the Vizier Bajazid Pasha, affected him the more, inasmuch as he had, at the instigation of some pashas, young men of his own age, rather scornfully over-ruled the objections which Bajazid had himself raised to the despatch of an untried force against the pretender. On the receipt of the news, struck with the enormous loss his empire had sustained, and recollecting that his own insistance had caused it, he turned to his friends, who were trying to cast the blame upon the absent, and told them that in the anger of the Almighty alone had they to seek the cause of the disaster: "It is our sins," he continued, "which have called forth the wrath of God. Be it our task to soften that wrath by prayers and tears. When the Creator is against us the creature can do nothing!" Setting an example to his nobles, he paid a visit to the great Shékh of Broussa—the Bokhari who had girded him with the sword of the Sultan on his accession, and who was to become still more famous in the years that were to follow—and implored his intercession with the God of Islam. For three days the Shékh lay prostrate before his Maker. In a trance, in which he fell on the third day, he heard the re-assuring voice of the Prophet: "The God of mercy," said the voice, "has heard the prayers of Amurath; tell him that Divine power will assure to him the victory." The Shékh arose, repeated the message to Amurath, girded his body with the sword used for the punishment of rebels, and bade him set out certain of victory. Amurath, as confident now of victory as he had been previously despondent, prepared at once to take the field.

In the enterprise the accomplishment of which lay before him, Amurath was immensely aided by the conduct of Mustapha. This pretender, after his easy victory in the marshy lands near Adrianople, had entered that city, and, deeming that the empire had been won, had given himself up entirely to pleasure. He had, however, at his side a man who now regarded Amurath with all the hate he had borne to his father, Muhammad, who had a soul above pleasure, who loved power, dominion, revenge. This man was

Schunëid, formerly lord of Smyrna. This noble had vainly tried to incite Mustapha to action so long as the political horizon appeared clear. But, armed at last with the information that Amurath was taking the field, he rushed into his presence and insisted upon the necessity of the pretender rousing himself to action. More than that, he pointed out that Amurath must not be allowed to land in Europe, that Mustapha must cross into Asia before his rival should be able to reach either Lampsacus or Scutari. Mustapha, now thoroughly on the alert, embarked his army on board the fleet at his disposal, landed at Lampsacus, and halted there three days, to receive the homage of the several pashas.

Meanwhile Amurath, confident of success, yet determined to neglect no precautions to gain it, had broken up from Broussa and marched to and encamped in a position in which his front was covered by the river of Ulubád, and the marshes which are formed by the overflow of the lake of the same name; his right by that lake and the marshes; his left by the sea. He made himself still more secure by breaking down the bridges across the river; and as, now, the only possible point on which he could be attacked was his right, and as to turn that required a march of three days, he felt he might securely wait the course of events.

Mustapha, meanwhile, approached, and encamped on the opposite side of the Ulubád river. He was urged by his adherents to challenge Fortune at once, and he had made every preparation to do so, when he was attacked by a violent bleeding of the nose. This affection lasted three days, and left him too weak to be capable of exertion. In the meanwhile Amurath had been joined by Michael Ogli—commander of the irregular cavalry which had gone over to Mustapha when appealed to before Adrianople—whom he had just released from the captivity in which he found him on his accession. Michael Ogli, after noting the positions occupied by the several corps of the pretender's army, proposed, and, with the consent of Amurath, put in execution a plan to entice the men formerly under him, who knew him and loved him, to make a great atonement for their action before Adrianople. Crossing the river during the night, he called to his men in a loud voice, to join him and to follow where he should lead them. The sound of the beloved voice had a magical effect upon the horsemen. Leaping into their saddles, they galloped to the spot whence it proceeded, and declared their willingness to do all that their chief might order. To his direction that they should join the army of the true Sultan they paid immediate and implicit obedience.

Mustapha's irregular horsemen were gained, but the irregular infantry remained true to him. Five thousand of them, indeed, resolved to avenge the desertion of their comrades by a night attack on Amurath's camp. The Sultan, however, outwitted them. Having been informed by a spy of their design, he despatched two thousand Janissaries, under Umur Bég, to fall upon their advanced guard as soon as it should have effected a crossing. Umur Bég carried out his orders, cut down the greater number of the enemy, and despatched the remainder prisoners to the camp.*

Meanwhile, encouraged by the successful action of Michael Ogli, a prominent noble in the Sultan's camp, Auf Pacha by name, despatched one secret messenger to Hamsa Bég, brother of the powerful Schunëid, promising the latter the governorship of Aïdin, if he would abandon Mustapha; another to Mustapha himself, warning him of the treachery of Schunëid, who, he said, had undertaken to deliver him to Amurath. The letter to Hamsa Bég worked wonders. That very night Schunëid, accompanied by sixty trusted followers laden with gold and silver, and leaving the lantern burning in his tent, fled in the direction of Aïdin.† When, in the morning, his flight became known, the troops of Mustapha, panic-stricken, dispersed in every direction. Mustapha himself

* One Janissary, conducting two of the enemy to the camp, happened to pass a butcher's shop where many sheeps' heads were exposed for sale. He offered to exchange, and did exchange, his two prisoners for one sheep's head. From this time dates, according to the Osmanli historians, the hatred which ever after existed between the Janissaries and the Ajabi, or irregular infantry.

† He held the Government of that place for two years, when his acts aroused the anger of Amurath. He rebelled again, and was strangled in 1425

fled, accompanied by only a few followers, to Lampsacus, whence he embarked for Gallipoli. He was followed by Amurath, at whose disposal the Genoese Governor of Phocæa, Adorno, had placed his fleet for the purpose. As Mustapha, from the ramparts of Gallipoli, beheld the fleet approaching, he sent a confidential messenger in a boat to the Genoese admiral to offer him fifty thousand ducats for the delivery of Amurath. Adorno unhesitatingly refused the proposal, and, as Mustapha's troops guarded the entrance into the harbour, he anchored outside it, and despatched his archers in twenty boats to effect a landing: Amurath followed with his array: the united forces attacked the defenders and put them to flight. Mustapha fled to Adrianople, where, collecting all the treasure upon which he could lay hand, he bent his course to Wallachia. On the way thither, he was betrayed by his own servants, taken prisoner, and eventually put to death. Amurath, after a stay of three days at Gallipoli, proceeded, accompanied by the Genoese, to Adrianople, whence the inhabitants poured forth to offer him welcome.

The Greek Emperor might well have beheld with some dismay the course of events. He had provoked a revolt against Amurath of such a nature that, had the man who was his tool been a man of real capacity and action, would have proved fatal to the son of the prince, reverence for whose last wishes had alone, he affirmed, inspired his action. It is true that he had apparently cast off that associate in the hour of that associate's triumph, but, in point of fact, it was the associate who had dismissed him. Then he had attempted to form a union with Amurath against the associate whom he had originally stirred up; but Amurath had deemed his terms inadmissible, and had refused them. And now Amurath ruled at Adrianople, the victor, without him, in spite of him! But for Manuel, indeed, Amurath might well argue, the trouble would not have arisen! Good reason, then, had the Emperor to dread the wrath he had provoked!

To appease, if it were possible, that wrath, Manuel promptly despatched to Adrianople ambassadors charged with professions of friendship. To those ambassadors Amurath delayed to grant an audience until his army was completely ready to march on Constantinople; then he sent them back to the Emperor with the message that he himself was following on their footsteps. A few days later he set out at the head of twenty thousand men (June 1422), and, after an uneventful march, encamped before the city, repulsed thence a second embassy despatched by Manuel, and, a few days later, under the influence partly of the irritation caused by the earlier action of Manuel, and partly of the exaltation caused by his recent successes, issued a proclamation in which he declared that the city and all its treasures should be given up to the conquerors. This unwise proclamation had a very remarkable effect. It drew to the camp of the Sultan all the adventurers and plunderers of the country. Men who lived by their wits, to whom toil was unknown, idlers, vagabonds, ruffians of every kind, some not possessed of sufficient money to purchase even arms, came swarming into its vicinity. It attracted also a different class of men. To partake in the promised plunder there crowded likewise the derweshes and the various orders of religious mendicants; the great Shékh Bokhari, the Shékh to whom the Prophet had appeared and promised victory over Mustapha to Amurath—was at their head. These demanded as their share the plunder of the churches and the religious houses. It would be difficult to exaggerate the excitement produced in the besieging army by the arrival of the Shékh Bokhari. Already he was a sacred personage. The vision referred to; the message of the Prophet; that message followed very promptly by its complete fulfilment; had made of him a power in the country such as it is difficult for those who have not lived in Eastern lands to form an adequate idea. When he entered the camp, riding on a white mule, people prostrated themselves before him; to touch his hand, his foot, even the bridle and the hoof of his mule, was to secure an amulet against evil. Shékh Bokhari seemed profoundly indifferent to these demonstrations. Absorbed apparently in contemplation of holier things he entered his tent, and set to work to endeavour to announce, from a study of his cabalistic books, the day and the hour when Constantinople should fall into the hands of the true believers.

After many days' meditation the inspiration came. The Shékh announced solemnly that on the 24th August (1422) at 1 o'clock in the afternoon he would mount his horse; that he would then wave his sword thrice round his head, shouting, at the same time, the war-cry; that, on the conclusion of that war-cry, the walls of Constantinople would fall.

The day and the hour arrived. The Shékh mounted his horse, and, escorted by five hundred dervishes, and preceded by a man carrying an enormous shield, advanced towards the city wall. Then he drew his sword from its sheath and went through the announced programme. The air immediately resounded with shouts from the Osmanli camp of "Allah!" and "Muhammad!" responded to from the walls of Constantinople by cries of "Christ!" and the "Blessed Virgin!"* as the Turkish army rushed to the assault. Fierce and terrible was the battle. The Muhammadan were excited by the promises which, they believed, had come from Heaven; in their minds victory was a foregone conclusion—it was within their grasp; death in the act of stretching out the hand to gain it meant Paradise—Paradise with all its glories immediate and eternal. The Greeks, on the other hand, had every reason to combat with the greatest obstinacy. Their all was at stake; their independence, their hearths and altars, their civilisation—all that makes life worth living. They fought, then, like men driven to the last extremity. The sun set on the terrible battle. Still it continued; there was no shrinking on either side. Suddenly, however, there was a cessation. The Osmanli fell back, stayed in their camp only long enough to burn their battering machines, and then beat a hasty retreat!

It is easy to understand that, in an age when a belief in the supernatural was widely spread, this retreat should be attributed

* "Christos! Panagia!" Panagia—the All-Holy One—the epithet given by the Greeks to the Virgin Mary.

to divine agency. The Greek historian* affirms that the discomfiture of the Osmánli was due to the sudden appearance on the walls of the assailed city, in the very height of the combat, of a Virgin, dressed in a violet robe, a light shining round her head; that, on seeing her, the Osmánli were struck by panic and fled. It is not necessary, however, to dive into the supernatural to account for the result. The fact is that the Sultan found the resistance to be more vigorous than he had anticipated. On the very day on which the assault was delivered he had received from the provinces information to the effect that another Mustapha, his own brother, then only thirteen years old, urged by Manuel, and supported by Elias Bég, had taken possession of Nissa. Not knowing how formidable might be this rival, he was unwilling to see his best army destroyed under the walls of Constantinople. At the moment, then, when success seemed, perhaps, more than doubtful he gave the order for the retreat, which, whilst it destroyed the pretensions of Shékh Bokhari, preserved for himself a still powerful army.

The second Mustapha proved less formidable than the first. Whilst the Greek Emperor gave him only a half-hearted support, Amurath purchased his chief adherent, Elias Bég. By Elias the young prince was betrayed into the hands of his brother's general, and by him executed. "When there are two Khálifs," said the Prophet, "to whom the world renders homage, it is necessary that one of them should die!"

A revolt of the Prince of Sinope and of Kastemuni was quelled with not greater difficulty. As one of the conditions of his forgiveness the conquered prince was forced to give his daughter in marriage to the Sultan.

In the midst of the festivities which followed this wedding (1424), the intelligence reached the Sultan of the death of the Greek Emperor, Manuel. Amurath did not allow the opportunity to slip by. Manuel had been a very capable ruler. He had, it is true, made mistakes; his greatest had been his behaviour towards

* Johannes Canano.

Amurath; but he had known how to repair them. His successor, John, possessed a character far more pliable. As the price of acknowledging his succession, Amurath wrested from this prince many cities and fortresses on the shores of the Black Sea, and wrung from him an agreement to pay an annual tribute of thirty thousand ducats. At the same time he renewed the expired treaties with Servia and Wallachia, and despatched an embassy to Ofen (Buda) to treat for a continuance, for two years, of the existing truce with King Sigismund, now become Emperor of Germany. The agreement was ratified by the exchange of presents of more than ordinary magnificence.

Towards the vassal princes who ruled under him, and the independent princes who neglected no opportunity to disturb him, Amurath displayed the same generous and magnanimous policy for which his father had been distinguished. The princes of Karamania and of Kermian more than once experienced the greatness of his character. He loved to forgive, when forgiveness was possible; to repair, by some splendid act of munificence, the evils which war had caused. The wise policy which accompanied that magnanimity was illustrated in the case of Salonica. The splendid position of that city, as the *entrepot* for the commerce of Greece, had ensured to it a long prosperity. That prosperity had been diminished neither by its capture in 1386, by Amurath I., and its subsequent recovery by the Greeks; nor by its conquest in 1394 by Bajazid, and somewhat later by Muhammad I., though on both occasions it was again lost to them. Struck down, Salonica always rose again. This place, which he should have guarded as the apple of his eye, the Emperor John had neglected. Wearied at last by the importunity by which he was assailed to provide some means for her defence, the Emperor had at last sold the city to the Venetians. Against a bargain of this character, Amurath protested. He would have left the city unmolested had John undertaken to maintain it, but he would not permit the interference of another Power. When, then, he concluded the treaties to which I have alluded with the other Christian Powers, he specially exempted Venice. Then, in

February of 1430, he broke up from Adrianople, and summoning Hamsa Bég from Asia Minor, laid siege to the town. During the short siege—the 26th February—there occurred an earthquake, in the midst of which some Turkish soldiers entered the town and endeavoured to persuade the Greek inhabitants to yield the place to the Sultan rather than hold it for a Latin master. A large number of the inhabitants declared themselves in favour of such a course, and this declaration, reaching the ears of the Venetians, inspired them with so great suspicion that they thenceforth insisted upon the attachment of men of other nations to every Greek guard. In vain did Amurath thrice invite a surrender on honourable terms, promising to spare the city, to give freedom to the inhabitants, and to allow free departure to the foreign garrison. The Venetians would not hear of capitulation, and the Greeks were controlled by the Venetians. At last the Sultan's patience was exhausted. With a rare magnanimity he sent word, on the last day of February, that he would storm the city on the morning of the day that was to follow. He kept his word. In spite of the brilliant defence of the obstinate Venetians, only fifteen hundred strong, his troops obtained possession of the defences before 9 o'clock. He was powerless to prevent the outrages and plundering which, in those times, and even in later times, always followed a successful storm. But he did his best, after the rage of his troops had been satisfied, to repair the evil. He allowed all the prisoners who had been taken to recover possession of their houses and property. To fill the houses left empty by death or other causes he transferred to Salonica the surplus population of the neighbouring town of Janitza. These naturally were Osmánli; they came in in such numbers as soon to constitute the majority. The conversion of the churches* into mosques followed as a matter of course. Thenceforth Salonica remained, as it still continues,

* One of these was the church which contained the remains of St. Demetrius. According to tradition there used constantly to drop from the coffin, which contained those remains, a balsamic oil, which used to effect marvellous cures. From the moment when the voice of the muezzin was heard from the roof of the Christian church the precious balsam ceased to exude.

one of the most important commercial cities of the Ottoman empire.*

The increase of the empire of the Osmánli in the direction of Europe continued under the steady rule of Amurath. In 1431 he obtained possession of the important town of Janina. Two years later he renewed for a further period the truce with Hungary, and, had Sigismund been as loyal as the Sultan, the events I am about to record would have found no place in history. But Sigismund was weak and disloyal; nor did he, in that respect, stand alone. The reader will find, to his regret, that in the transactions which form the history of the terrible war between Hungary and the Osmánli, the simple word of the Muhammadan was more worthy of trust than the plighted oath of the Christian.

Although Sigismund had agreed to a renewal of the truce with Amurath, and had accepted his presents,† neither he nor the Kral of Servia thought it beneath them to intrigue against the Sultan with the Muhammadan prince whose family the ancestors of Amurath had indeed despoiled, but to whom his own generosity had granted favourable terms of peace. The prince in question was Ibrahim Bég, Prince of Karamania. The cause of dispute with him was but slight. Ibrahim had stolen a valuable Arabian horse from a vassal of the Sultan. The vassal complained to his lord, who demanded the restitution of the animal. Incited by Sigismund and the Kral of Servia, Ibrahim refused. The Sultan sent, then, one army, and marched with another against the Karamanian prince. Him he easily conquered. Then he wished to punish the Christian allies who had stirred up the Karamanian to war. The Kral of Servia, George Brankowitch, succeeded in diverting the storm from his own country by declaring his readiness to perform a long-standing agreement—the giving to the Sultan the hand of his daughter

* Should the extension of the Austro-Hungarian empire replace this important city in the hands of a first-class European power it will attain, as the emporium of Europe, an importance to which it never aspired in its past.

† Amongst these, received by Sigismund in the cathedral of Basel, were twelve golden cups filled with pieces of gold, and cloths of silk embroidered with gold and covered with precious stones.

Marie; and to aid him with troops in his revenge upon Hungary. The revenge came first. Accompanied by the troops of Servia and by those of the Voivode of Wallachia, the Sultan crossed the Danube at Semendria, entered Transylvania, besieged, though fruitlessly, Hermannstadt, laid waste the country about Mediasch, sacked the town of Schässburg, burnt down the suburbs of Kronstadt, consumed for forty-five days the resources of the country, and returned with 70,000 prisoners. He had met with no opposition save at Hermannstadt.

The pleasures of the wedding succeeded to the toils of war. In the midst of the enjoyment of those pleasures, however, Amurath received information that the two vassal sovereigns who had accompanied him to the war, his father-in-law, the despot of Servia, and Drakul, the Voivode of Wallachia, were plotting against him. Both vassals, with their sons, were summoned to the Court of the Sultan to explain their conduct. Drakul obeyed, and after some delay, during which he renewed his oath of fidelity, was set free. But the reply of George Brankowitch could scarcely be misinterpreted. He fortified and re-victualled Semendria, the fortress which Amurath had demanded as a pledge of his fidelity, and placed it under the command of his elder son, Gregory; whilst, taking with him his younger son Lazarus, he fled to the Court of Albert, the successor of Sigismund, King of Hungary. Amurath then marched into Servia, took Semendria after a three months' siege (1439), and was about to lead his army against Nicopolis when information reached him that a Hungarian army, 24,000 strong, led by King Albert in person, was approaching the Lower Theiss. The Osmánli moved against the new enemy, who, panic-stricken, abandoned the field almost without a contest,* and were taken prisoners in such large numbers that five men were sold for five hundred aspres (fifteen francs), and a beautiful woman was exchanged for a pair of boots!

Whilst thus warring against Hungary, Amurath took great pains to extend his alliances in other directions. He exchanged letters

* *Geschichte Ungarns*, von Michael Horváth (German translation).

of friendship with the rulers of Egypt, of Karamania, and with Sháhrok, the grandson of Taimúr. He endeavoured, also, to establish diplomatic relations with Ladislas, King of Poland, whose brother Casimir was the rival of Albert, King of Hungary, for the possession of the crown of Bohemia. To this prince he offered his alliance on condition that he should break off all relations with Albert, and should support the pretensions of Casimir. The death of Albert at the close of the same year (1439), and the succession to the throne of Hungary of the very Ladislas with whom he had been negotiating, brought the correspondence to an abrupt conclusion. Amurath then proceeded to push the siege of Belgrad, the defence of which the Servian prince confided to the Hungarians.

We enter now upon the period which is called by the Hungarians, from the name of its hero, "the epoch of the Hunyadys."* John Hunyady, surnamed Corvinus, was the representative of an undistinguished family of the lesser nobility, or squirearchy, of lower Hungary. Born in a border province, subject to constant invasions, the young Hunyady had from his youth been bred up to arms. He soon displayed very remarkable qualities as a leader of men. He had come to the front in the war waged against the Hussites; but it was in defence of his country against the bands of predatory Osmánli that he had gained his greenest laurels. When the Hungarian army led by King Albert had disappeared before the presence of the Osmánli in the manner I have described, it was Hunyady alone who had rallied round himself a band of warriors who presented a bold front to the invader. For this service Albert had nominated Hunyady to the office of Ban of Servia. It was an empty title, for all Servia was falling fast into the hands of the Turks, and King Ladislas, when he succeeded Albert in the winter of 1440–1, had revoked it and confided to Hunyady the posts of Lord-Lieutenant of the county of Temeswar, and Captain-General of Belgrad, as well as the office, to be held jointly with Nicholas Ujlaky, of Voivode of Transylvania. Clothed with these offices,

* *Die Zeit der Hunyadys, von* 1439, *bis* 1490: Horváth.

Hunyady lost no time in showing himself competent to fulfil the duties devolving upon him. Collecting a body of Hungarian troops, he swept aside the several bands of Osmánli who invested the frontier, and boldly entered Transylvania. It is proper to relate that before this had happened the Sultan Amurath had been forced, by the obstinate resistance of its garrison during a period of six months, to raise the siege of Belgrad, saying, as he did so, that "sooner or later he would become its master." He retired then to his capital, leaving military control in Europe in the hands of his Master of the Horse, Meschid Bég by name, a man very ripe in years, but of some renown as a soldier. This general had, in March 1442, entered Transylvania from the side of Wallachia had beaten and nearly destroyed at Szt-Imre* the Transylvanian army led by the Bishop of the Diocese, George Lépes, and had laid siege to Hermannstadt. It was to relieve this important city, a city founded by Saxon colonists from North Germany, led by a certain Hermann, who still retain the customs and cherish the traditions of the land whence they came—that John Hunyady, accompanied by Simon Kemeny, dashed into Transylvania. Meschid Bég, as soon as he heard of his approach, broke up from his lines with the bulk of his forces to meet him. The battle engaged in the sight of the inhabitants of the beleaguered city. The fame of Hunyady had reached the leader of the Osmánli, and, knowing well his influence over his troops, and that his death would decide the battle, he had been at great pains to obtain an accurate description of the armour he was accustomed to wear, and of the horse he was wont to ride. This description he had made known to his boldest sipáhis, and he had impressed upon them the advantage which would accrue to the man who should bring to him, living or dead, the wearer of that armour and the rider of that horse. The matter had not been so secretly conducted but that it reached the Hungarian camp. The leading officers of the army, the chief alone excepted, were in despair. It was scarcely possible that their beloved general, whose passion it was to charge at the head

* Near the town of Karlsburg.

of hi roops, should escape destruction. Then Simon Kemeny stepped forward and said: "At all costs must John Hunyady be saved from this snare. The one course to ensure this is that he should exchange armour and horse with me." For a long time Hunyady refused to accept the sacrifice. At last, convinced of its necessity, he consented. The battle joined, I have said, on a plain within sight of Hermannstadt. The sipâhis, noting that the leader of the Hungarian right answered the description they had received, dashed furiously upon that wing, and, after a desperate defence, broke it, and drove it back in complete disorder, killing Simon Kemeny and five hundred heroes who had agreed to form his body-guard and who fought round him. Meanwhile Hunyady had routed the Turkish right led by Meschid Bég in person. Victory, however, was still hanging in the balance when the garrison of Hermannstadt sallied forth, broke through the Turkish lines, freed the prisoners they found there, and fell upon the rear of the Osmánli. Then the defeat was complete. The Osmánli fled with eager haste towards the Wallachian frontier. Hunyady did not lose a moment in pursuing them. His men cut down thousands, amongst them Meschid Bég and his son, in the disorderly flight. The country between Hermannstadt and the Wallachian frontier was strewn with corpses. It was calculated that the loss of the Osmánli amounted to 20,000. That of the victors did not exceed 3,000.

The Osmánli had fled into Wallachia, through the pass now known as the Rother Thurm pass—the pass of the Red Tower. Through that pass to the banks of the Danube, John Hunyady drove the remnants of the Osmánli army. Then, collecting his trophies, he loaded, with a portion, a van drawn by ten horses, placed on the top of all the heads of Meschid Bég and his son, and despatched it to George Brankowitch, the despot of Servia. Having done that, he gave rest to his army, whilst he himself prepared to meet the storm which he was confident the defeat of the Osmánli would certainly excite.

Amurath was absent, as presently to be related, in Asia Minor.

Unable, then, to take the field in person, he despatched Schaháb-u-dín Pasha, with an army eighty thousand strong, to deal with the daring Hungarian. With the boast that the very sight of his turban would cause the Christian host to flee, Schaháb-u-dín entered Transylvania. Hunyady had collected round him but fifteen thousand men, but, with this handful, he rushed to meet the invader. The two armies came in contact at Vasag (September 1442). Strange as it may seem, having regard to the disparity in numbers of the combatants, the battle was decided in the first half-hour. The terrible charge of Hunyady carried all before it. The front line of the Turks having been pierced, the whole army gave way. From that moment it became a slaughter rather than a fight. The number of killed was never counted. But five thousand men were taken prisoners, amongst them the Pasha who had commanded the defeated army.

This victory brought the religious element into the field. When the news of it reached Rome, Pope Eugene IV. summoned the Christian nations to a crusade against the unbelieving Osmánli. The better to carry out his purpose, he directed Cardinal Julian Cesarini to rouse the people of Germany and Bohemia. Julian was well-fitted for such a purpose. Audacious, unscrupulous, worldly, never having possessed, or, if he had once possessed, having lost the conscience of an honest man, he was just the instrument to carry out a policy which was founded neither upon justice nor upon public right. Seldom has an appeal to the ignorance and superstition of mankind been more successful. Since the battle of Nicopolis, there had not crowded into the Hungarian camp at Ofen men from so many different nationalities. Germans, Hungarians, Bohemians, Servians, Poles, and Wallachians formed the army, led by Ladislas, King of Hungary, with John Hunyady as his right hand. On the 22nd July the army broke up from Ofen, and marched on Semendria. At that place the Danube was crossed. Then John Hunyady was despatched in advance into Servia with a chosen corps of twelve thousand horsemen, whilst King Ladislas, whom Cardinal Julian accompanied, followed, at an interval of a two days' march, with the main body. The Osmánli were not ready to meet them. Three isolated columns were, indeed, marching on Nissa, on the Morara, with instructions to unite there and form the nucleus of an army. Hunyady was not slow to learn this fact. Forming his plans on the moment, he advanced with great celerity, caught the Ottoman columns separately (3rd November), and inflicted upon each singly a crushing defeat. It was an action quite in the style of the General Bonaparte of 1796, and deserving of the highest praise. The capture of Sofia followed this victory, whilst the remnants of the defeated columns took refuge in the passes of the Balkans. Hunyady then resolved to march on Philipopolis.

The march from Sofia to Philipopolis was, in the face of a determined enemy, a march full of danger. The range of the Balkans running from west to east, a distance of about two hundred and thirty miles from the lands whence spring the Isker and the Maritza, divide Bulgaria from Roumelia, and constitute the watershed between the Ægean Sea and the Danubian lands. The road from Sofia to Philipopolis crosses the westernmost part of the range. This part is very defensible. The Emperor Trajan had been so struck with its features that, aiding nature by art, he had, in the granite mountains, formed an almost impregnable mountain barrier, which, for many years, bore his name, and which in all the earlier histories is described as Trajan's Gates. The pass in which these gates were made is known as the Pass Succi. Trajan called the water power of the Isker to his aid in the defence position of the natural pass—the Pass Succi—which thus gained likewise the name of "the water pass"—whilst, by means of the depression caused by the rushing waters, he constituted, near Isladi, to the east of the original pass another strong defensive position—in fact, a second door or gate to the country beyond it. These two passes formed, at a height of about four thousand four hundred feet, the means for a very effective defence.

The Turkish commanders were well aware of the strength of this position, and when they heard of the defeat of their three columns near Nissa, and of the march of Hunyady upon Sofia, they determined to utilise to the utmost the advantages which the position in the Balkans would secure to them. They despatched thither, then, a working division to pile up stones, to fell trees, and to employ all the means in their power to render impossible ingress from the north into the passes. Nature came to aid them in these efforts. The winter had been a very severe one. Snow had fallen heavily —the river waters were frozen. Whilst the Osmánli, with all their resources behind them, well supplied with provisions and clothing, could rest in comparative ease, the invaders would have to clamber up slippery paths, exposed to a pitiless and continuous fire from the Turkish marksmen; after that, to storm, if they should reach so far, defences which had been made impregnable. Then, when their physical forces had been exhausted in the fruitless attempt, Amurath could send out his troops to avenge the defeats of Hermannstadt, of Vasag, and of Nissa.

Certainly the Turkish commanders neglected no precautions to produce this result. A day or two after the Hungarians had set out from Sofia, their main army reached the defensive position. With the exception of the Janissaries, who were with Sultan Amurath in Asia Minor, there were assembled to guard those passes the best troops of the Ottoman empire: the Sipáhis, who on many a field had by their splendid charges changed doubt into certainty; the Akindischi, or irregular cavalry; the Ajabi, or irregular infantry:— all were there, ready to avenge the defeat of their comrades by races whom they had held in supreme contempt.

Meanwhile Hunyady was approaching. The strength of this leader was akin to the strength which, in more modern days, was the strength, pre-eminently, of Frederic and of Napoleon. It lay in the absolute belief of his soldiers in, and their devotion to, himself. The same men who, under King Albert, had been so scared by the passage of the Danube by a Turkish force as to flee without fighting, were now ready, at a sign from their chief, to attempt the impossible, in the face of the best army the Osmánli could produce, and to attempt it with the conviction in their minds that whatever their chief should order they could and would accomplish.

It was on the 24th December that the vanguard of the Christian army ascended the steep slopes of the Balkans leading to Pass Succi. The way, naturally difficult, had been made still more so by the craft of their enemy. Every night the Osmánli had poured buckets of water down the steep incline, and this water, freezing, seemed doubly to secure their position. Still, in spite of this, in spite of every difficulty, the soldiers of Hunyady climbed upwards: they reached the western entrance to the gate, the water gate as it was called, the real Pass Succi. They reached it only to find it was impregnable. It was not to be stormed. After some vain attempts which only served to display the impossibility of the task he had undertaken, Hunyady drew off his men. Not, indeed, that he had renounced his aim; on that he was more than ever bent: it was to gain it by a new method. Hitherto he had made his attack on the true Succi Pass, that to the east. In that he had failed. But it occurred to him that the very fact of his attacking only one face would cause the other to be less guarded. He would at all events try it, and, to deceive the Osmánli, would act as though he were preparing for a new and more pronounced assault on Pass Succi. Withdrawing, then, to the rear his own chosen warriors, and instructing them to make their way round to the western pass, he brought up fresh troops to the front of the Succi Pass, and gave orders that they should busy themselves in preparing, in the sight of the enemy, an attack which, however, they were not to make, until they should have certain news of his success. This done he hurried to join his picked soldiers, led them to the western or Isladi pass. Evening was setting in as he made his assault. The entrance was not so difficult as that of the other pass, the frozen waters of the river furnished a better standing-ground, and, better still, the Osmánli were rather off their guard. Nevertheless the difficulties to be surmounted were enormous. If the assailants were terribly in earnest, the defenders were not less determined. It is on such occasions that the influence of a great man is divine. And, here, the great man was the leader of the Hungarians. Throughout that long Christmas eve, Hunyady displayed all the qualities which genius can command; his men fought as they had never fought before. Just as the day so revered in the Christian world dawned, he succeeded in planting the standard of the Cross on the crest of the blood-stained rock. Then, all was over. The Osmánli, beaten on the field they had chosen, fled in despair, to join, and, in joining, to communicate their dismay to a reserve force which was assembled at Jalovácz. Hunyady gave them no rest; joined by the King, Cardinal Julian, and the other commanders, he pushed through the Balkan range, descended its southern slopes, and found the Osmánli still at Jalovácz. It was the last day of the year 1443. The Christian leaders resolved to crown that day with a victory which had practically been gained before the battle joined. The allied host attacked the Osmánli. The battle, hotly contested for four hours, terminated then in the complete victory of the Christians. Among their prisoners were several of the most distinguished nobles of the Sultan's court.

Amurath had taken taken no part in any of these engagements. That very autumn the Prince of Karamania had broken the treaty and devastated the country as far as Angora. Against him Amurath had proceeded in person, had driven back his troops, and was making a triumphant progress through his dominions when the disasters I have recorded summoned him back to Europe. Anxious to secure peace on terms which should not be unfair, he offered to restore to the Voivode Drakul Wallachia, to George Brankowitch his two sons and the fortresses of Semendria, Chehir-Keni, and Krushovaz. Then he proposed terms to Hunyady. That warrior referred him to his sovereign. Negotiations then took place. Amurath was called upon to make great sacrifices. He made them frankly and without reserve. He agreed that Servia and the Herzegovina should be restored to George Brankowitch; that Wallachia should be under the overlordship of Hungary; that for the ransom of Mahmud Chelébi a ransom of seventy thousand ducats should be paid. These conditions were signed at Szegedin the 12th July 1444. Both parties solemnly swore to carry them truly into effect, the Turkish commissioners on the Korán, King Ladislas on the Bible. We have now to see how these oaths were kept.

Very shortly before the Peace of Szegedin had been signed a circumstance had occurred which had profoundly affected the Sultan. This circumstance was the death of his eldest and well-beloved son, Alla-u-dín, a prince of the highest promise. Under the stroke Amurath bowed, broken-hearted. Only forty years old, he had already drunk deep of the cup of misery. He had seen an uncle, real or pretended, the first Mustapha, supported by Manuel, perish fighting against him. Another Mustapha, also a relative, in revolt, had similarly, in spite of his expressed order, lost his life. And now, his eldest son, the apple of his eye, had been taken. His next son, Muhammad, was but twelve years old, and there were other sons still younger. Amurath had a great affection for his children, and he could not but picture to himself what their fate would be if he were to die before measures had been taken to secure their lives. After deep meditation, tired of glory, careless of pomp, yearning after the "lettered leisure" which has been the dream of so many statesmen, Amurath came to the conclusion that he could secure every end by resigning in favour of the son who had now become the eldest. In this way alone, he felt, would the succession be assured without bloodshed, and the lives of his children preserved. He carried this idea into practice, made over the empire a few days after the signature of the Peace of Szegedin to his son Muhammad, and retaining for himself only the administration of Menteschе, Sarukhan, and Aïdin, retired to Magnesia.

The news of the abdication of Amurath and the transfer of the Sultanat to an untried boy in his fourteenth year, came to the Christian leaders assembled at Ofen at a very critical moment. Ten days before they had sworn in the most solemn manner to maintain a treaty for ten years with the Osmánli. Two days before they had received letters from Cardinal Condolinieri, Admiral of the Papal fleet in the Hellespont, and from John Paleologus, the Emperor of Constantinople, pointing out that the time was singularly favourable for suddenly attacking the Turkish possessions in Europe, inasmuch as the Sultan, confident of peace, had returned to Asia. Upon the top of this came the news that Amurath had abdicated, and that a child reigned in his stead! A revulsion of opinion immediately ensued. The solemn oaths on the Bible, taken only ten days before, were forgotten. The Papal Legate, Cardinal Julian Cesarini, who had borne a prominent part in that sacred ceremony, promptly shifted his colours, and adjured the King, in the name of the Holy Trinity, of the Virgin Mary, and of St. Stephen, to break his plighted oath. Cardinal Julian was supported by the representative at Ofen, of Venice, and by all those who placed expediency in a higher scale than honour. It is to be regretted that John Hunyady himself, whose name and character were till then without spot, was, after long hesitation, not proof against the promise of the title of King of Bulgaria if he would only give his support to the war party. The policy prevailed, and it was resolved to invade the possessions of the Osmánli on the 1st September.

Meanwhile the "infidels"—as the followers of the Prophet* were styled by the professing Christians, who had but just decided that perjury, when coupled with prospective advantage, was not a crime —had performed their part of the contract, and had evacuated all the countries which, by the treaty, they had undertaken to restore. Amurath was in his calm retirement in Magnesia; the young Muhammad was learning how to govern under the tutelage of the able men whom his father had trained. Not a single cloud was discernible on the horizon, when suddenly, on the 1st September,

* I spent thirty-five years of my life in constant association with Muhammadans, and I can aver I never once found a man of that religion who would break a promise accompanied by the placing of the right hand on the region of the heart. By that action the man pledged his honour, and I never knew it forfeited. In the country where I write this—in Hungary—I find the same experiences. The word of a Turk, accompanied by the action I have described, is regarded as safe as a Christian bond

less than two months after the treaty of Szegedin had been sworn to, the Hungarian army, only ten thousand strong, led by King Ladislas in person, accompanied by Cardinal Julian, crossed the Danube at Orsowa.

As had ever been the case, John Hunyady commanded the vanguard. But it was not alone on their army, which was small, nor on the prowess of Hunyady, that the leaders of the Christian host counted. I have mentioned in an earlier part of this chapter,* the name of George Castriota, better known as Skándar Bég, and who figures as the hero of Lord Beaconsfield's beautiful story, *The Rise of Iskander*. At the period at which we have arrived, George Castriota was the ruling Prince of Epirus. The youngest son of Ivan (John) Castriota, the youthful George, had been delivered by his father, with his three brothers, as a hostage to Amurath when that Sultan conquered Epirus. Brought up as a Muhammadan in the Sultan's palace at Adrianople; remarkable not less for the vigour of his body and the charm of his manner, than for the qualities of his mind, George Castriota so conciliated confidence and regard that, at the early age of eighteen, the Sultan entrusted him with an independent command. The favour with which he was regarded rose with his fortunes. His many acts of personal daring, alike on the field and in private quarrels, so won the regard of Amurath that he gave the young Epirote the title of Skándar Bég, or Prince Alexander. Up to the year 1443, Skándar Bég followed with fidelity the fortunes of his Muhammadan lord. But in that year, a year known as the year of the five victories of Hunyady, his faith in the supremacy of the Crescent faltered, and, always in heart an Epirote, he deemed that the hour of the deliverance of his country had arrived. After the three defeats of the Osmánli army by Hunyady in the autumn of 1443, followed by the capture of Sofia, Skándar Bég, then twenty-nine years old, quitted his colours, wrung from the Sultan's Secretary of State an order to the Governor of Croja to deliver over that city to the bearer of the despatch,

* Page 135 *note*.

reached, on the seventh day, at the head of three hundred men who had joined him during his flight, the upper range of the Dibra—the range which, running parallel to the Adriatic, covers, to the east, northern Albania. On this height he left his three hundred followers, with a like number of mountaineers who had declared for him, and proceeded alone to the gates of Croja. The governor of that city at once recognised the validity of the order presented by Skándar Bég, and delivered over to him his command. That night the Epirote chieftain opened the gates of the city to his six hundred, who proceeded, in the name of liberty, to massacre the sleeping Turkish garrison. Only those of them escaped with their lives who made a personal appeal to Skándar Bég, and to that appeal the young Epirote listened favourably only when the supplicant would agree to renounce the faith of Islam. The slaughter of the Turkish garrison of Croja was the signal for the slaughter of the Turkish garrisons in all the neighbouring towns and military posts. Between the southern range of the Dibra and the beginning of the northern range of the Tomoros, runs the road from Macedonia to Epirus. In the depression there existing, Skándar Bég posted himself with two thousand men, and summoned there to meet him the leading men of the cities of Epirus to consult with him as to the measures necessary to be taken for the freedom of their common country. The appeal was freely responded to. From all parts of Epirus there streamed to the appointed spot well-armed men with their retinues, in numbers calculated at twelve thousand. Prominent amongst the active adherents of the cause were the nephew of Skándar Bég, Hámsa, who had fled with him from the Turkish camp, and Moses Galento. The result of the conference was the resolution to recover promptly all Epirus by force of arms. Three thousand men were at once despatched under Galento to take Petrella, called by the Turks Arnaud-Belgrad, a small, but, from its commanding position, a most important military post. Galento took Petrella, granting terms to the garrison—a policy which was wise as well as merciful, for it encouraged the garrisons of other strong places to retire on similar conditions. It thus happened that within thirty days the patriots had occupied all the strong places in the land, and Skándar Bég was lord of Epirus. The memorable Christmas day, 25th December 1443, on which John Hunyady stormed the passes of the Balkans was celebrated by Skándar Bég, now became once more George Castriota, by his own solemn re-entry into the Church of Christ!

The Turks, however, did not so easily renounce Epirus. Ali Pasha, a brave but over-confident commander, was despatched with an army forty thousand strong, to recover that dependency from the man whom every honest Osmánli denounced as a deserter, a traitor, and a renegade. George Castriota—for it seems proper to call him by the name which he had himself resumed—was equal to the occasion. As, after the slaughter of the garrison at Croja, he had appealed to the patriotism of the Epirotes, so now, after he had freed Epirus, he appealed to the patriotism and religious fervour of his Christian neighbours. Again was the appeal well responded to. There came, to fight under his banners, each with his contingent, the lords of the two Albanias, of Sissus, of Musachi, of Dayna, of Drivast, of Montenegro, and others of lesser celebrity. In an incredibly short space of time, Castriota had at his disposal a force of fifteen thousand well-armed and devoted soldiers. With these he took post on the lower Dibra about forty-five miles from Croja. Here, in the spring of 1444, he was attacked by, and completely defeated, Ali Pasha. Never had the Turks been more thoroughly beaten. It is said that they left twenty-two thousand dead on the field of battle. Certainly two thousand of them were taken prisoners: they lost twenty-four standards, and their army dissolved.

The abdication of Amurath and the Peace of Szegedin seemed to secure to George Castriota the independence of his native land. He now occupied a very important position, and when the Christian lords and prelates assembled at Ofen came to the conclusion that perjury, when practised against an unbeliever, was lawful, they recognised that position by appealing to him for assistance.

Castriota promised to march with 30,000 men to co-operate with them. For the moment, however, circumstances—to be related presently—prevented him from performing that promise.

It is time to return now to the crusading army, 10,000 strong, which had just crossed the Danube at Orsowa, under the command of King Ladislas, its vanguard led by John Hunyady. The army marched without hindrance to Nicopolis. There it was to have been joined by the reinforcements led by Drakul, Voivode of Wallachia. Drakul indeed appeared, but it was to protest with all his eloquence against the undertaking of so great a war with means so insufficient. "The hunting-retinue of the Sultan," he declared in the council of war assembled to discuss further proceedings, "is larger than your army." Hunyady, enraged at Drakul's opposition, gave him the lie direct. Thereupon the Voivode drew his sword upon the Hungarian, and there is no saying to what extent his embittered feelings would have driven him, but that he was promptly disarmed and placed under close arrest. Ultimately Drakul had to compound for his freedom by a promise to contribute a considerable sum of money, and 4,000 men, led by his son, to the expedition.

From Nicopolis the leaders of the crusade issued a manifesto announcing their intention to free Europe from the presence of the Osmânli, and warning the Muhammadan inhabitants of Sumen, Mahoratz, Petretz, Varna, and Galata to quit their several abodes and to proceed forthwith to Gallipoli, whence they would be able to obtain easy shipment to Asia. Some obeyed; but the majority, not yet doubting the power of Islam, remained. The army, meanwhile, had continued its march. Skirting the slopes of the Balkans, it had traversed the plains of Bulgaria, devastating and destroying all that it met on its path—even to the Greek churches of the people whom they professed to wish to free from the Osmânli yoke. The towns of Sumen and Petretz, which refused to open their gates, were taken by storm, and their inhabitants put to death. At length, increased by the Wallachian contingent and other reinforcements to double its original strength, the army reached

Varna. Thence it had begun its march towards Gallipoli when its leaders were startled by the information that Amurath had eluded the Papal fleet which was to have barred to him the passage of the Bosphorus, and was marching towards them. They had but time to reach the vicinity of Varna when they learned that Amurath was within two miles of them.

It becomes us now to inquire how it was that Amurath, who had abdicated the sovereignty and retired to seclusion in Magnesia, came to lead his warriors against the Christian invader. That illustrious prince, greater even as a man than as a prince, had divested himself of all his dignities, confiding in the solemn and sacred promises of his Christian enemies—promises which seemed to ensure ten years of peace, a time sufficiently long to enable his son Muhammad to attain the years of wisdom. In thus acting Amurath had no after-thought. His first and main desire was to secure the succession without bloodshed, and save the lives of his younger children; his second to devote his life to those intellectual pursuits which possessed for him a more than ordinary fascination. Yet—strange freak of Destiny!—scarcely had he reached the place he had chosen for his retirement than the success of the schemes, to secure which he had chosen to abdicate, was imperilled by the perjury of his Christian enemies! His part of the contract he had performed religiously and scrupulously. He had diminished the extent of his empire to secure peace to the remaining portion. A great modern statesman, a statesman who still lives and who holds in his hands the scales which regulate the balance of power in Europe—the illustrious Prince Bismarck—has declared that a State which begins to recede has taken the first step towards its fall. Certainly history points out that no nation has ever made a recession, by whatever pure motives it may have been actuated, without awaking the greed of its neighbours.* We have seen in the instance before us

* Witness the receding step taken by the present Government with respect to the Transvaal. Of that step the German colonisation of West Africa and the action of the same Power relative to the islands in the Australasian seas are the direct consequences.

how it was that the abdication of Amurath decided the perjured crusaders to attempt to drive the Turk out of Europe.

Still, when the news reached him of the action of his Christian enemies, Amurath made no sign. He had abdicated; his son must now display the qualities which were in him. Far otherwise was it with the councillors of the young Muhammad. These men, able as they were, had always leaned on their sovereign. His had been the brain that inspired, his the clear cool courage which had ever animated. Without that strong centre support, without that counsel which never had failed to hit the true point of the discussion, these men were divided and helpless. They could not lean upon a boy in his fourteenth year, promising as he might be. They hastened, then, (the Grand Vizier, Khálil Pasha, at their head,) to Magnesia, sought an interview with their late sovereign, laid before him the difficulties and dangers of the situation, and implored him to resume the authority which he had resigned.

Amurath was very unwilling. He listened patiently to the arguments urged by the ministers of his son, and then answered them thus: "You have a master; it is for him to defend you. What! do you envy me the repose I have so well merited after all I have done and suffered for the empire?" But the ministers were not content with this reply. They continued their solicitations. At length Amurath yielded. Resuming his authority, he raised an army of forty thousand men, not only bought off the commanders of the Genoese and Venetian vessels, but hired their vessels of war as transport ships,* crossed to Galipoli in one night, marched thence on Varna, and reached the vicinity of the crusaders' army on the 6th November.

In the councils of that camp opinions were divided. The Cardinal Legate, the Bishops of Erlau and Waradin, were all for intrenching their camp and awaiting the enemy's attack. The

* Engel, vol. iii. pp. 79, 80. Amurath paid at the rate of a ducat per man transported to Galipoli. Engel adds: "The cupidity of some Venetian and Genoese galley-captains decided that night, for centuries, the fate of the Turkish empire and its neighbours."

Osmánli, they argued, outnumbered them in the proportion of two to one; the position lent itself to defence; their right occupying a chain of hills, their left bound solidly to it by the centre; it could be stormed only by the display of qualities which had not been very conspicuous amongst the Osmánli in the campaign of the preceding year. An attack on their part, on the contrary, would expose their army, in case they should not succeed, to absolute destruction. These arguments were combated by John Hunyady with all the force of his impetuous nature. He accepted the appeal to the experience of the preceding year, as proving that a vigorous attack by a daring body of Christian warriors, well led, and confiding in their leader, could not fail. His voice carried all before it, and it was decided on the evening of the 9th November—for the Sultan, when once in sight of the Christian host, had halted for four days—to attack the army of the Sultan the following morning.

Amurath, on his side, was well content to meet the attack. Under his own personal leading his troops had never failed him, and he felt confident that could he induce them to stand firm against the first impetuous attack, he would make the Christians repent bitterly their broken faith. He ranged his army in three lines. In the front line were the Asiatic troops, mostly infantry, led by Karad Shah, Beglerbég of Asia Minor. Then came the Sipáhis, under the command of Tusa Khán, Beglerbég of Rumelia. The reserve Amurath commanded in person. There he stood, surrounded by his Janissaries, their front covered by a palisaded ditch. Close by his side he had planted in the ground a lance, bearing on its point the broken treaty, a memento to his soldiers that they were supporting the cause of fidelity to a plighted oath against perjury.

On the side of the Christians the arrangements for the battle had devolved, by consent, upon John Hunyady. That leader, in a reconnaissance made early in the morning, had noted the position of the Osmánli, and their great superiority in numbers. He resolved, then, in conformity with his usual tactics, to break their strongest defence with masses of cavalry, then to take instantaneously the unbroken wings, held meanwhile by his infantry, in flank and rear. In this view he massed all his cavalry, except five hundred,* without guns, on his left, near the lake opposite to the post occupied by the Béglerbég of Rumelia. On the hills on the right he posted the bulk of his infantry, led by the Bishop of Grosswarden. The guns, supported by a very thin line of footmen, occupied the space between the two wings. In the second line, commanding the reserve, stood King Ladislas, at the head of five hundred chosen Polish horsemen; by his side Prince Stephen of Báthory. Close behind each wing and the centre were the whole baggage-waggons, so ranged that in the case of a repulse the infantry might re-form behind them, and under their shelter offer a resistance sufficiently prolonged to enable the cavalry to rally.

Just before Hunyady ordered the attack, a storm of wind from the lake blew with such force across the plain that all the Hungarian standards, that of the King excepted, were blown to pieces. The violence of the wind lasted but a few minutes; then Hunyady, not giving his men time to draw discouraging omens from the occurrence, charged with all his force the soldiers of Asia. The shock was so terrible that not all the exertions of Karad Sháh could induce his men to stand firm. They gave way as the cornstalks yield to the reaper, and, before many minutes had passed, the place which they had occupied with so much display in the early morning knew them no more! It seemed as though Hunyady would justify the proud boast he had made in the council-room. One third of his foes had disappeared without causing him any appreciable loss. He was re-ordering his horsemen to attack the second line, the once dreaded Sipáhis, when his eagle eye missed the four thousand Wallachians whom the son of Drakul had led to the first charge. Those Wallachians, true to their plundering

* In front were five squadrons of his own Hungarians; then the Wallachians, commanded by the son of Drakul; then a troop of Hungarians; then the horsemen of Bishop Simon Bozgon; then the Croatian squadrons; then Cardinal Julian's crusaders.

instincts, had no sooner recognised the defeat of the first line of the Osmánli, than they had swept out of the line and, making a long detour, had fallen upon the baggage of the Turkish army. But Hunyady did not know this. His memory reverted at once to the scene at Nicopolis—to the day when Drakul had drawn upon him his sword, and, placed then under restraint, had been compelled against his will to agree to despatch a contingent of Wallachians to co-operate with the invaders. In the disappearance of the Wallachians, then, Hunyady saw, or thought he saw, a concerted action between father and son, a revenge for the insult of Nicopolis. Believing that they had simply returned sullenly to camp, he ordered his remaining troops to halt whilst he rode back to force the missing troopers to retake their place in the line.

The pause in the attack saved the Osmánli. Looking at the turn the battle afterwards took, it is scarcely to be doubted that had the defeat of the Asiatic troops been followed by an immediate assault on the Sipáhis, they would not have held their ground.* But the absence of Hunyady was—regarding it as an incident in the lifetime of a battle—a very long absence. He sought, in vain, the Wallachians; then, being in the camp, he lost more time by riding to the King to inform him of his success, and to beg him on no account to quit the ground upon which he then stood, as, in the event of the new attack not succeeding, it was behind the unbroken horsemen of the King's body-guard that he would re-form his squadrons. Ladislas promised to comply.

Then John Hunyady returned, and led his men to charge the second line of the enemy. But the Osmánli were no longer the men whom he had left suddenly terrified by the defeat of their first line. The delay had given them confidence. The very leader of the beaten line, Karad Sháh, had ridden to the Sultan to reassure him, and the Sultan had communicated his own confidence to his men. The Sipáhis, then, met the charge of the Hungarians

* Von Hammer says that they had been already thrown into disorder by the Wallachians, but his statement is scarcely borne out by the authorities on whom he bases his narrative. The account given by Engel (vol. iii. pp. 81–3) is far more trustworthy.

with a confidence not inferior to their own. Never were the qualities for which they were famous more conspicuously displayed. Driven back by the impetuosity and the weight of the first attack to the very edge of the palisaded ditch behind which the Janissaries were massed, they rallied then, and, forcing back their enemy, recovered the ground they had yielded. The day seemed almost lost for Hunyady, when a chance shot laid low the leader of the Sipáhis, the Beglerbég, Tara Khán ; the Hungarians profited by the sudden dismay caused by his fall, charged, as King Charles would have charged at Naseby, and snatched out of the fire the victory over the enemy's second line! The Sipáhis in their turn fled in disorder. There remained now only Sultán Amurath and the Janissaries. But before he could attack these, Hunyady felt himself compelled to clear the field completely of the Sipáhis. He followed them up, then, with vigour; perhaps, indeed, too far, looking to what followed. How the battle would have gone had King Ladislas kept the promise he had made to the great Hungarian, may perhaps be doubtful, for the Janissaries were yet unbroken, and they were fighting for and under the eyes of their beloved sovereign. But King Ladislas did not keep that promise. The Polish horsemen who formed his body-guard had noticed with increasing impatience the defeat and dispersion of the Sipáhis. Feeling then no doubt but that the battle was gained, and dreading lest they, by their inaction, should be excluded from all share of the plunder, they urged upon the King not to allow Hunyady and the Hungarians to appropriate all the honours of the day, but to charge at their head and by that charge decide the victory. For one moment, Ladislas, mindful of his promise to Hunyady, hesitated. But the hope that it might fall to him, the King of Poland and of Hungary, to vanquish the great Amurath in personal encounter, overcame every scruple. Before, then, Hunyady had returned from the pursuit of the Sipáhis. Ladislas gave the order to gallop to the battle-field, and to charge.

We must return for a few moments to Amurath. That illustrious man had beheld with many fluctuations of feeling the progress of the battle. The defeat of his Asiatic troops—the long pause which followed—the splendid gallantry and all but decisive success of his Sipáhis—their final overthrow—all these events, following quickly one upon another, had caused his mind to alternate from despair to hope, from hope again to despair. When at last the defeat of the Sipáhis had become absolute, and he saw himself and his Janissaries exposed to the full brunt of the attack of the victorious army, he, for a moment, gave himself up for lost. Only, however, for a moment. The sight of the stalwart forms and resolute countenances of his Janissaries, the confident expression of his A'gas, and, it may have been, a glance at the violated treaty on the lance by his side, reassured him. Little time, however, had he for reflection. The palisaded ditch had not stopped the Polish lancers. In another moment those gallant men, led by their King, had joined battle with his Janissaries. A terrible hand-to-hand combat now ensued. In the midst of the desperate fight a Janissary struck the King's horse with his axe, and brought it and the rider to the ground. In another second the King's head was severed from his body, and placed on a pole to be set beside the treaty he had violated. The sight of their sovereign's head thus paraded struck the Polish horsemen with dismay. They paused in their attack; the Janissaries renewed their onslaught with terrible effect. Of the five hundred Poles who had crossed the palisaded ditch, only two returned! *

Hunyady, meanwhile, having driven the Sipáhis far from the field, had returned to the place where he had left the King. To his surprise, instead of the King, he found his long lost Wallachians laden with plunder! These were too satisfied with the result of the battle, as it had affected them personally, to risk any more hard blows. Despite, then, of the entreaties of Hunyady, they set out at once with all their booty for their own country. The soldiers of Hunyady were too exhausted with their long combat to venture to

* Engel, vol. iii. p. 83. He adds: "Stephen Báthory also saved the King's standard." But Báthory was a Transylvanian.

affront the victorious Janissaries. That night they, too, with Hunyady at their head, abandoned the field and followed the Wallachians. The Sultan ventured upon no pursuit that night. But early the following morning he stormed the position still held by a remnant of the enemy's infantry, and defended by the baggage-waggons.

Such was the famous battle of Varna; such the Muhammadan answer to the perjury of Christian knights and nobles! It may not be uninteresting to note that in the battle and in the storm of the baggage-waggons, the following eminent perjurors lost their lives. Besides King Ladislas, Cardinal Julian, the Bishops of Erlau and Grosswardein, and Stephen Báthory—the same who had saved the King's standard from the Janissaries. The loss on both sides was very great. In the actual battle that of the Osmânli exceeded that of the assailants; but defeat in those days was fatal to an army. Hunyady had so small a following when he reached the Wallachian capital, that Drakul, mindful of his old grudge, had no difficulty in making him prisoner.*

Amurath was satisfied with having saved his country. He cared not for the position he had regained at the price of the repose which he prized above all earthly possessions. Still in the perfection of manhood, still blessed with the affection of his people, Amurath resigned once again his high office to return to the beautiful gardens and the sunny sky of Magnesia. Scarcely, however, had he resumed the happy life of comparative seclusion than he was once more summoned to save the country. Suddenly there appeared, to break that seclusion, Sarudsche Pasha, confidentially despatched by the Grand Vizier and the two Bégler Bégs, Ufghar and Ishak Pasha, to announce the revolt of the Janissaries. These formidable warriors, taking advantage of the terrible confusion caused in Adrianople by a fire which had consumed the market-place, had quitted their quarters, pillaged the city, and, in open defiance of the authority of the young Sultan, had taken their post on the

* He was forced by his own magnates to release him.

dominating hill of Butschul. The chief eunuch, against whom they had a grudge, had escaped their vengeance by a miracle. So threatening did their attitude become, that the Grand Vizier, considering any means permissible for the moment, offered to increase their pay by half an asper (a centime and a half) per day if they would return to their duty. The offer was accepted, but Khâlil Pasha and the two Bégler Bégs, sensible of the danger of the principle they had been forced to put into action, despatched, as we have seen, Sarudsche Pasha to Amurath, with instructions to set before him the state of affairs in the clearest manner, and to implore him, as he had once before saved the State from an alien foe, to rescue it now from unlicensed military rule. The clear mind of Amurath recognised at once all the dangers of the situation. Without a moment's hesitation he renounced his life of ease, and re-assuming the sceptre, returned to Adrianople. There, his very presence sufficed to bring the Janissaries on their knees. The young Sultan, whom the Grand Vizier had persuaded to quit the city for a few days' hunting, found, on his return, the palace occupied by his father. He at once, with apparent willingness, proceeded to Magnesia to assume the governorship which his father had resigned to re-ascend the throne.

Meanwhile the proceedings in the Grecian peninsula of George Castriota, the revolted child of the Sultan's own training, had been causing considerable alarm in the provinces still occupied by the Osmânli. To repress the continued aggressions of this patriotic warrior, the Sultan invaded Greece at the head of an army sixty thousand strong, took possession of the isthmus and the city of Corinth, and forced the princes of the Peloponesus to pay him tribute as their liege lord. He then advanced into Epirus, and, his army being now increased to a hundred thousand men, laid siege to Croja. In that city Castriota had left a garrison of four thousand chosen men, whilst he himself at the head of forty thousand should throw himself upon the communications of the invader. His plan, conducted with singular ability, succeeded to perfection. The garrison held the city, whilst Castriota rendered the position of the besiegers so difficult that Amurath was forced to retire. A second attempt, made the following year, was attended with a similar result.

Perhaps it was the non-success of the Ottoman arms in Epirus which prompted John Hunyady to resume the policy which had been so fatally foiled at Varna. Whilst the Sultan was besieging Croja for the second time, the Hungarian hero made every preparation to invade the dominions of the Osmánli. In spite of the personal opposition of Ulric Cillay—a rival candidate for the office of Ban of Slavonia—and of George Brankowitz, Despot of Servia, who wished to live in peace with his powerful neighbour, Hunyady raised an army of twenty-four thousand men—the finest and best equipped force he had ever commanded—and, to revenge himself of Brankowitz, entered (August 1448) and laid waste Servia. George, in despair, appealed to the Sultán, and Amurath, renouncing for a second time his attempt on Croja, hastened with an army a hundred and fifty thousand strong to his assistance. Hearing of the approach of the Sultan, Hunyady quitted Servia and took up a position on the famous plain of Kosowa.

To the readers of this chapter that plain is already known.* Fifty-nine years before, Amurath I., in the hour of victory, had been murdered there by the Servian, Michael Kibilowitch. The circumstances connected with that deed had not been of a nature to inspire the Hungarian leader with any supernatural confidence, for, despite the death of Amurath, the defeat of the Christain host had been crushing. Yet his determination to fight a battle on the Amselfeld was fixed and irrevocable. Eight successive messengers did the peace-loving Amurath despatch to Hunyady, each bearing a proposal to treat. The Sultan would have accepted any terms which would have secured from aggression the allies who had been true to him. But Hunyady was immovable in his resolve. Confident that the Sultan was very impressionable, he no sooner learned that Amurath had crossed the Sitnitza—an operation which lasted three days—than he released a Turkish soldier whom he had taken prisoner, and who had been greatly

* Page 134.

struck by the prowess and confidence of the Hungarian array, and allowed him to return to his own people. The relation of the released soldier so affected Amurath that, without waiting to reflect, he gave orders to recross the Sitnitza. Emboldened by this move, and eager to follow up his success, Hunyady, not waiting for the reinforcements which George Castriota was bringing to his aid, broke up from his position, crossed the little river near Brod, and marched against the enemy. That same evening (16th October) he ranged his army for the attack which he was determined to deliver on the morrow.

Hunyady had drawn up his army in the following manner. On the right were posted his own chosen Hungarians, and the Szeklers—the representatives of a small Hungarian colony which occupied the district of Haromszék in Transylvania; on the left the Wallachians; in the centre the Germans, the Bohemians, and the Transylvanians. The Osmánli were formed as at Varna: the Asiatic troops forming the first line, the Sipáhis the second, the Janissaries, covered by a palisaded ditch immediately behind which were ranged the camels, the third.

Hunyady began the attack on the 18th in his accustomed manner. He led the thirty-eight squadrons which formed his entire cavalry to a brilliant and impetuous charge. The Asiatic troops of the Sultan, in obedience to a preconcerted plan, moved rapidly to their right, to leave the field open to the sweeping dash of the Sipáhis. The shock between the most famous horsemen in Christian Europe and the warriors whose name had already begun to strike terror far and wide beyond the borders of their country, was terrific. For a moment the superior numbers of the Osmánli seemed to carry all before them; but Hunyady supplied the place of numbers. Whilst his inspiring presence animated his men, his cool calm judgment detected the smallest mistake on the part of his foe. He was always at the right spot at the right moment. But the odds against him were enormous. Never had his attack been met with greater courage and determination. Scarcely had he repulsed the Sipáhis on one wing than their great superiority in men enabled them to appear in force on the other; nor was it until after many pauses, many rallyings, many fierce bids of the Osmánli for victory, that, late in the afternoon, he succeeded in driving them, broken and beaten, before him.

For a few minutes it seemed as though he might gain on the spot a great victory, when, all at once, with firm step and in strong array, the Janissaries, emerging from their intrenchment, marched boldly to the front and stopped the pursuit. Behind them the Sipáhis rallied, and, gathering renewed courage, dashed round either flank on their surprised enemy. Again was the combat renewed. But it was now no longer equal. To the number and renewed ardour of the Sipáhis were now added the steadiness and valour of the Janissaries, fighting under the eye of their Sovereign. The combined action of the two produced its natural effect. With the shades of evening, Hunyady was glad to lead back his horsemen to the position they had occupied in the morning.

That night Hunyady held a council of war. At that council a Karamanian noble, who, Muhammadan though he was, had from hatred of Osmánli supremacy followed the Hungarian camp, urged that, as it was clear that the action of the Janissaries had alone prevented Hunyady from gaining a decisive victory, a night attack should be made upon those formidable warriors. Hunyady approved and followed this counsel. But the gallant men who formed the body-guard of Sultan Amurath were not to be surprised. They met the attack when it came with so much firmness and courage that just before the dawn of the 19th Hunyady was glad to draw back, baffled but in unbroken order, to his camp.

He resumed the attack, however, on the Osmánli before the sun rose. This time he launched his full strength against the Asiatic troops of the Sultan, men less confident than their brethren when pitted against Europeans, and whose defeat, he felt certain, would not fail to cause discouragement and disorder in the ranks of the Sipáhis. But Amurath displayed on this occasion the military clear-sightedness for which he was renowned. No sooner had he seen the cavalry of Hunyady well engaged with his Asiatic soldiers than he despatched his Thessalian cavalry to attack the Hungarians in the rear. To meet this new enemy Hunyady had to form a double front—the second front facing in a direction opposite to that taken by the first. This caused him great perplexity. It necessitated his presence on the spot where his own men were combating. Whilst he was thus engaged the leader of the Wallachian contingent, hopeless of success, had sent to the Grand Vizier to propose terms. These were readily granted, and before Hunyady could release his own men from their false position the Wallachians had gone over bodily to the enemy. Hunyady was so discouraged by this desertion that he gathered round himself his most trusty warriors, and ordering his centre—the Germans, Bohemians, and Transylvanians—to the front to hold the Janissaries, fled from the field! This was the closing act of the second day's battle.

Ignorant of the flight of Hunyady the Osmánli began the third day's combat by an attack on the Hungarian camp, which was defended by the baggage wagons with guns at intervals, and occupied by the troops of the three nationalities who had been sent forward the previous evening to cover the flight of the Commander-in-Chief. At the outset Amurath conducted the attack with caution; but no sooner had he become aware that John Hunyady had fled than he launched his whole army, the guns covering its front, on the weak intrenchment. The Germans and Bohemians and the Transylvanians of Saxon blood—for the contingent from that country had been recruited mainly from Hermannstadt—defended the camp with all the stubbornness inherent in their race, nor was it until the shades of evening were falling that the death—with his face still to the foe—of the last German hero gave Amurath possession of the entire field of battle. The slaughter on both sides had been great. Nine thousand Hungarians, two thousand Germans and Bohemians had fallen to rise no more. Nor had the Wallachians profited by their treachery. Taken in the confusion for enemies, they had been treated as such by the Osmánli and had been killed to the number of six thousand. A considerable number of Hungarian cavalry, who had been engaged against the Asiatic troops of the Sultan, had been taken prisoners. There remained only three or four thousand to be accounted for, and of these the bulk had followed Hunyady in his flight to Servia. On the side of the Osmánli the loss in men had been still greater, but their victory had been complete.

The death of the Greek Emperor, John Paleologus, the following year, seemed likely to cause a contest for the possession of the waning throne. John had died childless. His natural successor was his brother Constantine, but the younger brother Demetrius, under the pretext that he alone had been born in the purple, claimed the inheritance. Constantine referred the matter to the decision of the Sultan, and Amurath conferred on the seventh ruling member of the family of Paleologus the sceptre which his son Muhammad was destined to break in his hands. The same year he had the happiness of celebrating the nuptials of that son with a princess of Sulkador, the daughter of the Turkoman ruler of that country.

Fifteen months later, 6th February 1451, the great Sultan, who had reigned gloriously for thirty years, who had twice, in the prime of life, abdicated, and resumed power only to save his country from the ruin which threatened it, was struck by apoplexy. He recovered his consciousness, made his will, and despatched an express to his son Muhammad, then at Magnesia. Three days after the first attack, the 9th February 1451, he expired. He was then only in his forty-seventh year.

Amurath contrasts favourably with all the Christian monarchs who were his contemporaries. Though engaged in many wars, he was a lover of peace. His word was to him more than a bond. He had a merciful disposition, a sympathising and affectionate nature. The only charge that has been brought against him was his fondness for the pleasures of the harem and the table; it is admitted, however, that he never allowed these to interfere with his duties as ruler. He was pious and charitable. Whenever he took a city he built in it a large mosque, a smaller mosque, an Imárat, a Madrasah or college for learning, and a caravansarai. He built the famous mosque of Adrianople, known as the mosque with the three galleries. Near this mosque he erected another building—a school of the traditions of the Prophet, the Turkish name for which it bore—and richly endowed it. He was the first of the Ottoman princes who constructed long bridges: that crossing the marshes between Salonica and Venishahr; another at Eskené with a hundred and sixty-one arches; and a third at Angora, the product of the toll of which constituted a revenue for the poor of Mecca and Medina, may be specially cited. Under the reign of Amurath II., moreover, poetry, jurisprudence, and the study of theology made marked progress. Amurath did much likewise for the re-organisation and discipline of the Osmánli army.

Muhammad, known in history as Muhammad II., was at Magnesia when he heard of his father's death. He proceeded with all haste to Gallipoli, and thence to Adrianople. Resolved to have no future rival in his path, he signalised his arrival there by the putting to death of his brother Ahmad, though he was still a child. The fact was that though he had been received with the greatest enthusiasm by all classes, though not the shadow of a shade of opposition had stood between himself and his accession to the throne of his father, Muhammad could not forget that he was but the son of a slave, whilst the mother of Ahmad was a princess of Sinopé. Not content with the murder of the young prince, Muhammad forced the mother to espouse, on the spot, a slave named Ishak. He would have treated his other stepmother, the daughter of George Brankowitch of Servia, in a similar manner, but that he dreaded lest such action should rouse that prince to hostility. He accordingly sent her back with every mark of honour to her own country. To greet him on his accession Muhammad received and replied graciously to embassies from the rulers of Servia, of Constantinople, of Ragusa, of Wallachia, of Genoa, of Rhodes, and of Hungary. With all of these a further truce for three years was arranged. From Karamania alone, his most southern neighbour in Asia, issued hostile sounds.

Ibrahim Bég, ruler of that principality, judging, from the fact that the dangers of the Ottoman Empire had twice forced the young Sultan to make way for a stronger will, that Muhammad might not now be able to cope with a formidable opposition, invaded Aïdin and Menteschê. A demonstration made by the Governor of Asia Minor, followed by the departure of Muhammad himself for the seat of war, promptly changed, however, the ideas of Ibrahim Bég and forced him to sue for peace. Muhammad granted that peace on easy terms.

The heart of the new Sultan was, indeed, at the present moment occupied by but one ambition. He had long felt, and every hour of his rule had confirmed the feeling, that, before anything else, the question of the continuance of the Greek rule at Constantinople must be settled. Muhammad entertained dreams of conquest outstripping, in the vastness of their range, the boldest which his ancestors had dared to conceive, and, to make of those dreams realities, the magnificent situation of Constantinople was necessary to him. Unscrupulous, energetic, allowing no sentiment to interfere with the gratification of his plans, careless of the rights and of the blood of others, Muhammad would have forcibly solved the question even if the behaviour of the Court of Constantinople had been guided by prudence. But that Court managed to irritate him in a matter on which he was most sensitive. There was residing in the city of the Eastern Cæsars an Ottoman prince, Orkhán by name, grandson of the Sulaimán who had been his own grandfather's rival. In the first moment of delight which had attended his accession, Muhammad had promised the Byzantine ambassador to pay yearly three hundred thousand aspers (rather less than four thousand francs) for the support of that prince. When the ministers of the Greek Emperor found that, either from accident or from design, the promised payments were not made, they allowed threats to escape them, that unless double the amount were at once disbursed, they would give facilities to Orkhán to pose as a rival candidate for the dignity and office of Sultan! Muhammad neither forgot nor forgave this injudicious utterance. Even the peace-loving Grand

Vizier, a friend of the Greeks, the Khâlil Pasha who had twice invited Amurath to re-ascend the throne, but who had yet been retained by Muhammad in his office—even Khâlil saw in this action the death-warrant of the Byzantine Empire. "If," he declared to the members of the Greek embassy, when the threatening words were repeated to him, "if Constantinople escapes the daring and wild impetuosity of my sovereign, qualities which I know that he possesses, then shall I believe that God protects your lies and your deceit. Fools! Scarcely is the ink dry of the treaty we have made with you than you come to Asia to frighten us with your customary bogey! But we are not weak and inexperienced children. If you can do anything, do it. If you wish to make Orkhán ruler of Thrace, make him so; if you would call the Hungarians across the Danube, call them; if you would recover your lost territories, try to recover them. But it is as well that you should know that you will succeed in none of these attempts: more than this—that what you now possess will be taken from you! I shall repeat our conversation to my master, and what he wills will happen!"

The events just recorded had happened whilst Muhammad was in Asia Minor, prosecuting his campaign against Ibrahim Bég, of Karamania, and had contributed much to secure favourable terms for that prince. Muhammad hastened back towards Adrianople, resolving to reserve his final dealing with the Greek question till he should arrive at that capital. But on approaching Broussa, an event occurred which roused all his anger. Outside that city the Janissaries met him in tumultuous array, loudly demanding a present. Dissembling his anger he ordered ten girdles of gold pieces to be disbursed to them.* A few days later, however, he sent for the A'ga of the Janissaries, boxed his ears, caused him to be bastinadoed, removed him from his office, and authorised his successor, Mustafa Bég, to introduce such changes in the organisation of the body-guard as would prevent a renewal of the

* This act formed a precedent, under succeeding Sultans, the amount being continually increased.

demand. Arrived at Gallipoli, Muhammad issued orders for the collection of materials to build on the Bosphorus a castle which should secure the position of an army besieging Constantinople. The Greek Emperor was terrified beyond expression on learning of this design, and used all the means in his power to propitiate his powerful neighbour. But Muhammad was not to be moved from his design. The work, begun the 21st March 1452, was completed three months later. It was a marvel of strength, the thickness of the walls being twenty-five feet, and that of the fortified towers thirty feet!

In vain did the Greek Emperor profess contrition; vainly did he humiliate himself so far as to send daily costly wines and viands for the consumption of the Sultan. In the eyes of that prince, the cup of Byzantine iniquity was full to overflowing. The "fiat" had gone forth. Constantinople was to become the seat of empire for the descendants of the Turkoman refugee from the fury of Chingiz Khán!

On the 21st June 1452 the Greek Emperor, hopeless of accommodation, closed all the gates of Constantinople, and imprisoned all the Turks in the city except a few eunuchs from the Sultan's harem. These he sent to Muhammad, with the following message: "The Emperor places the fate of the city in the hands of God. Its gates, in consequence of the broken treaty, he has closed, and its inhabitants he will defend to the best of his ability, unless the Almighty should please to dispose the heart of the Sultan to peaceful views." The Sultan, disdaining to employ excuse or pretext, at once declared war. For the moment he contented himself with pushing on his preparations for the siege. The 28th August following he spent three days in narrowly inspecting the fortifications of the city. He returned the 1st September to Adrianople. The next seven months he passed in feverish activity, himself superintending and even preparing the plans for the siege. He could talk and think of nothing else. He could not even sleep, so completely had that thought mastered every other. At length everything was in readiness. On Friday, the 6th April, the

Sultan appeared at the head of an army of two hundred and fifty thousand men before the doomed city, whilst his fleet, consisting of three hundred galleys and two hundred smaller vessels, blockaded it on the side of the sea.

The city of Constantinople had, up to the fifteenth century, long enjoyed a pre-eminence far surpassing that of the other European capitals, Rome alone excepted. So far back as 660 B.C. there had been built, on the ground now occupied by the old seraglio and the large garden attaching to it, the city of Byzantium. The magnificent situation of the city caused its possession to be disputed for many years by the Persians, the Spartans, and the Athenians. Three hundred years before the Christian era it was besieged, when garrisoned by the Athenians, by Philip of Macedon. Students of ancient history will easily recall to mind the legend which caused the defenders to assume as the peculiar design of the city the shape of the crescent. They will recollect how, on one dark night, the besiegers were creeping to the walls, in the hope of surprising the citizens in their sleep, when suddenly the rising half-moon revealed their presence to the startled Athenians, who, thus roused, were able to repel the assault. The design of the crescent adopted from that time forth to stamp the coinage, was a heritage which attached itself to the life of Byzantium. It were long to relate how the city with the other Greek portions of the Macedonian Empire, fell later under the sceptre of Rome, and how, in the year 330 A.D. Constantine the Great re-named it after himself, and made it the joint-capital of the known world. In the middle of the sixth century the Emperor Justinian contributed much to its embellishment. We next hear of Constantinople as being involved in the troubles which accompanied the dissolution of the Roman Empire; we read how, in 616 A.D., it was besieged by the Persian King, Chosroes; ten years later, by the Avas; in 668, by the Arabs, who, however, were repulsed by the employment of Greek fire. A similar result followed the siege of the city by the same wild people, in the years 716–18. During the ninth, tenth, and eleventh centuries the city was often hard pressed by a people who came from the southern parts of Russia. In the year 1204 Constantinople was conquered by the Crusaders, who founded in it a Latin Empire, which, however, fifty-seven years later, came under the sway of Byzantine lords. The city was the seat of government of a member of a ruling family of this race when it was besieged by Amurath II. in 1422. The representative of the same family, Constantine Paleologos, occupied the supreme position there when, thirty-one years later, the son of Amurath II. appeared before its walls, under the circumstances I have attempted to describe.

The Constantinople of 1453, the Istambol or Stamboul of the Turks, lay on a three-cornered tongue of land on the south-western point of the Bosphorus, formed to the north by a small re-entering bay, the Golden Horn, eating from that point into the land to a distance of about four miles, and to the south by the sea of Marmara. On the third or westernmost side of this triangle, which in fact is its base, the strip on which Constantinople is built is joined to the mainland of Rumelia. From the base, where it is about three miles broad, it extends eastward between the two waters I have mentioned to a point where the Golden Horn, the Bosphorus, and the sea of Marmara meet, and the tongue of land terminates in a well-rounded point. On this is the true Constantinople, looking to the south on the Sea of Marmara, and towards the outlet of that sea on the Dardanelles; to the east on the Bosphorus, marked out by its seven snake-like windings, by its seven hills, one rising at every winding, by its seven currents and counter-currents, having beyond its northern mouth the stormy Black Sea. The same winding stream at its western exit, or rather at its head, bending westward, forms the safest harbour possible to conceive, the most roomy, the most suitable for anchoring in, the most secure from storm and wind. This, from its form and its natural wealth, was called "The Golden Horn." This harbour forms one side, the shores of the Sea of Marmara another side, and the land walls a third side of the city. On the two first sides—that of the harbour and the Sea of Marmara—it was covered by a simple wall without a ditch, on the land side by a double high wall with high towers and two deep ditches. On each of the three angles of this triangle, the walk round which would require three hours, there rose a fortified castle. On the one end of the harbour, on the point both sides of which were washed by the sea, formerly called the Acropolis, stood the castle of St. Demetrius, by the church of the same name. Thus it happened that this suburb was called likewise the suburb of St. Demetrius. Similarly the sea was called the arm of St. George, from the fact that the church built on its shore had been so named. On the second angle of the triangle, at the further end of the harbour, and on its furthest point, stood the Kynegion, an amphitheatre for the display of contests between animals. Behind this was a large palace, used by the later Greek emperors in preference to any other, and whose name (the Blachernen) is now employed to designate the Greek quarter in that neighbourhood. On the third angle of the city, that is, on the further end of the land wall, there rose the fortress of the five towers, called the Kyklobion or Pentapyrgion, but known later as the Seven Towers. After we have noted the three angles of the triangle I propose to explore the sides of the angles, keeping only on their outside and leaving untouched the palaces and churches, which, however famous, do not enter into the narrative of the siege. On the shore turning towards the Sea of Marmara—on one end of which was the Acropolis, on the other the Five Towers—there were, between the two, two artificial havens, now filled in and enclosed within the walls. One of these was called the Eleutherian or Theodorian, afterwards the Wlangabostan, the other the Julian or Sophian, later the Kadrighalimali, or Gallery haven. The last was in the quarter Kondoskale, the short landing-place, formerly called the Heptaskalon or seven-stepped landing-place. The sides of both havens were adorned with palaces, the one in the Eleutherinian, the other in the Sophian style. Between the sea-girt angle of St. Demetrius and the Sophian or Julian haven, stood the lower imperial palace, the Bakolion, so called from the figures, cut in stone, of a lion and an ox. Between the Eleutherian or Theodorian haven (Wlangabostan) and the

Kyklobion or Five Towers stood the palace Psamatia, near the gate still known by that name.

I now turn to the city gates. On the second side of the city facing the water, that, namely, of which the Acropolis formed one point and the Blachernen quarter the other, there are now fourteen gates opening on the harbour. At the time of the siege, however, there were but five. Of these the first, and furthest outside, is the gate Kynegion, now the gate of the menagerie; then came the Xyloporta, or wooden gate, not to be confounded with the existing wooden gate on the side of the harbour, known as the Odun Kapusi; the third is the great gate of the palace of the Blachernen; the fourth the Petra gate, in the Greek quarter of the present day; finally, the Oraca, now the gate of the Jewish, or fish-market. On the lower side of the palace of the Blachernen was the bricked-up door of a subterranean passage. This, called then KERKOPORTA, or gate of fluted wood, is identical with the gate known in the present day as the Xyloporta.* It had been bricked-up by order of Isaac Comnenus, in dread lest the prophecy that the Emperor Frederic should enter the city by that gate might be fulfilled. Passing over the remainder of the side on which was the harbour, and the Five Towers on the sea face, we come to the gates on the land face. These must be minutely examined, because it was to this side that the whole power of the besiegers was directed, and each gate was specially indicated as a point of attack. On the land side, then, the first is the Charsish gate, called also Kaligaria, and now known to the Turks as the Egri Kapa, or crooked gate. The second, known in the present day as the gate of Adrianople, was called Myriandri, or Polyandri, signifying the gate of a thousand or of many men. In the siege of the city by the Avas the fiercest fight had taken place between this gate and that nearest to it. In the Turkish siege, however, the most desperate battle was fought between the third

* "Die *Κερκοπορτα* des Ducas xxxix. page 159, ist augenscheinlich das Thor *ξυλοκερκου* des Nicetas (ii. 4), und also eines und dasselbe mit der heutigen Xyloporta."—*Vide* Hammer, 516, and *note* page 667.

gate and the centre one, near the gate of St. Romanus, called also Top Kapusi, or the Cannon gate. On this gate are two towers, that of St. Romanus and of Bagdad. We come next to the gate the road out of which leads first to the Well Palace, now called Baliklu, then by way of Khegium (Chekmedshe) to Selymbria (Siliori). It bore, on that account, the triple name of the Well gate, the Khegium gate, and the Selymbria gate, which last name it now possesses. The fifth and last gate on this side mentioned in the history of the siege is the Golden gate, the portal through which the triumphal bands were wont to enter, and which, adorned with statues and portraits in half-relief, was the great gate of the city. It remained, however, walled up by reason of a prophecy that the Latins would enter the city through it; nor did the defenders, not following the action taken with respect to the Kerkoporta, to be presently related, remove the bricks during the siege,

To return to the actual siege. We have seen how Sultan Muhammad, leading an army said to have been two hundred and fifty thousand strong, encamped before the city on the 6th April. His army was in every respect well-appointed. Fourteen batteries of artillery were forthwith planted to play upon the land face of the city. Prominent among the field-pieces was a famous cannon, the most gigantic ever used in the history of sieges. To draw this huge piece fifty oxen were employed, whilst two hundred men walked on either side of it to keep it steady in position, whilst fifty waggoners and two hundred pioneers cleared the way in front. Muhammad himself, who had his tent pitched behind the hill in front of the Charsis, or Kaligaria gate, was terribly disappointed to find this famous piece failing almost at the outset. To load it required two hours, and at the eighth discharge it burst, killing the Hungarian engineer who had cast it. The piece was, however, repaired, and, though it was at first unskilfully worked, it became very efficient under the direction of an ambassador from John Hunyady, who paid an opportune visit to the Osmánli camp!

The lines of the besiegers stretched from the original Xyloporta on the one side to the Golden gate on the other. In front of the face ending with the latter were ranged a hundred thousand men; fifty thousand assailed that terminating with the palace of the Blachernen, whilst a hundred thousand horsemen were drawn up in the rear. In the centre stood the Sultan with his body-guard of fifteen thousand Janissaries, whilst Saganos Pacha was posted with a few thousand men on the height behind Galata on the side of the harbour opposite to that connected with the city. The fleet, the composition of which has already been mentioned, was commanded by Balta Ogli, a Bulgarian renegade.

Before describing more particularly the events of the siege, it is well to glance at the means at the disposal of the besieged. The armed and disciplined Greeks occupying the city numbered no more than four thousand nine hundred and seventy-three. To these may be added two thousand permanent foreigners, and from three to five hundred Genoese despatched to the aid of the sorely threatened city.* The Venetians at the last moment had held back their fleet: the armed ships at the disposal of the Emperor consisted, therefore, of only three Genoese, one Spanish, one French, three from Candia, and five others. In guns they were equally deficient. Those that they possessed, though smaller than those of the besiegers, were yet too heavy for the city walls, which shook and trembled under each discharge.

Still, despite their inferior provision, the besieged were animated by that resolution, that courage of despair, which goes so far to supply the lack of material. What though the fire from the fourteen Osmánli batteries battered down a portion of the walls: during the night that followed, the zeal and energy of the besieged filled up the breaches with timber and with fascines. Did the Sultan bring miners from Servia to work their way under the walls, the Greeks brought into play their famous fire, and drove them from their subterranean strongholds. By the daring and skilful employment of this fire they succeeded in burning to ashes a machine

* To the Greeks may be added about a thousand armed for the occasion, whilst from various sources the number of the foreigners was raised to three thousand.

which had destroyed the tower of St. Romanus, wringing from the Sultan himself an exclamation of wonder and admiration!

Nor did the superiority of the Turkish fleet make itself very signally manifest. On the 15th April that fleet had appeared before the city. Five days later five ships, four of them Genoese and one a Greek, which had been delayed by contrary winds at Chios, appeared in the Sea of Marmara. A hundred and fifty Turkish vessels at once sailed to prevent the entrance into the harbour of these new arrivals. It was a clear bright day; the sea was calm; the walls of the city were crowded with spectators; the Sultan, mounted on his horse, watched with eager glance from the European shore the proceedings of the combatants. There seemed no possible doubt as to the issue. It was a fight of a hundred and fifty ships against five!

But it was in numbers only that the Turkish fleet was superior. Of their hundred and fifty ships, eighteen only were galleys of war, and these were manned by inexperienced and untrained landsmen. To meet their adversaries the entire squadron had to run the gauntlet of the city walls, whence there rained upon them arrows, Greek fire, and masses of rock. The Genoese, on the other hand, trained to sea warfare, fought with all the coolness of experienced warriors: nor was the captain of the Greek vessel one whit, in that respect, behind his colleagues. The five ships caused terrible havoc amongst the inexperienced Osmánli. Many of their vessels ran against each other and became disabled; two caught fire. The battle seemed lost for the superior numbers. The Sultan, who had watched the fight with ever-rising fury and indignation, could at this point contain himself no longer. Digging the spurs into his horse, he dashed into the water to encourage his sailors, distant from him but a stone's throw, to renew their efforts. The Turkish sailors tried to respond to the call. In vain, however, and Muhammad had to undergo the pain of seeing the five Christian vessels sail unharmed, like Shadrach, Meshed, and Abednego in the burning fiery furnace, through his disordered fleet into the harbour!

A proud, impetuous, self-willed, unrestrained, and unscrupulous man was Muhammad II. The sight which he had witnessed roused all his fury, and that fury he turned upon the renegade admiral, to whose cowardice he attributed the defeat. Scarcely, in the first outburst of his anger, could he be prevented from having him roasted. The punishment he did award was terrible—degrading more to the executioner than to the victim. Forcing four slaves to hold the admiral stretched out flat on the ground, the Sultan himself administered to him a hundred blows with a heavy club, whilst one of his soldiers (an Ajábi) crushed in with a stone the eyes and cheeks of the victim!

At a council of war, held by the Sultan after this, for him, untoward event, the Grand Vizier, Khálil Pasha, the same, be it remembered, who had twice in the lifetime of Sultan Amurath been the means of relegating Muhammad into retirement, gave his voice in favour of abandoning the siege. The second vizier, Saganos Pacha, brother-in-law and favourite of the Sultan; the Mullá, Muhammad Kuráni, the preceptor of his younger years; and the Shékh Akshem-ú-dín, who had excited the Osmánli troops by the eloquence of his preaching; took, however, the opposite view and urged the Sultan to persevere. Whilst the discussion was at its height, Muhammad asked the three combative councillors how they would propose to break through the chain which defended the harbour, how otherwise they would devise a plan for forcing an entrance into the city, and how, whilst it were attacked on one side, they would assail it on the other. The three councillors, puzzled, remained silent. Upon this the Sultan disclosed his plan. The same spirit of enterprise, writes von Hammer, whom I have closely followed in the account of the siege, which had made him spur his horse into the sea, urged him to propose to traverse the dry land in ships. His idea was to transport a certain number of his ships from the shores of the Bosphorus, where they lay anchored in front of Berchishtash, across the land into the inner part of the harbour. The distance to be traversed did not exceed five miles, but the ground was uneven and undulating. To overcome this difficulty the Sultan caused the entire way to be planked. He then had the planks greased with the fat of oxen and of rams. This having been accomplished, he set his fleet in movement. In the space of a single night seventy-two galleys, each with two ranks of rowers, and others with three or five ranks, with sails set to catch the wind, the pilot in front, the captain on the poop, drums beating and trumpets sounding, were transported from the shores of the Bosphorus across hills and valleys into the harbour. When the day broke, the defenders beheld to their dismay a Turkish fleet of more than seventy sail, descended as it were from Heaven, and anchored in the harbour opposite to the city walls!

The commander of the Genoese galley, Johannes Longus, a member of the noble family of the Gustiniani, resolved to make an attempt to burn the Turkish fleet at nightfall. His plan, however, was disclosed to the Osmánli by the Genoese of Galata, who throughout the siege employed all their artifices to profit themselves by serving alternately both the contending parties. Ignorant of this treachery Gustiniani started on his enterprise. The Osmánli allowed him to approach within a certain distance, and then directed at his ship a discharge so well directed that it sank, and a hundred and fifty of his choicest sailors were drowned. The Turks repaid the Genoese of Galata for their assistance by sinking one of their richest merchant vessels. The prisoners they had taken during the night they put to death before the walls of the city in sight of the besieged.

Having thus baffled the Genoese admiral the impetuous Muhammad proceeded to take the fullest advantage of the position he had so skilfully gained. He first constructed a kind of pontoon bridge across the harbour. This bridge was composed of large casks securely fastened to each other, and stretching from the shore to the ships which were anchored immediately under the walls of the city. Across these casks he caused to be fixed boards, fifteen feet broad —so that five soldiers could march abreast—three hundred long, and strong enough to bear the weight of guns. The defenders, viewing with dismay the proceedings of their enemy, made a serious attempt to render them abortive. For this purpose they commissioned a Venetian named Jacob Kok to destroy by fire alike the bridge and the ships. Kok manned four fast rowing-boats with forty chosen men, took with him Greek fire and other necessary materials, and, leaving two of his crew on the bridge, set out at nightfall on his errand. He succeeded in setting fire only to one galley before the Osmânli who were on the look-out sank his boats by hurling on them huge masses of stone. The one galley was burned, but the fire kindled on the bridge was extinguished before any serious damage had been caused. The next morning the Osmânli executed the prisoners they had made in this attempt in the sight of the defenders, who, not to be behindhand, slew two hundred and sixty Turkish prisoners and exposed their heads on the battlements.

The unsuccessful attempt made by Kok caused considerable ill-feeling between the Venetian and the Genoese section of the defenders, the latter attributing the failure to the bad-leading of the Venetian. Unmindful of the help rendered him at the time of the attempt of Gustiniani by the Genoese of Galata the Sultan now proceeded to erect a battery on the hill of St. Theodore, on the same side of the harbour as Galata, with the view to destroy the Greek and Genoese vessels lying at anchor before that suburb. Vainly did the Genoese urge him to spare their peaceful merchant ships. Replying that they were not merchant ships but pirates which had come to help his enemy, the Sultan ordered the battery to open fire. The first discharge sank one ship ; the others, however, cutting their cables, took post close to the houses of the town, which completely covered them. The batteries continued their fire indeed for a considerable time, but it is recorded that a hundred and fifty rounds succeeded in killing only a pretty woman —and that she was struck, not by the ball itself, but by a stone which the ball had displaced !

The siege had now lasted forty-six days, and still Constantinople held out. Her defences had, however, been considerably damaged. A continuous fire from the land side had destroyed four of the towers of the Kyklobion (the Five Towers), and had effected a wide breach in the defences near the gate of St. Romanus. The ditch in front of it had been filled in with the débris of the crumbling wall. Under these circumstances the Sultan, inspired either by merciful considerations or by a desire to be assured by the evidence of a trustworthy eye-witness of the exact state of the city, directed his brother-in-law, Isfendiar Ogli, to seek an interview with the Emperor. Isfendiar was to pose not as an ambassador, but as a well-wisher of the Greeks, who should urge the Emperor to spare the lives of his subjects by the surrender of a city which could no longer be defended. Taken before the Emperor in council, Isfendiar acquitted himself of his mission. In that supreme hour Constantine Paleologus proved himself worthy of the traditions of his house. He would indeed thank God, he said, if the heart of the Sultan, like the hearts of his ancestors under similar circumstances, were really bent on peace. He would remind him, however, that no Sultan who had besieged Constantinople had reigned long afterwards. He was willing to accept peace on the condition of paying tribute, but he would rather die than yield the city.

Muhammad received this answer on the 24th May. He at once announced to his troops that, five days later, the 29th, he would lead them to the storm. Summoning then to his tent the divisional commanders, he promised them the plunder of the city, with the exception of the walls and the public buildings, which he reserved to himself. The hearts of the soldiers were gladdened when their commanders announced to them this gift of their sovereign. The colonels of the Janissaries pledged themselves to victory, whilst they begged from their Sultan the release of their comrades who had been imprisoned after the unlucky sea-fight. The request was granted, and the whole army broke out into tumultuous rejoicings. Heralds went through the camp promising high posts to those who should first mount the walls, and death to those who should flee. Dervishes streamed through the camp, calling upon the soldiers in the name of the Prophet and of his standard-bearer to plant the banners of Islam on the walls of Constantinople. Soon after night had set in, the signal was given, by sound of trumpet, for a general illumination. On the masts of the ships and on the centre-poles of the tents, on the shores of the Bosphorus, on the heights behind Galata, in the centre of the harbour, along the lines of circumvallation, from the ports opposite the Blachernen to those fronting the Golden Gate, appeared illuminations lighting up the city walls and displaying the joy of the besiegers. The city itself seemed all but surrounded by the fiery half-moon which was thus suddenly formed, whilst from the camp there rose cries, repeated over and over again during the night: "there is but one God, and Muhammad is his prophet"; "there is but one God and none other like unto Him." The besieged beheld this illumination with changing feelings. At first they believed that the Turkish vessels in the harbour were on fire. Soon, however, the rejoicings in the camp, the dancing of the dervishes, the tumultuous shouts announcing assured victory, made them aware of the fatal truth. Dismay then filled their hearts. To the shouts of triumph of their enemies they responded with passages for deliverance: "Kyrieleison! Kyrieleison! Turn, O Lord, thy threatenings from us and save us from the hands of our enemy!"

The certainty that the storm was resolved upon, that there would be no quarter, that their only safety lay in the repulse of their enemy, rekindled the waning resolution of the defenders. Gustiniani spent the night of the illumination in filling up and repairing the breaches. He built a new wall of fascines to replace the walls about the gate of St. Romanus which had crumbled under the enemy's guns. Had everyone behaved like this noble man, the defence might have been successful. Little supported by the Greeks, refused by Lucas Notarias, Admiral of the Greek fleet, the guns he had asked for, Gustiniani, aided by his seven brave companions, Careto, Bochiardi, Fornari, Selvatico, Galetusio, Cataneo, and John the Illyrian, repaired the breaches, made sorties, watched and baffled the enemy's movements—behaved, in fact, as the heroic defenders of Arkât, of Prague, of Saragossa, and of Lakhnao behaved in the years that were to follow. The Sultan, who noticed his prowess, was wont to exclaim: "What would I not give to have Gustiniani on my side!" In vain did he attempt to win him over. He found him as proof against gold as against steel. The efforts of the Genoese hero were rendered enormously difficult by the condition of the city walls at the commencement of the siege. The Emperor had indeed, before that period, provided the money for their repair, but he had entrusted it to the monks, and these, instead of employing it for the purpose for which it had been given, had buried it!

The storm might now ensue any day, and it behoved the defenders to be prepared to meet it. The city walls were manned accordingly. It was naturally to be expected that the principal attack should be directed against the gate of St. Romanus—the centre of the wall on the land side. At that gate, then, the Emperor Constantine took post, having at his side Gustiniani and three hundred chosen Genoese, and Don Francis of Toledo. The gate next to it, that of the Thousand Men, now that of Adrianople, was guarded by the following of the brothers Paul and Antony Troilus Bochiardi of Genoa, men who during the siege had repaired at their own expense the heavy damages caused there by the Osmánli guns, and were now prepared to defend the post to the very last. The gate next to it on the land side, the Charsis gate, was confided to Theodorus of Caristos, and to the German, John Grant, the former a famous archer, the latter a noted artillerist. At the furthest end of the harbour side, at the gate of Kynegion or the Menagérie gate, as far as the church of St. Demetrius, was posted the Cardinal Isodorus—a man who, sent from Rome to effect a union between the Latin and Greek churches, had devoted his purse and his person to the defence of the Christian cause. The Venetian, Jerome Minotto, guarded with his men the palace of the Blachernen, whilst the Italian, Jerome, watched the lower seaside; the Genoese, Captain Leonard of Langasco, the Xyloporta—the wooden-gate—and the tower of Anemas. The defence of the entire remaining length of the harbour side was confided to the High Admiral, Lucas Notarias. Between the point of the

Acropolis and the lighthouse stood Gabriel Trevisano with four hundred Venetian nobles, whilst Andreas Dinio, the galley-captain, lay with his galleys before the entrance of the harbour. The Spanish Consul, Pedro Giuliani, defended the sea walls from the palace Bukoleon as far as Kontoskalion, and the Venetian, the defender of the Golden gate, from the same along the seaside as far as Psamatia. Between the Golden gate and the gate of Selymbria, or the well-palace, the defence was entrusted to the noble Genoese, Maurice Cataneo, who had already distinguished himself in the sea-fight before the harbour, whilst the learned Greek, Theophilus Paleologus, commanded between Selymbria and the gate of St. Romanus. To Demetrius Paleologus and Nicholas Gudelli were assigned no fixed posts: they were required to go the round of all, and to render help where it was required.

Meanwhile, confidence was fast waning in the camp of the Osmánli. Rumours of approaching assistance from Italy and from Hungary to the beleaguered city had considerably damped the ardour which the promises and martial instincts of Muhammad had so signally evoked. The three days which immediately followed the famous illumination were days of decreasing confidence and of increasing alarm. On the evening of the third day, however, the sudden appearance of a meteor darting through the heavens over the city somewhat revived the hopes of the credulous Osmánli. They regarded it as a sign of the wrath of the God of the Christians against His worshippers. Muhammad, who, partly under the pressure of the Grand Vizier, Khálil Pasha, and partly under the dread of the consequences of failure, had begun also to vacillate, summoned now, 27th May, another Council of War. At this meeting the peaceful recommendations of Khálil Pasha were overborne by the war party, and the decision of the previous council was confirmed. On the following morning the Sultan gave his orders for the assault. Two strong columns were to assail the city on the land side; eighty galleys were formed in line between the Wooden gate and the gate of Platœa; the remainder of the fleet was formed, from the entrance of the harbour (the gate Oraia, now the Fishmarket gate) round the point of St. Demetrius, past the sea-gate of Hodegetria to the harbour of Blanka (now Wlanga bostan)—an enormus crescent. The Osmánli columns on the side of the Golden gate were one hundred thousand strong; on the left side of the camp fifty thousand; the supports numbered one hundred thousand. In their centre stood Muhammad with his body-guard of fifteen thousand Janissaries. As night fell, the arrangements alike for assault and for defence were completed. The leaders on both sides did their utmost to inspire their men with confidence. The Emperor, after taking the sacrament, rode round the defences, speaking words of encouragement to the soldiers. It was only when the cry of the early morning cock heralded the approach of day that he had concluded his rounds and reached his post at the gate of St. Romanus. A very few minutes later the sound of the approach of the stormers gave the signal that the battle for the possession of the finest city in the world had begun. Contrary to ordinary custom, ancient and modern, the contest was not heralded by the discharge of cannon. To weary the defenders by keeping them constantly on the alert, giving them no time to rest, Muhammad, with the grey dawn of the morning, had despatched, as forerunners of the real assault, a storming party composed of recruits and invalids. But as the day broke, the fight engaged on both sides of the city. If the assault was terrible, the defence was stern and heroic. The losses of the Osmánli greatly exceeded that of the assailed, sheltered behind their walls. The noise was terrific, the shouts of the men, and the exhortation of their leaders being mingled with the music of horns, of trumpets, and of kettle-drums, shortly to be drowned by the roar of cannon. At the close of two hours of fierce combat the assailants had made no progress; the defence was successful at all points, and it seemed as though Muhammad would be baffled. In vain did the Sultan employ executioners with whips and iron rods to flog his soldiers to renewed efforts; vainly, in his terrible anger, did he threaten them with his iron club. The stones hurled from the battlements, the destruction-spreading Greek fire, availed more than the rods of his executioners and his own threats. On every side the assailants were repulsed, and the day would have closed as a day of triumph for the assailed Christians, but for the consequences of an extraordinary act of gross negligence on their part.

I have already spoken of the splendid part which John Gustiniani had taken in the long defence. I have related how, on the day of the assault, he had been posted with his three hundred Genoese at the gate of St. Romanus, to fight under the eyes of the Emperor Constantine. Everything had gone well at this most important post. The Emperor, seated on his horse, had encouraged the Genoese soldiers alike by his words and by his example. Suddenly a ball or an arrow, it is not certain which, struck Gustiniani in the fleshy part of the arm or of the leg, some say of both. Turning to the Emperor, he begged him to continue firm in the defence whilst he proceeded on board to have his wounds dressed. Constantine begged him not to quit the defence. But Gustiniani, not to be withheld, hastened on board a vessel, exclaiming to the Emperor, in reply to his inquiry as he quitted him, that he was going "to the place to which the God of the Turks had opened a way for him," and, ascending the side of a Genoese ship, sailed for Galata. The extraordinary desertion of their captain produced so great a confusion and dismay amongst the men that the effects were soon apparent to the enemy; and Muhammad, whose sharp glance nothing escaped, took advantage of it to induce the Osmánli to make one charge more for victory. But this attack, though made with an almost superhuman energy, had been already repulsed, when there occurred the SURPRISE, the consequence of the negligence I have spoken of, which decided the fate of the city.

In the description of the gates of the city I have referred to one known then as the KERKOPORTA, or gate of fluted wood, which had been bricked up by Isaac Comnenus to prevent the fulfilment of the prophecy that the Emperor Frederic would enter the city through it. Unhappily, a very few days before the assault, the

Emperor, to facilitate a sortie from a quarter whence no sortie would be expected by the Osmánli, had caused the bricks which barred up the entrance to be removed. No such movement had been attempted, and so accustomed had the defenders become to regard Kerkoporta as a gate which was not attackable, that its defence had not been considered in the general scheme. The Emperor had forgotten, and those in whom he confided had forgotten, to provide a living substitute for the bricks which had been removed. This combined negligence and forgetfulness proved fatal to the defenders of Constantinople. Their valour had with great difficulty repulsed the assault which the flight of Gustiniani had provoked—the last, the most desperate, the most nearly successful of all the assaults prior to the surprise; the attack of the Osmánli had failed; no efforts of Muhammad could have brought another man to the breach; he would have had to retreat, as, on a similar occasion, his father before him had retreated; when suddenly all was changed — changed by a SURPRISE, the work of but fifty men!

The defence at the gate of St. Romanus had practically succeeded, and the besieged were rejoicing in their victory, when suddenly a rumour reached them that the Turks had penetrated within the city. The rumour was so far true in that fifty Turks, stealing through the undefended Kerkoporta, had attacked the defenders of the palace of the Blachernen in the rear. The number of the assailing body was contemptible, and a little presence of mind would have taken its measure and baffled its purpose. But the danger of a surprise consists in the disturbance it causes in the mental vision of the men surprised. That two, or three, or twenty Turks should attack the defenders in the rear was sufficient to induce the belief amongst them all that the bulk of the Osmánli army had penetrated within the walls; to expel from their minds coolness and the power of reflection, and to replace those qualities by panic and dismay. So it was on this occasion. Fruitless was it that a few men, cooler than their fellows, turned upon the fifty penetrating Turks and slew them to a man. The moral effect caused by their ingress had before that done its work. At every post except at those on the harbour side the garrison had become demoralised. We have seen how rumour, based on the fact that fifty Osmánli had penetrated within the city walls, had quickly reached the victorious defenders of St. Romanus and had persuaded them that the city was in the power of the enemy. At once, without a moment's reflection, there succeeded to the exhilarating joy of victory overwhelming panic! In vain did a few daring and devoted men, the learned and not less brave than learned Theophilus Paleologus, Don Francisco Toledo, and John the Illyrian, whose cooler brains had not been deceived, exert themselves to the utmost to maintain order and reanimate their men. Under the influence of the panic positions from which the Turkish army had been repulsed were abandoned in disorder. The Sultan was not slow to notice the disorder and to explain its cause to his soldiers. The men whom neither reproaches nor iron clubs could have induced to stir one foot to another assault, were persuaded to dash forward with alacrity by the information that some of their own men were within the city and were appropriating to themselves the spoil destined for the whole army. Even then, had the defence been vigorous, the assault would have met the fate of its forerunners. But, this time, there was practically no defence. Panic—the panic caused by the surprise effected by fifty men—had done its work but too well. The Emperor Constantine, who throughout that long assault had shown himself worthy of the high position he occupied, who had never faltered, but, remaining cool and undisturbed, had repaired the flight of Gustiniani, and shown himself a hero as well as an Emperor, felt in this fatal hour that the surprise had given victory to the assailants. True, however, to himself, to his high position and lofty lineage, and feeling that a noble death was preferable to a degraded life, he led those who would yet follow him in a last charge against the stormers. When his followers shrank back he turned to them reproachfully with the words: "Is there, then, no Christian who will kill me?" then, dashing forward, fell under the swords of the Osmánli. The defence had long since been practically abandoned. Through the gate Charsis

—the crooked gate—over a bridge formed of the dead bodies of the defenders—the Osmánli dashed in. The survivors of the garrison were intent only on reaching the still untaken defences on the harbour side. Many succeeded in reaching the ships, but a still greater number were baffled by the superstition of the guardians, who locked the gate of egress and threw the keys into the sea. This remainder then took refuge in the great church of St. Sofia, where after a short delay they surrendered to the victors.

In this manner Constantinople was captured. The surprise of the Kerkoporta created a panic which gave the city to the Osmánli. The sudden appearance of fifty men behind the defences produced the result which the sacrifice of thousands in front of them had failed to obtain. It was nothing that those fifty men were killed. The disturbance of the defenders' minds, and the consequent panic, had made their work memorable for ever in history. The Surprise which thus produced the fall of Constantinople is in all respects worthy of study. Not only does it emphasise the truth that great results may spring from very little causes, but it inculcates the necessity of attention to every detail, even to the very smallest. It shows moreover that unless a surprise be met with coolness and self-possession it may prevail over a victory already achieved, and may render useless defences from which an enemy had been repulsed. Had only twenty men been posted to protect the Kerkoporta the enormous army of Muhammad, collected with so much effort and with so many circumstances of pomp, would have been dealt a blow under which the empire of Othman would have reeled. The Surprise by fifty men turned the decisive victory already practically achieved into one of the most decisive defeats of which History gives record!

CHAPTER V.

INNSBRUCK.*

THE league of Schmalkalden was the protest of certain Protestant princes of Germany against the high-handed course adopted by the Emperor Charles V. at the Diet of Augsburg. On the 25th of June 1530, the document called the Augsburg Confession, a document signed by princes so considerable as John, Elector of Saxony; George, Margrave of Brandenburg; Ernest, Duke of Lüneburg; Philip, Landgrave of Hesse; Wolfgang, Prince of Anhalt; by the representatives, likewise, of the free cities of Nuremburg and Reutlingen; had been presented, and read in the Latin and German languages, to the Emperor and the Diet. The Emperor had replied by moving the Diet to issue decrees which, refusing liberty of conscience, prevented the possibility of reconciliation, and by then dismissing it (November 1530). A few weeks later (22nd December) the Elector, John, of Saxony and his son, John Frederic, the Landgrave, Philip of Hesse, and a few other Protestant princes, met at the town of Schmalkalden in Cassel, and in the parlour of the still existing inn called the Krone (Crown) drew up the heads of an agreement for the common defence of their

* My principal authorities for the events recorded in this chapter are: Ranke's *Deutsche Geschichte im Zeitalter der Reformation*; Robertson's *Charles V.*; Dolce's *Vita di Charles V.*; *The Memoirs of Sir James Melvil of Halhill*; Arnold's *Vita Mauritii, electoris Saxoniæ*; Schleukert's *Moritz Churfürst von Sachsen*; Langern's *Moritz Churfürst von Sachsen*: Reutter's *Geschichte Herrn Moritzen.*

religious and civil rights. Ten days—from the 22nd to the 31st of December—were spent in the deliberations. The princes and nobles then separated, only to meet again in greater number on the 29th of March of the following year. The five days following were spent in finally arranging and signing the articles which constituted the agreement known to the world as the Schmalkalden League.* To this league there gave adhesion the Elector of Saxony and the Landgrave of Hesse, the Prince Wolfgang of Anhalt, the Dukes Philip, Ernest, and Francis, of Brunswick and Lüneburg, the Counts of Mansfeld, the cities of Magdeburg, Bremen, Lübeck, Strassburg, Lindau, Constance, Memmingen, Biberach, Isny, Reutlingen, and Ulm; a little later, Esslingen, Brunswick, Göttingen, Einbeck, and Goslar. John of Saxony and Philip of Hesse were nominated chiefs of the League, which was constituted for six years; its final articles were settled at Frankfurt in December 1531. The power of the League soon made itself felt. Charles V. was engaged in a desperate war with the Turks. The League refused him all assistance. Supported by the Duke of Bavaria and the King of France, they declined likewise to recognise the election of his brother Ferdinand as King of the Romans.† On the 22nd of January 1532 the League signed a treaty of alliance with Frederic I., King of Denmark. Its growing power forced at length the Emperor to treat. Pressed on all sides, and dreading lest it should obtain the active support of the Kings of France and England, he ratified, the 2nd of August 1532, at Ratisbon, a peace, which, signed by the Protestant princes on the 23rd of July preceding, and known as the Peace of Nuremburg, gave him at least time. So skilfully had he arranged its clauses, that in reality he conceded nothing, and the treaty secured to the Protestants the rights only which they already possessed, whilst, as I have said, it gave to the Emperor time—the time to meet the other dangers which stood in the way of his carrying out the policy

* Schmalkaldischer Bund: called generally, in our language, "The League of Smalcald," the German name having been anglicised.

† He had been elected the 5th January 1531.

which lay nearest to his heart—the repression of the civil and religious liberties of Germany. So great was the store he set upon this treaty in that especial respect, that between the years 1534 and 1545 he renewed it six times. The sudden peace, however, which Charles concluded with the King of France at Crespy (18th September 1544), and the convocation of the Council of Trent (1545), roused the Protestant princes from the fools' paradise in which they had permitted themselves to slumber. Supported by the Pope, by his brother Ferdinand, and, strongest support of all, by Maurice of Saxony, the representative of the Catholic branch of the reigning electoral house and first cousin of the then Elector, John Frederic, Charles launched against the Protestant princes an edict placing them at the ban of the Empire; then, marching against them with an army drawn from all parts of Europe, he surprised and totally defeated their leaders at the battle of Mühlberg (24th April 1547). Amongst the prisoners taken at that battle was John Frederic, Elector of Saxony, son and successor of the Elector, John, whom we have seen one of the founders of the League of Schmalkalden, and who had died in 1532. To mark alike his sense of the constant opposition displayed by John Frederic and his father to the imperial plans, and of the invaluable assistance rendered him by his cousin Maurice, the Emperor transferred on the spot the electoral dignity from the former, whom he caused to be tried and condemned to death, to the latter, and commissioned Maurice to carry out to its bitter end the execution of the law launched against the defeated princes. Before I proceed to narrate the other acts by which the Emperor followed up his victory, it is fit that we should glance at the antecedents and the character of the new Elector of Saxony.

The eldest son of Duke Henry the Pious and of his wife, a daughter of Duke Magnus of Mecklenburg, Maurice of Saxony was born at Freiberg the 21st March 1521. In 1539, he embraced, at Torgau, the doctrines of Luther, and, eighteen months later, the 9th January 1541, he married the daughter of Landgrave Philip of Hesse. The same year his father died, leaving a will, in which he devised his domains in equal portions to Maurice and his younger brother Augustus. The Elector of Saxony, however, the John Frederic whom he subsequently supplanted, insisted upon Maurice taking the entire inheritance. He soon had cause to regret his insistance. An attempt made by the Elector to levy the tax imposed in the territories subject to the Duke of Saxony for the purpose of carrying on the war against the Turks, without having previously obtained the sanction of the Duke, roused the anger of Maurice, who raised an army to prevent this infraction of his rights. The intervention of the Landgrave Philip and of Luther brought about, however, an accommodation, and, during the same year, 1542, Maurice joined the imperial army in Hungary. There he so distinguished himself that the Emperor proposed to him to follow his banner in the war which he declared the year following against the King of France. The successful campaign of 1544, in which Maurice bore a distinguished part, terminated, the 18th September that year, in the Peace of Crespy, and Maurice was then free to turn his attention to the relations between the Emperor and the princes adhering to the Reformation, then extremely strained. Despite the favour with which he was regarded by the Emperor, Maurice supported the leaders of the Schmalkalden League in their action for the defence of the reformed religion, although he carefully abstained, partly from policy, partly to avoid the drain of money which contribution to its funds would have entailed upon his resources, from openly joining it. Contact with its leading members very quickly satisfied him that their cause was already lost; and his quick intelligence, sharpened by his boundless ambition, indicated to him that his fortunes would best be served by making common cause with the Emperor. As Charles was as eager for such a result as was Maurice, it took but little time to arrive at an understanding, and in a secret treaty, signed at Ratisbon the 19th of June 1546, Maurice, openly nominated by the Emperor Protector of the duchies of Magdeburg and Halberstadt, was secretly assured the electoral dignity and the hereditary possessions held by his cousin, John Frederic, on the sole condition of rendering efficient assistance to the Emperor. For a while, dreading the effect which was being produced by the rumour then industriously spread that the sole object of the Emperor was to extirpate Protestantism, Maurice held back; but, in October of the same year, armed with the powers which he had deemed necessary, he presented to the estates of his duchy an imperial decree guaranteeing the maintenance of Lutheranism within its borders, and obtained from them in return the means necessary for an invasion of the electoral territories. He then entered Saxony, rapidly occupied the towns and strong places of the electorate with the exception of Wittenberg, Gotha, and Eisenach, and received everywhere the homage due to the Elector. This sudden invasion of Saxony produced a decisive effect on the campaign. The army of the princes of the Reformed League was at the moment holding the Emperor in check on the banks of the Danube. The news that Saxony had been lost forced its leaders to retreat and divide their forces, each eager to protect his own. John Frederic, returning to Saxony, did indeed force Maurice to evacuate all he had gained, and even occupied his duchy. But the price at which this apparent advantage was gained was the ruin of the cause which he had at heart. The division of the Protestant forces gave the Emperor time to assemble his veterans from all parts of Europe. On the 27th March 1547 he was joined, at Eger on the frontiers of Bohemia, by Maurice; marched thence, at the head of twenty-seven thousand men, against the isolated John Frederic, forced him to fall back on Wittenberg; then, pressing on, reached, on the 24th April, the Elbe, in front of Mühlberg, which John Frederic had only quitted a few hours before. In pursuit of him, Maurice and the Duke of Alba crossed the river at the head of four thousand light cavalry, caught the main body of the Saxon army at the village of Kossdorf, and held it till the main body of the imperialists arrived. The defeat of John Frederic was then crushing and complete, and Maurice, who had been twenty hours on horseback, who had twice been nearly killed, returned to camp to find that his cousin, placed long previously at the ban of the empire, was a prisoner, and that he was Elector of Saxony in reality as well as in name!*

The victory at Mühlberg confirmed Charles V. in his resolution to extirpate civil and religious freedom from Germany. Even had he been the great man he believed himself to be, he would not have succeeded. But Charles, though resolute and determined, wanted that breadth of view which enables a man to grasp all the bearings of a difficult and delicate situation. He believed in the power of persecution, and, never caring to consider the evil effect on the general plan of a crooked course, always seized it, if that course seemed to his essentially narrow mind to offer an immediate advantage. Never did he give more complete evidence of this want of breadth of view than after the victory of Mühlberg. Not content with treating an Elector of the Empire as a common criminal, he displayed towards the Landgrave, Philip of Hesse, a bad faith which enormously weakened his moral influence throughout Germany. Seeing that the fate of the reformed party had been decided at Mühlberg, Philip, in April 1547, applied to his son-in-law, Maurice of Saxony, and to Joachim II. of Brandenburg, to make terms for him with the Emperor. Charles duly empowered the two Electors to arrange with the Landgrave conditions of capitulation. Philip signed the conditions and surrendered, whereupon Charles, acting upon the advice of his minister, Granvelle,† consigned him to prison, and kept him there for five years, upon the pretext that he refused to recognise the validity of the Council of Trent. This act of bad faith not only roused a general indignation throughout Germany, and even in Europe, but it contributed, it can scarcely be doubted, to alienate Maurice. It is

* John Frederic was brought to trial, and on the 10th May was sentenced to death. On the 19th, however, he purchased his life by signing an act by which he renounced the electoral dignity in favour of Maurice, on the sole condition that the latter should guarantee a pension of fifty thousand florins to his son. He remained, however, a prisoner of the Emperor till the period of the surprise of Innsbruck. He died in 1553.

† Better known as Cardinal Granvelle; wonderfully painted by Motley in his *Rise of the Dutch Republic.*

certain that his representations to the Emperor on the subject were couched in language which showed that he regarded the detention of Philip as a wrong done to himself.

But Charles proceeded further in his impolitic course. Convoking a diet at Augsburg, he proceeded to the city which might, in one sense, be termed the cradle of the Reformation, took possession violently of all its churches, caused them to be purified from the Lutheran contamination, and re-established in them the exercise of the rites of the Roman Catholic faith. He then presented to the diet a decree known as the Interim, a declaration conceding the use of wine at the administration of the Communion, the marriage of priests, and some trifling changes considered by the Protestants to be of importance, and sought by means of it to reconcile the two religions. Although this compromise was disapproved alike by Catholics and Protestants, Charles had influence enough to cause the diet to accept it. But it remained, virtually, a dead letter, and when Charles ventured to lay before the same diet a scheme for the federation of Germany, a scheme which would have rendered the imperial power predominant, not even his influence could induce its members to consider it. Whilst Maurice of Saxony had used all his persuasive power to cause the failure of the imperial scheme of federation, he had supported the Emperor in his endeavours to carry through the scheme of religious compromise. He did even more. Returning to Saxony, he used all the arts of which he was a consummate master, to induce his subjects to accept it. To this end he re-established the University of Wittenberg, and displayed so much deference towards the Protestant leaders, that Melancthon was induced to write a letter in which he accepted the principle of the compromise. The result, after many discussions, was the acceptance by the Reformers of Saxony of a formula of agreement, which, under the name of the Interim of Leipzig,* came to be generally accepted.

* The Interim of Leipzig was accepted, by the Landtag (diet) held in that town, on the 22nd of December 1548. Whilst it recognised the supreme authority of the Pope and of the Bishops, so long as that authority was not misused, it brought back into the Lutheran rites many Catholic customs which had been dropped, and

Maurice had not beheld unconcerned the efforts of the Emperor to limit the power of the princes of Germany, and, if he had seemed to aid him in his religious policy, he had displayed so much conciliation in carrying it into effect that he had really strengthened the bases of his power. Charles, meanwhile, had proceeded to the Low Countries to proclaim there his son Philip as his successor; had returned thence to hold at Augsburg a second diet to enforce the Interim, and thence had moved, sick and weary, suffering from gout and disappointment, attended by but a handful of troops, to Innsbruck, there to watch, and, as much as possible, to influence, the proceedings of the Council of Trent, whilst, by means of Maurice, he kept his hold on Germany. To that end he had commissioned Maurice to march against the city of Magdeburg, which he had placed at the ban of the Empire.

Maurice, the most astute politician in Europe, a man who possessed to perfection the art of so concealing his own designs that it was impossible even to suspect them, had read the Emperor's heart, and had resolved not only to defeat his plans, but to make himself the arbiter of the situation. Whilst, then, in pursuance of the commission given him, he marched against Magdeburg (November 1550), he carefully considered how he might utilise the opportunity. Magdeburg was strong, and Maurice had no intention, for the moment, of endeavouring to overcome that strength. His mind was soon made up. He resolved to break with the Emperor, but to use his own time for the rupture, so that the blow, when he did deal it, should be decisive. Meanwhile, he kept up appearances. He hemmed in Magdeburg, and defeated a relieving force led by Count Heideck, a partisan of the extreme Reformers, and took Heideck prisoner. But it was remarked by those about him that, a few days later, Heideck was released, and became one of his most trusted advisers. From the first his emissaries had travelled far and wide. Very soon he came to an under-

declared generally that forms and ceremonies might, at will, be disregarded. The larger form was drawn up by Melancthon, Bugenhagen, and Major. The fiery spirits of both religions refused to accept it. It disappeared before the more general scheme of toleration promulgated in the Peace of Passau, 31st July. 1552 standing with John, Margrave of Culmbach, with John Albert, Duke of Mecklenburg, and with William, son of the imprisoned Landgrave of Hesse. In concert with these, he (May 1551) despatched agents to France and England to negotiate with the Courts of Paris and London a secret treaty against the Emperor. This mission so far succeeded that on the 5th October 1551 Henry II. of France concluded with the representative of the confederate princes a treaty by which, becoming protector of the Protestant princes, he agreed to pay them two hundred and forty thousand crowns monthly for the first three, and sixty thousand crowns for the subsequent, months of the alliance; the princes engaging, in return, not to oppose the intention formed by Henry of becoming imperial vicar for the bishoprics of Metz, Toul, Verdun, and Cambray. Whilst this negotiation was pending, Maurice still lay before Magdeburg, making a show only of pressing the siege, whilst he augmented and disciplined his forces. But no sooner did he learn that the negotiations at Paris had terminated favourably than he entered into correspondence with Albert, Count of Mansfeldt, who had the chief command in the besieged city, and, by means of Count Heideck, now entirely in his confidence, communicated to him the scheme he had formed for the release of his father-in-law, the assurance of liberty of conscience, and the limitation of the imperial power. A complete understanding was thus arrived at. Magdeburg was to accept terms similar to those which the Emperor had granted to other Protestant cities, but Maurice pledged his word that its fortifications should not be destroyed, and that freedom of conscience and the enjoyment of their ancient immunities should be secured to the inhabitants. The better to bind Maurice, the magistrates went so far as to elect him a Burgrave, an office which gave him a very ample jurisdiction in the city and its dependencies. These secret terms having been settled, Magdeburg surrendered, and Maurice entered the city with all the pomp due to a conqueror (9th November 1551).

The task entrusted by Charles to Maurice was now completed, and it became the latter to disband his army. But this was just the course he was determined not to adopt. The army raised and maintained with so much pains was necessary for the carrying out of the plans which he had concerted with his allies. The question he had to solve was how still to deceive the Emperor; how to retain his army and yet to produce upon the mind of Charles the impression that it had been disbanded. Maurice, an adept in dissimulation, was equal to the occasion. No sooner had he entered Magdeburg than he dismissed his Saxon soldiers, having arranged that they were to re-assemble at the first warning he should give them. As to the mercenaries, alike his own and those who had defended Magdeburg, he assembled them in the plain outside the city, paid them a portion of the arrears due to them, and ostentatiously dismissed them. The discharged soldiers enlisted on the spot under the banner of George, Duke of Mecklenburg,* a partisan devoted to himself, who bound himself to employ them and to pay them the portion of the arrears still owing to them. The arrangement was common enough in those days, and the Emperor, believing that Duke George had hired the troops to become sole master of the territories then comprised under the name of Mecklenburg, was completely taken in.

But it was necessary to maintain still longer the game of deception. The severe winter still kept Maurice from openly acting, yet the treaty with France was about to be signed, and it was scarcely to be expected but that some rumour of it would reach the ears of a sovereign who maintained spies in every capital. But Maurice had not in vain attended the councils of that sovereign. He knew well the essential narrowness of his mind, the patience with which he would pursue his autocratic aims to impose his will on Germany, alike in religious and secular matters. Maurice, a far greater man intellectually, thought himself then justified in meeting deceit by deceit, in beating the Emperor at his own game and with his own weapons. Knowing that it lay much at the Emperor's

* The several divisions of the duchy of Mecklenburg were united, in 1471, under one family, whose members ruled in common. It was not till 3rd July 1611, and, more specifically, the 3rd March 1621, that the formal division into the Güstrow and Schwerin branches was accomplished.

heart that the reformed states of Germany should recognise the authority of the Council of Trent, Maurice began suddenly to evince the greatest zeal in that direction; appointed Melancthon and other divines to prepare a confession to lay before that Council, and, by his ostentatious example, induced other states and cities which had adhered to the reformed doctrines to pursue a similar course. This astute policy singularly favoured the ultimate views of Maurice. The subject of whether or not a safe conduct should be granted to the Protestant deputies induced between the Pope and the Emperor a dispute which so entirely absorbed the attention of the latter that he had no leisure to watch narrowly the movements of his ambitious vassal.

Maurice was now almost ready for action. Before, however, absolutely declaring himself, he deemed it politic, with a view to strengthen his position with the princes of Germany, to make one more formal demand for the release of his father-in-law, the Landgrave of Hesse. The efforts of the ambassadors, despatched conjointly by himself and the Elector of Brandenburg on this errand, though supported by the representatives of Elector Palatine, by the Dukes of Würtemberg, Mecklenburg, and Zweibrücken, and other minor princes, as well as by letters from the King of Denmark, the Duke of Bavaria, the Dukes of Lüneburg, and even by the Emperor's brother, the King of the Romans, failed. Charles haughtily announced that he would communicate his intentions to Maurice on the arrival of the latter at Innsbruck, and that until then he declined to discuss the subject.

Morally strengthened by the refusal of the Emperor even to discuss the redress of a great wrong, Maurice now prepared to throw off the mask. Sending a skilful agent to Paris to warn Henry of his intentions and to remind him of the treaty, he issued orders to his Saxon troops to assemble, and sent an express to George of Mecklenburg to hold his mercenaries, who were quartered in Thüringen, ready to march at a moment's notice. When everything was in readiness, he had still one task of further deception to accomplish. This was to rid himself of that particular minister by his side whom he knew to be in the pay of the Emperor, and, at the same time, by that act to confirm the credulous confidence of Charles!

Maurice managed this delicate matter with extreme cleverness. Announcing to his councillors his immediate departure for Innsbruck to render to the Emperor the visit which had been so much talked of, he selected to accompany him the minister who, he knew, had been placed near his person as a spy upon his actions. Upon the day fixed he set out, accompanied by that minister, having previously secretly issued orders that his Saxon troops should join the mercenaries under George of Mecklenburg in Thüringen. After journeying five or six days in the direction of Innsbruck, Maurice feigned sickness, and, halting where he was, directed the inconvenient minister to continue his journey alone, to make his excuses to the Emperor, and to assure him he would follow in a few days. As soon as the minister was well out of sight, Maurice mounted his horse, rode to join the camp at Thüringen, issued a manifesto, and set out by forced marches towards Augsburg.

The manifesto issued by Maurice gave evidence of the bold purpose and resolute nature of the man who dictated it. Going straight to the point, it announced three reasons for taking up arms —to secure the reformed religion threatened with immediate destruction; to maintain the constitution and laws of the Empire, and thus to save Germany from the dictation of an absolute monarch; to deliver Landgrave Philip of Hesse from a long and unjust imprisonment. A document which appealed alike to patriotism and to the passion for religious freedom, which by its demand protested against a breach of faith which had done as much as any act of his reign to damage Charles V. in the estimation of the German people, could not fail to have a great effect. From the moment of its appearance the success of Maurice was the ardent hope of every patriot.

The very day of the issue of the manifesto saw Maurice set out at the head of an army twenty-two thousand strong. No obstacle hindered his progress, not a voice was raised to endeavour to induce the inhabitants of a single town to shut its gates. Marching as the deliverer of the people, he was received all along the route with acclaim. His immediate direction was Augsburg. There he was sure of a welcome, whilst from that free city, or from any point on his way thither, he could always shape his course to meet the event of the hour.

Charles, meanwhile, had been living in a fool's paradise at Innsbruck. Having permitted his attention to be entirely absorbed by the matter relating to the Council of Trent, he had allowed suspicions which had been excited to be too easily allayed. The craft of Maurice had, in fact, known how to explain satisfactorily every act which might have borne a doubtful appearance. Thus it was that, tormented with gout, his time and attention occupied mainly by religious disputations, Charles had lain at Innsbruck, with a body of soldiers hardly strong enough to guard his person, without money, without credit, relying upon the prince who had thus suddenly and without warning turned against him!

It was a situation sufficient to try the nerves and intellectual power of the strongest. Amongst such, in 1552, Charles V. certainly did not count. Though but fifty-two years old, his inordinate gluttony had ruined his digestive powers. The mind which had been unable rightly to comprehend the glorious vista which had been disclosed by the victory of Pavia, had become neither enlarged nor strengthened. He was still dreaming only of enforcing his own personal views in the matter of religion on the peoples over whom he ruled, and of increasing his own power by the gradual undermining of their rights and liberties. His mind was absorbed in this congenial task, when the manifesto of Maurice roused him to look the danger in the face.

What was he to do, this master of half Europe, without men, without money, without credit, thus surprised in the mountains of Tirol? The first thought which came to him was flight towards the Low Countries, but a brief examination proved that the advance of Maurice had rendered flight in that direction impossible. He then fell back upon the resource of gaining time. The only means of gaining time was negotiation. Unwilling, even in that dark hour, to compromise his own dignity by treating with one whom he had ever regarded as a subject, and whom he now hated as a rebel, Charles requested his brother Ferdinand to enter into negotiations on his behalf. The envoys of Ferdinand reached Maurice in Franconia. Feeling himself master of the situation, and confident that he could direct the negotiation to his own advantage, the Elector accepted the proposition, made over the command of his army to George of Mecklenburg, with instructions to push on to Augsburg, and proceeded to Linz on the Danube, there to meet the King of the Romans.

In proceeding to Linz the one thought of Maurice was to retain the sick Emperor at Innsbruck whilst he should propose terms, which, not acceptable at the moment, might yet serve as the basis for future negotiation. He completely outwitted Ferdinand. That prince, unable to accept the impossible terms offered, saw, or thought he saw, in the pacific protestations of Maurice a disposition which would ultimately yield to reason. He, therefore, agreed that a second interview, each party having full powers to conclude, should take place at Passau on the 26th May, and last till the 10th June, and that between those dates an absolute truce should exist between the contending parties.

It was the evening of the 6th May. Calculating that he could reach his army on the 9th and have yet sixteen days before him to work the surprise which he contemplated, Maurice agreed to and signed the terms offered, and set out the next morning to rejoin his army. He caught it on the 9th at the little town of Gundelfingen on the Brenz, about forty miles from Augsburg. There were yet sixteen days of action before him, for the truce only began on the 26th, and, within that time, much might be accomplished. Charles still lay at Innsbruck, tormented by gout, lulled by the agreement of Linz into a feeling that he was secure and having at his disposal but a slender guard. The temptation was too strong for Maurice. By a bold march he could so terrify the Emperor as to enable him to dictate terms at Passau!

Maurice then gave the orders to push on with all speed. That same day he crossed the Danube at Offingen, and pushing on thence by forced marches reached the town of Füssen, a distance of about a hundred and twenty miles, on the 18th!

To the modern traveller, the pretty town of Füssen is well known as one of the gates into western Tirol. The road thence by Reutte, the ruined castle of Ehrenberg, Lermoos, and Nassereit to Innsbruck is full of attractions for the lover of beautiful scenery. Romantically situated at the foot of a lofty range, on the left bank of the rapid Lech, deriving its name from the chasms and gorges (fauces) caused by that river, it covered the defile which, traversing the Kniepass, formed the outwork to the strong castle of Ehrenberg, a little beyond Reutte. The magnificent road, hewn out of the solid rock, so easy to the traveller of the present day, did not then exist.* The importance of its situation, as constituting, so to speak, the glacis of the fortress of the mountains of Tirol, was in those days highly estimated.

Füssen had a garrison of eight hundred men. Surrounded as the town was in those days with high loop-holed walls, it had been easy for these to defend it sufficiently long to warn the Emperor of the danger which was approaching him. But either they had no heart in the cause, or they were badly led, for they gave away almost immediately before the fierce attack which Maurice, a consummate general, well aware of the importance of the place, launched against them. It had been better in every way for the cause of the Emperor had they stood their ground to be slain, for they fled only to communicate their terror to their comrades who guarded the Kniepass, and to cause them to abandon a position which, well-defended, would have been hard to force.

Maurice was now master of Füssen, and Füssen was but sixty-nine miles from Innsbruck. Between him and that place

* It was constructed towards the end of the eighteenth century by order of Joseph II.

there was, immediately in front of him, the Kniepass, then the picturesquely situated town of Reutte, and some two miles beyond that the Ehrenbergerklause, guarded by the strong castle of Ehrenberg, efficiently garrisoned and deemed impregnable.

Under ordinary circumstances, and against an ordinary leader, Ehrenberg had been impregnable. But against the most consummate soldier of the sixteenth century even that strong fortress was powerless. Maurice had not stayed at Füssen one single second longer than was absolutely necessary. Knowing the difficulties immediately before him, and aware that the utmost promptitude alone could enable him to overcome them, he pushed on instantly, traversed the Kniepass unmolested by its defenders, whom his success at Füssen had terrified, reached Reutte, and, staying there only long enough to glean information regarding the mountain paths from a shepherd who came to him, dashed on against the impregnable fortress.

From Reutte, on the right bank of the Lech, the road ascends to the pass called the Ehrenbergerklause, to the west of which, guarding it, on the summit of an isolated pine-clad hill, upwards of three thousand feet above the level of the sea, stood then the castle, as there stand now the ruins, of Ehrenberg. Though it was so strong, alike in its position and its defences, that even a small force could easily have maintained it, the Emperor, sensible of its importance, had placed in it a very efficient garrison. An attack upon it by the ordinary routes, even had that attack been led by Maurice, could scarcely have succeeded, and it was the conviction of this truth which made the ever-impetuous leader halt a few minutes at Reutte to endeavour to obtain information as to the mountain paths which must be known to those whose avocations led them daily on to the green alps which afforded pasture for cattle.

As he was thus halting, there approached him the shepherd of whom I have spoken. This man, hearing the reason why the troops, who had marched hitherto without caring to take a rest, had halted in the streets of his native town, ran quickly to the handsome warrior, sitting with impatient look on his horse, from whom all seemed to take their commands, and intimated that he had something important to communicate. Very recently, he stated, a goat had strayed from the flock which he was tending on the mountain: he had hurried after the animal, and after a long chase had recovered it: but, during that chase, he had come upon a path, utterly unknown to him before, which led to the summit of the hill upon which stood the castle of Ehrenberg. He was ready, he added, to guide the soldiers to that path.

At this information the heart of Maurice leapt with joy. Though the shades of evening were falling, he would not delay another minute. Turning to Duke George of Mecklenburg, he directed him to take with him a body of men upon whom he could absolutely depend, and follow the guide, whilst he himself would approach by the ordinary road, and attack when he should hear the signal agreed upon. The order to advance was then given: the troops pushed forward. In about an hour Maurice began to make a demonstration, which served to keep the attention of the garrison directed towards him. Meanwhile Duke George and his men had commenced the difficult ascent which would give them a commanding position on the other side. When he gathered that they had reached it, Maurice began his assault. For a moment, but only for a moment, the garrison showed great earnestness in resisting it. Suddenly, however, the rumour that a body of the enemy's troops had already effected an entrance on the other side, spread a panic amongst them. They at once laid down their arms and surrendered. Maurice had captured the strongest fortress in Tirol almost without bloodshed!

The daring leader was now within sixty miles of Innsbruck.* The writers of the time call it a two days' march. It is possible that, by using great expedition, Maurice might have accomplished the journey within that time, but, considering the nature of the country, it does not seem to me probable. In the case of

* The distance from Reutte to Innsbruck is fifty-nine miles, over a country the first two-thirds of which were then very difficult for soldiers.

his having been able to do so, he himself would have been the bearer of the news of his own success to the gouty Emperor. What was there to save Charles from becoming his prisoner? His last stronghold had been captured: he lay in fancied security at Innsbruck; the way to Innsbruck lay open; no one could traverse it so quickly as Maurice and his lightly-equipped horsemen!

But, in war, there sometimes occur circumstances against which it is impossible to provide, which it is impossible to foresee, and which, slight in themselves, are capable of changing the fate of events. It was simply a happy inspiration which brought Dessaix in the very nick of time to the field of Marengo; the refusal of the higher officers under him to obey his orders robbed Masséna of victory at Fuentes d'Onoro; * the defection of the Saxons affected very considerably the result of the three days' battle at Leipzig; the heavy rain of the night before retarded by three hours the commencement of the battle of Waterloo; and here, just as the capture of the Emperor seemed absolutely certain, there occurred one of those mishaps which, by causing the delay of some hours, afforded him the opportunity to escape!

The troops which Maurice had led against the fort of Ehrenberg were the mercenaries whom he had employed against Magdeburg, and who had afterwards enlisted under George of Mecklenburg. They were men who lived on war, men ready to strike in any cause, and to serve under any leader who should secure to them plunder and money. The terms of their engagement differed widely from those which prevail in the present day. Instead of receiving regular pay, they lived on the towns near which they were quartered, and depredations at the expense of the industrial portion of the community were commonly regarded as coming within their rights. The storming of a town or of a fortress gave them a claim to pillage for three days, or to a compensation in money, carefully calculated, as the price of abstention. It was this which made

* "Had Masséna not been blind," said the Duke of Wellington, "he would have beaten me at Fuentes." Masséna was not blind, but the divisional officers under him were jealous of him, and refused to obey his orders.

them regard that dangerous service as one of the great prizes of their profession.

Before these men Ehrenberg had fallen. But it had not fallen in a manner which brought it under the unwritten law which allowed them compensation for abstention from pillage. Ehrenberg had not been stormed. It had surrendered upon terms. Clearly, then, the mercenaries were not entitled to any money compensation. But their life, ever since they had quitted Thüringen, had been a long series of disappointments. There had not been a single storming. Every city, every town, every village, except Füssen—and Füssen was then but little more than a village—had opened its gates to the man universally regarded as the deliverer. To achieve this result they had made forced marches, had forded rivers, had partaken sparingly of food. And now, when they had marched to the storm of a fortress, that fortress had disappointed them by a sudden surrender. It was too hard to be borne longer. There was yet a wearisome march before them, a march likely to terminate by another surrender of a town. And that surrender would terminate the campaign. Clearly, then, unless they could bring the capture of Ehrenberg within the terms of the understanding which allowed them compensation for assault, their long marches would produce but a barren result!

Reasoning in this manner, the mercenaries struck, at the critical moment, for payment for the capture of Ehrenberg. Not only did they refuse to march, but they broke out into open mutiny, and threatened the lives of their officers unless their terms were complied with. Even had Maurice wished it ever so much, compliance was impossible. Armies did not march in those days with large treasure-chests at the disposal of the commander. But he showed himself here, as on every occasion, a born leader of men. When he found expostulation unavailing, he assumed a high hand. Gathering around him a few officers and men on whom he could depend, he penetrated, despite of hard words, of hostile gestures, even of threatened violence, into the very midst of the mutineers, and seized the ringleaders. For a time his life, and the lives of those about him, were in imminent danger. But the boldness which could not quail bore him safely through the storm. Still holding his prisoners, he told the men that they had no right, and they knew that they had no right, to the money they demanded, but that, if they would march quietly with him, not only would their present conduct be condoned, but he would give them large opportunities for enriching themselves.

The bold language of a threatened man to those who threaten him, and who are inwardly conscious that they are wrong and he is right, never fails in its effect. The words of Maurice quelled the mutiny. The hope of future booty made the men, easily impressed, as eager now to march as they had been before unwilling. All fell into their ranks, and the army once more set out.

But several precious hours had been lost. It was conjectured, indeed, even by statesmen of the period, that Maurice was not unwilling to allow the Emperor time to escape. "I do not yet possess," he, it is said, stated to his confidants, "a cage large enough to hold a bird of that size." This conjecture derives strength from the movements of Maurice after the mutiny had been quelled. If it were true that he could have accomplished the distance, fifty-nine miles, two-thirds of which led him across rough mountain paths, in two days, then he displayed a want of dash foreign to his nature. For, whilst he quitted Reutte on the 19th, it is certain that he reached Innsbruck only on the morning of the 23rd. I am inclined to believe that the conjecture I have referred to embodies the exact truth. Whilst the capture of Charles would only have embarrassed Maurice, the fact that he had compelled the Emperor to flee in terror from Innsbruck, on the eve of the meeting at Passau, would, on the other hand, secure to him all the ascendancy he required, would enable him to enforce the conditions which he had set forth in his manifesto.*

* On this point I venture, with all respect, to differ from Robertson. That writer states: "Maurice entered Inspruck a few hours after the Emperor and his attendants had left it; and, enraged that the prey should escape out of his hands when he was just ready to seize it, he pursued them some miles." This sentence contains, I think, many errors. Maurice stormed Ehrenberg the evening of the 18th May; the mutiny detained him at Reutte a great part of the following morning; he

Meanwhile Charles lay impassive, suffering in body, exercising his mind how decisions on theological questions might produce political results, at Innsbruck. He had confirmed the proceedings of his brother Ferdinand with respect to the meeting to be held at Passau, and he had resolved to make no concessions which would imperil the success of the policy he had steadily pursued—the fettering of the consciences and the restriction of the liberties of the German princes and people. He was resting, tormented by gout, in this mood, when, late in the evening of the 19th May, a messenger reached him with the information that the army of the Elector of Saxony was in rapid motion; that it had captured the only defensible post which covered Innsbruck, and that it was in full march for the capital of Tirol!

No one understood better than Charles the firm and resolute character of Maurice. Arguing from his own ideas of policy, he arrived at once at the conclusion that it was his own person which the Elector desired to secure. He gave orders, then, for immediate departure. So great were his sufferings from gout that he could bear only the motion of a litter. His litter, then, was brought to the door, and by the light of torches, accompanied by his courtiers and attendants, some on horses, some on foot, and by the ex-Elector of Saxony, John Frederic, the man whom he had deposed to make way for the very Maurice who was now marching against him, he took his way across the Brenner to the point now known as Franzensfeste, and, diverging thence to the left, proceeded in very miserable plight along the Pusterthal to Villach. Even there, he scarcely thought himself secure!

certainly did not leave that town till the afternoon of the 19th. Now, Charles left Innsbruck the evening of the same day. Looking at the distance, fifty-nine miles, and the nature of the country, it is impossible that Maurice could have reached Innsbruck "a few hours after the Emperor had left it." But admitting, for the sake of argument, that he did, and that he pursued him. Charles was carried in a litter; he was so ill that he could bear only easy motion. Carried over rough mountain paths, he could not have travelled at a greater rate than two or three miles an hour. If he had been vigorously pursued, then, within a few hours of his departure he must have been captured. But this part of the story is a myth. Maurice wished to frighten, not to capture, the Emperor, and he so timed his march from Reutte to Innsbruck as to ensure his escape.

Maurice, meanwhile, marching leisurely, entered Innsbruck the morning of the 23rd. Made acquainted at once with the details attending the Emperor's flight, he despatched a few men to make a show of pursuit. But it was only a show. He had gained all he wanted—a moral power which would enable him to dominate the proceedings at Passau. To compensate his soldiers for their disappointment at Ehrenberg, he abandoned to them the baggage belonging to the Emperor and his ministers. He protected only the property of the King of the Romans.

Three days later (26th May) the Peace Congress met at Passau. Besides Maurice—who, directing his army into Bavaria, had hastened thither from Innsbruck—and Ferdinand, King of the Romans, there were present other imperial ambassadors; the Duke of Bavaria, the Bishops of Salzburg and Aichstadt, representatives of all the Electors, of the free cities, and of the most considerable princes of Germany. From the very first day Maurice assumed the ascendancy which devolved upon him by reason of the position his great achievement had gained for him. With commendable prudence he limited his demand to the granting of the three concessions upon which he had insisted in his manifesto: the enlargement of the Landgrave of Hesse; the maintenance of the Constitution and laws of the Empire, and the security of liberty of conscience. In the end, after many negotiations, including a visit made by Ferdinand to Villach, these terms were granted in principle by the Emperor, and the agreement of Passau (Passauer Vertrag*),

* The actual terms of this treaty were as follows: that before the 12th August the confederates should lay down their arms and disband their forces; that on or before that day the Landgrave of Hesse should be set at liberty, and conveyed in safety to his castle of Rheinfels (the ruin, near St Goar, still exists an object of interest to the traveller); that a Diet should be held within six months, to deliberate as to the most effectual method of preventing religious disputes in the future; that in the meanwhile the adherents of the Confession of Augsburg should be allowed the free exercise of their religion; that the same freedom should be granted to the Catholics; that Protestants and Catholics should be admitted indiscriminately to seats in the Imperial Chamber, which should administer justice impartially between both parties; that if the Diet, to be convened, did not settle religious disputes, the agreement of Passau was to remain in force for ever; that none of the confederates should be punished for their action in the war; that the consideration of the

signed the 31st July 1552, became the result of the surprise of Innsbruck!

Imperial encroachments on the Constitution of the Empire should be remitted to the approaching Diet.

At that Diet, held at Augsburg in the autumn of 1555, there was signed, on the 25th of September, the famous peace called "the Peace of Religion," the natural outcome of the agreement signed at Passau. It is true that peace decided nothing permanently. The real Peace of Religion was not made till the Treaty of Westphalia was signed, nearly a hundred years later, 24th October 1648.

CHAPTER VI.

FORT DUQUESNE.*

The second Treaty of Aix-la-Chapelle, concluded between Great Britain, France, Holland, Germany, Spain, and Genoa, and signed by the British Plenipotentiaries the 7th October 1748, had given no promise of a lasting peace to Europe. Though it had renewed and confirmed the Treaties of Westphalia in 1648, of Minwegen in 1678 and 1679, of Ryswick in 1697, of Utrecht in 1713, of Baden in 1714, of the Triple Alliance in 1717, of the Quadruple Alliance in 1718, and of Vienna in 1738, the elements which had burst forth on the death of the Emperor Charles VI. were still smouldering. The Empress Queen was brooding over the loss of Silesia, whilst the interests of France and England, alike in Asia and America, still continued to jar.

On those continents, indeed, the peace had been almost nominal. How the normal state of hostilities affected the destinies of the two countries in Asia, I have told in another work.† In the New World, the feeling was equally rancorous, and the action still more pronounced. The peninsula known as Nova Scotia, dis-

* My chief authorities for this chapter are (1) Bancroft's *History of the United States*; (2) Marshall's *Life of Washington*; (3) Walpole's *Memoirs of George II.*; (4) Mante's *History of the late War in North America*; (5) Sargent's *General Braddock*; (6) Parkman's *Montcalm and Wolfe.* To this last excellent and interesting work I am specially indebted.

† *History of the French in India.* 1868. London: Longmans

covered by Cabot in 1497, visited and named Acadia by Verrazzani in 1524; colonised and settled and renamed Nova Scotia by William Alexander in the reign of James I., had, after several changes of masters, been finally conceded to England by the Peace of Utrecht. The conditions of that treaty had, I have said, been confirmed by the Peace of Aix-la-Chapelle, but the limits of the English colony had been extremely ill-defined; and although the territories in dispute had never been explored, the question of the right to possess them had caused ill-feeling and acrimonious correspondence between the Courts of St. James and Versailles. In Canada and New England the feeling was not one whit better. There, too, the question of the boundary-line gave rise to continued bickerings. And as in India the rivalry of the native lords of the soil gave the European adventurers of both nations the opportunity of fighting without being at war; so, in North America, any hostile raid made by a party of wild Indians on the one nation was attributed to the deliberate purpose and prompting of the other. Nor is there reason to suppose that the recriminations, in this respect, largely dealt in, were always unfounded. Ever since the restoration of Canada by the English in 1632, that colony had remained French, and it had been the aim of the able men who were periodically sent from France to govern it, to extend its dimensions in a southerly direction One design, above all others, had animated them all. Long sensible of the advantage which would accrue to France, in case of war, by a connection of her settlements on the St. Lawrence with those on the Missisippi, they had designed to erect for this purpose a chain of forts outside the French territories, and had even taken some steps in that direction by building, at Presqu'isle, where Erie now stands, a fort called Fort Presqu'isle, and another at the further end of the creek now known as French Creek, called Fort le Bœuf. Their action had been viewed with jealousy and alarm by the English colonists, and when, in 1754, a fort, called from the French Governor, the Marquis Duquesne, Fort Duquesne, was built on the Ohio, at the confluence of the Monongahela and Alleghany rivers, on the foundations of a stockade taken from the English,* the indignation and alarm of these—the indignation and alarm especially of the men of Pennsylvania and Virginia—knew no bounds. Well might the French action produce such a result. The establishment of the contemplated line of forts would absolutely close the markets of the West to the English traders !

The capture, in a time of peace, by the colonists of one nation of a stockade, unfinished though it was, guarded by the soldiers of a rival nation, was undoubtedly an act of war, and as an act of war it was regarded by the Lieutenant-Governor of Virginia, Robert Dinwiddie. To understand the exact circumstances with which he was called upon to deal at this conjuncture, it will be necessary to trace his action during the years preceding it.

Robert Dinwiddie had been clerk in a Government office in the West Indies ; then Surveyor of Customs in the Old Dominions, and though past sixty, he was the most watchful sentinel against French aggression, and its most strenuous opponent.† He was fortunate in having at his right hand a young man of great ability. George Washington had already displayed qualities of no common order. Born in Virginia, the 22nd February 1732, the younger representative of an ancient family, two members of which had emigrated in 1657, he had received the ordinary education suitable to a planter. Especially distinguished as an athlete, he had, when only sixteen years old, attracted the attention of Lord Thomas Fairfax, whose cousin his brother Lawrence had married, and become the companion of his sports. In 1748 Lord Thomas commissioned his young friend to visit an unexplored portion of his vast domains, which, comprised between the Rappahannock and Potomac rivers, extended as far as the Alleghany mountains. The excursion, which took him into the Shenandoah valley, lasted five weeks, and

* Ensign Ward and forty men were at work upon it when the French, some five hundred strong, summoned him to surrender. He complied, and re-crossed the mountains to report his mishap to Washington, then second in command, under Lieutenant-Governor Dinwiddie and Colonel Fry, of the Virginian regiment, and who had selected the position on the Ohio.

† Parkman's *Montcalm and Wolfe*, vol. i., page 137.

his report of it so enchanted his patron that Lord Thomas himself proceeded to the Blue Mountains, having first procured for Washington the office of Surveyor of Public Lands. For three years Washington performed the duties of this office. They gave him very full opportunities to study the country. His constant journeys brought him much into the society of Lord Thomas Fairfax. The intercourse was extremely profitable to him, for from the lips of that lord he learned more about the past and present history of England than any books could have taught him.

Whilst George Washington was engaged in his interesting duties, many of his countrymen, amongst them his elder brother Lawrence, were dreaming of extending the borders of the English colony from the Alleghany mountains to the banks of the Ohio. Some of them actually formed a company to carry out this idea. This design brought the English in contact with the French—for they, too, had put forward claims to the valley of the Ohio—and the Governor of Canada, the Marquis de la Gallissonnière,* promptly despatched, in 1749, a small body of soldiers † under a very famous colonist, Celoron de Bienville, to obstruct, as much as possible, the progress of the English pioneers. This happened, be it remembered, when the two nations were at peace in Europe, for the Treaty of Aix-la-Chapelle had been signed only the year before.

In the view of possible hostilities the English governor of Virginia proceeded, 1751, to divide that province into military districts, each of which had at its head an adjutant-general whose duty it should be to superintend the organisation of the militia. One of these appointments, which carried with them the rank of major, was conferred upon George Washington, then but nineteen

* Rolland Michel Barrin, Marquis de la Gallissonnière, was humpbacked, and of small stature. When he first showed himself to the Indians in Canada, they exclaimed to him: "You must, indeed, possess a mind of great capacity; for otherwise our Father beyond the seas would not have sent a man with so small and so deformed a body to rule over us." He was a sailor, the same who forced the English fleet, under Byng, to retire from before Minorca.

† The force consisted of fourteen officers and cadets, twenty soldiers, a hundred and eighty Canadians, and a band of Indians.—Parkman.

years old. Washington, with his customary energy, applied himself zealously to his new duties. From an officer named Muse, he received lessons in the practice of artillery; another taught him fencing. In both these sciences the scholar soon surpassed his masters. From his duties, however, he was suddenly called away in the winter of 1751–2, to accompany his brother Lawrence, then dying from consumption, to Barbadoes. From that island the brother returned, only to die, in July 1752.

During their absence, the situation beyond the Alleghany mountains had become greatly involved. The French, better versed in the kind of diplomacy which most influenced the Indian tribes than their rivals, had enlisted the support of the most formidable of these, and their progress had caused considerable alarm in Virginia, in Maryland, and in Pennsylvania. Of the first of these provinces Robert Dinwiddie was then governor, and he, a man of capacity and action, resolved to send at once to the countries contested by the rival colonists a man upon whom he could absolutely depend. He selected George Washington. Washington set out for the disputed territory at the end of October 1753. He was accompanied by Jacob van Braam, the man who had taught him fencing, and whose chief qualification for his new employment was a slight knowledge of French; by an intrepid trader named Christopher Gist; by Davison, to interpret to the Indians; and by four woodsmen as servants, With these men he traversed the valley of the Monongahela, to the point of junction of that river with the Alleghany. The confluence of these two rivers forms the Ohio. Pursuing his journey along the traders' path, he reached, the 7th December, the Indian town of Venango. Here he came upon a trading-house, originally English, but which had been seized and was then occupied by the French. After enjoying the hospitality offered him by his rivals—a hospitality somewhat disfigured by the strenuous attempt, which very nearly succeeded, to entice away the Indians whom he had enlisted—Washington and his companions pushed on for four days till they reached Fort le Bœuf—the wooden stockade I have already mentioned. To the commandant of that stockade Washington delivered a letter, with which he had been furnished by Dinwiddie, requiring the peaceful departure of himself and the men he commanded from lands "so notoriously known to be the property of the Crown of Great Britain." Washington soon discovered that the demand, unless it were sustained by force of arms, would not be entertained. He, therefore, determined to return. Eager to make his report to Dinwiddie, and not caring to endure the slow monotony of a regular march, he quitted the bulk of his companions, and, accompanied only by Gist, traversed the footpaths which led back to Virginia. He arrived at Williamsburg, after many hazardous adventures, the 10th January 1754.

During his absence Dinwiddie had met the Assembly of Virginia, convoked to vote the grant of money necessary to meet the emergency. The Assembly, to avenge itself on the Governor for his refusal to attend to their wishes in a smaller matter, had refused the grant. That was the situation when Washington returned and made his report. It was a difficult one. The Governor had a keen sense of the danger, but he had not the money to supply the means which would have averted it. Robert Dinwiddie, however, belonged to the category of men whose firmness, whose self-reliance, and whose energy have made the Empire of Great Britain. Denied even a penny by the Assembly which should have supported him, Dinwiddie took it upon himself to order two hundred of the militia on the service. Washington was to have commanded, with a trader named Trent as his lieutenant. But before the men were ready to set out the Assembly again met (February 1754), and voted ten thousand pounds for the service. Whilst Trent was then sent on in advance to build a fort on the forks of the Ohio, the number of the detachment was raised to six hundred, and the command was offered to Washington. With characteristic modesty he preferred that the command should be bestowed upon an English gentleman, Mr. Joshua Fry, then settled in the colony, whilst he accepted the second position, with the rank of lieutenant-colonel. Washington set out with about a hundred and sixty of the men whom he had raised—"poor whites, brave, but hard to discipline, without tents, ill-armed, and ragged as Falstaff's recruits"*—the 2nd April, to hasten to the support of Trent. But before he had gone far the mortifying intelligence reached him that Trent, having reached the point on the forks of the Ohio which he had indicated, and having begun there the erection of a fort, had left Ensign Ward and forty men to complete it, whilst, for some unexplained reason, he had fallen back with the remainder; that Ward had been driven, 17th April, from his post by a vastly superior number of Frenchmen; and that these latter had begun, on the same spot, the creation of a fort on a much larger scale—a fort which, on its subsequent completion, was named, in honour of the then Governor of Canada, FORT DUQUESNE. Washington received this information the 23rd May, when he and his handful of men were on the Youghiogany—a mountainous and marshy region, always exposed to an attack from the French or their Indian allies. He was told, at the same time, that a combined force was on its march against him. The information only whetted his resolution to give his men a fair chance. He marched without delay to a vast glade in the prairies known as the Great Meadows, and, intrenching himself there, sent out scouts to seek news of the enemy's movements. They brought him none; but two days later, the 25th, his old comrade, Gist, who had made a settlement on the further side of Laurel Hill, arrived in his camp with the information that some thirty Frenchmen had endeavoured to ransack his house the day before. This intelligence decided Washington. He at once despatched a body of seventy-five men to seek for and surprise the enemy. Their search was vain, and the small force returned to camp weary and dejected. Hardly had they arrived when Washington received a message from an Indian chief to the effect that the hiding-place of the French had been discovered. Half doubting the truth, and fearing lest the information might be intended to decoy him from his camp, he left in it the bulk of his soldiers and started himself, at the head of but forty men, in the

* Parkman.

evening, to ascertain the truth. The night was dark and rainy, the way difficult and uncertain, and the sun was rising when he discovered the enemy. The discovery was simultaneous. The French grasped their weapons, and a fire from both sides ensued. The first shot from the English laid low the commander of the French party, Coulon de Jumonville. After a short fight, and the loss of nine of their men, the surviving twenty-one surrendered. One had previously escaped. The English had one man killed and three wounded.

The prisoners declaimed against the hostile act, declaring that their commander had been but the bearer of a summons from the commandant of Fort Duquesne requiring the English to quit French territory; but the papers found upon the dead officer sufficiently showed that the party was a reconnoitring party, and, as such, was obnoxious to attack. Be that as it may, the shots fired on this occasion were the prelude to a war which extended to both hemispheres, and which terminated in the expulsion of the French from North America.

Washington returned to his encampment, sent his prisoners to Dinwiddie, and sent to Colonel Fry for reinforcements. A few days later he received information that Colonel Fry was dead, and that the reinforcements, composed of the other moiety of the Virginian regiment, were on their way to join him. Whilst waiting for these he threw up for his protection a rough stockade, which he happily called Fort Necessity. Here he was joined very shortly by the moiety of his regiment, by his Indian allies, and by an independent company from North Carolina, commanded by Captain Mackay, an officer who held his commission from the Crown.

The absurd and inflated pretension which led an officer holding his commission from the Crown to decline to recognise the authority of a local officer of superior rank who held his only from a governor appointed by the Crown—the gross absurdity of which it required a retreat from Lexington and a battle of Bunker's Hill to expose—broke out on this occasion. The royal captain, as he styled himself, would take no orders from the colonial colonel.

Finding Captain Mackay, then, rather a hindrance than a help, Washington left him at Fort Necessity and marched towards Gist's settlement. His force numbered about three hundred men of English blood.

Washington was intrenching himself in his new position when the news reached him that a formidable force, led by Captain de Villiers, was on its march from Fort Duquesne to avenge the death of Jumonville. He at once sent for Mackay. Mackay arrived the 28th June. After some discussion it was resolved, as the actual intrenchment was commanded by the neighbouring heights, to fall back on the Grand Meadows, and, if possible, still further. The force fell back accordingly, and reached its old position the 1st July. The great fatigue endured by the men rendered further retreat impossible. Here, on the 3rd, they were attacked by five hundred troops of the French army, supported by a large body of Indians. After a contest of nine hours, in which the French from the superiority of their ammunition had all the advantage, Villiers called upon Washington to capitulate. Washington at first declined even to treat; but when the offer was renewed, and he recognised that he had no food for his men, and that his ammunition was nearly spent, he hesitated no longer. That night he signed terms by virtue of which his troops were allowed to march out with the honours of war. On the other hand he agreed to restore the prisoners taken in the affair with Jumonville, and to allow two officers to remain as hostages with the French force until those prisoners should reach Fort Duquesne.

Rarely has a man suffered more from the want of knowledge of a foreign tongue than did Washington on this occasion. In the capitulation which he signed the death of Jumonville was qualified as an assassination (*assassinat*). Washington understood no French, and Van Braam, who acted as interpreter, apparently anxious that the surrender should go through, translated the word to mean, not "assassination," but "death." The admission that he had assassinated Jumonville gave a handle to his enemies of which they did not fail to make use.

The retreat of the English was difficult and depressing. The Indians had killed all the horses; the sick and wounded had thus to be carried, and the bulk of the baggage to be left behind. To intensify the sufferings of the men, aggravated as they were at the outset by the threatened attack of the Indians, there came the overwhelming thought that the surrender of Fort Necessity had given the country as far as the Alleghanies to their hated rivals.

To none, unless we except Washington himself, was the disappointment greater than to Robert Dinwiddie. Great as it was, it was immensely aggravated when, on summoning the burgesses, they declined, unless he would give way in a manner which he regarded as unconstitutional, to vote even the small sum of twenty thousand pounds to provide means to repel French aggression. In indignant terms he reported to the Home Government "their intolerable obstinacy and disobedience to His Majesty's commands," and recommended that Parliament should tax them half-a-crown a head.* Subsequently, though still in 1754, when the number of the French on the Ohio had considerably increased, he wrote again in the most urgent terms to Lord Granville on the subject of the stubborn obstinacy of the colonists in the presence of an overwhelming danger. For a long time the Government was as careless of colonial demands as its successors showed themselves in 1884. But at length there was an awakening. Dread of French ambition, stimulated by Dinwiddie's representations, supported more or less by those of the other governors, and sharpened by the events occurring under their own eyes, roused at last the Home Government to decisive action. On the 14th November, in his speech to Parliament, the King announced his intention to "protect those possessions which constitute one great source of the 'wealth' of his subjects." Two regiments, each five hundred strong, to be raised by enlistment in Virginia to seven hundred, had been previously ordered to America. To command them and all the royal and colonial forces on that continent, Major-General Braddock, a favourite of the Duke of Cumberland, was appointed.

* Parkman.

A great American writer* has thus tersely described the new commander-in-chief: "A man in fortunes desperate, in manners brutal, in temper despotic; obstinate and intrepid; expert in the niceties of a review; harsh in discipline." The portrait is based upon revelations made by Horace Walpole in his gossiping letters to Sir Horace Mann, and, if the details in these letters may be absolutely trusted, it cannot be pronounced unfaithful. When all has been said, however, that can be said against the moral character of Edward Braddock, there stands out, in bold relief, the admission that he was beloved by those who served under him. "You never knew me," he replied to a general officer, his senior in rank, who asked him how long he had divested himself of the brutality and insolence of his manners, "you never knew me insolent to my inferiors. It is only to such rude men as yourself that I behave with the spirit which I think they deserve." This reply reveals the man. Self-confident, even aggressive in demeanour, he held his head erect, fearing no one, gaining the love of his subordinates and the ill-will of his equals. Self-confidence is either a strength or a snare. It is a strength when the possessor of it is well-instructed in details, capable, endowed with a calm and cool judgment; it is a snare when he has none of these qualities. To Braddock, going to a new country, it was sure to be the latter. However, he was a brave man. "Whatever were his failings," writes Mr. Parkman in his admirable history,† "he feared nothing, and his fidelity and honour in the discharge of public trusts were never questioned."

Braddock, accompanied by the two regiments I have spoken of, reached Alexandria, in Virginia, the end of March, and on the 14th April following he met in council there Robert Dinwiddie and other provincial governors. The most considerable of these, not even excepting Dinwiddie, was Shirley, Governor of Massachusetts, a man who, though a lawyer by profession, thirsted for military glory, and who believed that the safety of the colonies

* Bancroft's *History of the United States.*

† *Montcalm and Wolfe*, vol. i. p. 191; Macmillan.

was involved in the expulsion of the French from every part of the continent. He had already pressed his ideas upon the Home Government, and had obtained its sanction to attack the French fort of Beauséjour, at the head of the Bay of Fundy, and Crown Point, the key of Lake Champlain. The son of Shirley was appointed secretary to Braddock.

With two such men as Dinwiddie and Shirley to support him in the Council, Braddock found his task easy. It was resolved to attack the French simultaneously on four points. Shirley, leading two regiments newly raised in the colonies, was to march against Niagara; Colonel Monckton, an officer of merit in the King's army, leading a band of men raised in New England, was to march against Fort Beauséjour and complete the conquest of Nova Scotia; Colonel William Johnson, an officer very capable and possessing great influence over the Indians, was to command a colonial force raised in New England, New Jersey, and New York, and capture Crown Point; whilst to himself Braddock reserved the command of the expedition which was to wrest Fort Duquesne from the hands of the French! It is with the last enterprise that I have to deal in this chapter.

The distance from Alexandria to Fort Duquesne was about two hundred and twenty-six miles. The track thither, as far as the point where Wills' Creek joins the Potomac, on which was a rude stockade named Fort Cumberland, about a hundred miles from Alexandria, was comparatively open. Then followed nearly seventy miles of mountains covered with primeval forest—the main Alleghany, the Meadow Mountain, the Great Savage Mountain—to a place called the Little Meadows. From the Little Meadows was a rough road of nearly sixty miles to Turtle Creek, a stream entering the Monongahela, between nine and ten miles from Fort Duquesne.

With the determination to undertake this expedition the difficulties of Braddock began. His first endeavour was to arrange for funds for the prosecution of the campaign; but here he had to meet the same difficulties which Robert Dinwiddie had encountered before him. In the council of governors the opinion was unanimous that unless they were coerced by the Home Government the Assemblies would not allow the control of their funds to pass out of their own hands.* Nor did Braddock find greater readiness to furnish supplies for the army. The colonists ridiculed the idea of there being any danger of French aggression, and, in Pennsylvania especially, they displayed so much apathy and indifference that even Washington recorded his opinion that "they ought to be chastised." Contracts were made only to be broken; and it seems difficult to imagine how the army could have been brought into a condition to move at all, but for the intervention of a man whose name remains ineffaceably inscribed in the annals of his country's history.

The *Deus ex machinâ* was Benjamin Franklin. Braddock had, on the breaking up of the Council, proceeded with his troops to Fort Cumberland, and had made that place the base of his forward movement. He was there fretting and fuming, growing old with anxiety regarding the wagons and supplies which, promised always, never arrived, when Franklin, then Postmaster-General of Pennsylvania—the most recalcitrant of all the provinces—and who possessed within it considerable influence, visited him. Born at Boston in 1706—then, therefore, in his fiftieth year—Franklin already enjoyed, by means of his scientific discoveries, a European reputation. He was a member of the Royal Society of London—having been specially admitted without payment of the usual fee—an associate of the Academy of the Sciences at Paris, whilst the universities of Oxford and of Edinburgh had conferred upon him those high honours which are bestowed only upon the illustrious. But Franklin was more even than a scientist. To a common-sense which was never at fault he united a combined power of vigorous thought and decisive action which has never been surpassed.† The arrival of such a man produced a marvellous

* The Council shortly afterwards broke up.

† "Indigent," says M. Mignet in his admirable life of this illustrious man, "he made his way by work to wealth; ignorant, he raised himself by study to science; unknown, he obtained by his discoveries as well as by his services, by the greatness

influence. Dining daily with the General, he had an opportunity towards the close of his stay of inspecting the return of the wagons which had been brought in. To his surprise he found that their number amounted only to twenty-five, and that not all of these were serviceable. Braddock, in despair, was on the point of renouncing the expedition when a remark made by Franklin, as to the capability of the farmers of Pennsylvania to supply his wants, drew from him an earnest request to the speaker that he would return thither and use his influence to induce them to help him. Franklin cheerfully assented, returned to Pennsylvania, issued an address to the farmers appealing to their interests and their fears, and in a fortnight despatched to Braddock a sufficient provision of wagons and horses to enable him to advance.*

It was only on the 10th June that Braddock's force was able to march from Fort Cumberland. He had with him two thousand two hundred men. The two regiments he had brought with him from England, Halket's and Dunbar's, had been increased by enlistment in Virginia to a strength of seven hundred each; he had, besides, four hundred and fifty Virginians equally formed into nine companies, thirty sailors from the fleet, and some three hundred and twenty gunners and axemen. Besides the horses for the guns, he had collected, thanks to Franklin, some six hundred baggage horses and about a hundred and eighty wagons. Of Indians he had but few. The combined neglect and contempt of the colonists had alienated the tribes—the Iroquois, the Delawares, and the Shawanoes—whose cordial co-operation would have sufficed to avert the calamity.

Preceded by three hundred axemen to clear the road, its flanks strongly protected by skirmishers, and the woods on either side examined by scouts, the little force, with its train of

of his ideas and by the extent of the benefits he conferred, the admiration of Europe and the gratitude of America. Franklin possessed alike genius and virtue, prosperity and glory. His life, always happy, is the finest justification of the laws of Providence."

* Braddock declared that Franklin was "almost the only instance of ability and honesty I have known in these provinces." *Vide* Parkman's *Montcalm and Wolfe*, vol. i. p. 199.

baggage and guns making a length of four miles, set out on the 10th June to accomplish the nearly seventy miles of combined forest and mountain which lay between Fort Cumberland and the Little Meadows. That place was reached without any sign of opposition on the ninth day (18th June). But the toils of the journey had been great, and the usual accompaniments of fatigue in a malarious country, fever and dysentery, had already made their presence felt. To the anxious mind of the General the doubt had more than once occurred whether the continuance of a march at a rate so slow would not give time to the enemy to strengthen the garrison and defences of the fort, the capture of which was the object of his expedition. In this difficulty he consulted, on his arrival at the Little Meadows, George Washington, then serving on his staff as aide-de-camp.* Washington we have already seen, knew the country, and, appreciating the importance of despatch, urged the General to push on with a body of chosen troops. Braddock followed his advice, and, leaving Colonel Dunbar to command at the Little Meadows, pushed on, on the 19th, with twelve hundred chosen men (besides officers), a few guns, thirty wagons and a large number of pack-horses. Amongst the chosen men were the Virginians.

Still the progress was slow. Instead of displaying the rashness which has been commonly attributed to him, Braddock marched with a caution sufficient to counteract all the advantages in the hope of obtaining which Washington had proffered his advice,† and it was not till the 8th July that he reached a point on the Monongahela, at a short distance beyond which Turtle Creek joins that river, ten miles, by the direct path, from Fort Duquesne. This path, however, led through a defile so perilous

* The other aides-de-camp were Robert Orme, who wrote an account of the expedition, and Roger Morris.

† "The prospect," wrote Washington to his brother, "conveyed infinite delight to my mind, though I was excessively ill at the time. Bnt this prospect was soon clouded, and my hopes brought very low indeed, when I found that instead of pushing on with vigour without regarding a little rough road, they were halting to level every molehill, and to erect bridges over every brook, by which means we were four days getting twelve miles."—Parkman, Bancroft.

that, to avoid it, Braddock, on the morning of the 9th, forded the Monongahela, with the intention of re-crossing it about half a mile below the creek.

I have already recounted how the French had, only in the year immediately preceding, driven the English from the unfinished stockade they were creating on the Ohio, at the confluence of the Monongahela and Alleghany rivers, and how, on the site chosen by their rivals, they had erected a fort which, in honour of their Governor, they had called Fort Duquesne. The fort was simply a square of four bastions, with the water close on two sides, and the other two protected by ravelin, ditch, glacis, and covered way. On these sides the forest, to a distance of a little more than a hundred yards, had been cleared away, but the fort lay under the disadvantage of being commanded by a wooded hill beyond the Monongahela. The armament consisted of a number of small cannon mounted on the bastions.

The garrison had originally consisted of the men, about five hundred in number, partly Frenchmen partly Canadians, who, the year previous, led by the Marquis Contrecœur, had driven Ensign Ward and his soldiers from the incipient stockade. Contrecœur still held the command, having under him three captains, de Beaujeu, Dumas, and de Lignères, and a garrison increased by some Canadians and by eight hundred Indian warriors. These were encamped on the cleared space between the fort and the forest.

Contrecœur had not been unmoved by the information, which had reached him in the month of June, of the march of the English. Hitherto he had confined himself to summoning all the Indians who were likely to respond to his call, and to sending out occasional scouts to bring information. But when, on the 7th July, the Chevalier de la Perade, who had been despatched upon this duty, returned with the information that the English were not far distant, and when, on the day following, two other scouts, the brothers Normanville, not only confirmed this information but specified the actual spot where they had seen the enemy, he deemed that the moment for some decisive action had arrived. He at once, then, summoned his captains to a consultation. The reader will recollect that the day was the 8th July, the day on which Braddock was approaching Turtle Creek. The time available for the garrison to carry out any plan which their leaders might arrange was therefore short. The deliberations, however, were long. Contrecœur, with the weight of responsibility on his shoulders—a weight which always finds out the weak point of a weak man—with the knowledge that the fort, commanded by the hill beyond the Monongahela, could not offer a successful defence against an assault which should be at once vigorous and skilful, was quite ready to accept and approve any proposition which might be made to fall back and save the garrison. His captains were men of a different stamp. One of them especially, de Beaujeu—the bearer of one of the greatest names in the history of France—strongly urged the advisability of holding the fort with the bulk of the white troops, whilst with the remainder, and the greater part of the Indians, the effect of an ambush and surprise were tried on the advancing enemy. After a long debate, Contrecœur was brought round to the views of his subordinate, and de Beaujeu was commissioned to propose the plan to the Indians. To his intense annoyance they refused to adopt it. "Do you want to die, my father?" exclaimed the chiefs; "and to sacrifice us besides?" With that reply constantly repeated, de Beaujeu was forced for that day to be content, and when the night of the 8th July closed the conviction prevailed that the resolution of the Indians was not to be shaken.

Nor did night bring them better counsel. When, in the early dawn, de Beaujeu, still eager for action, repeated his proposition, the Indians still refused. The gallant Frenchman, earnest and not despairing, made then an impassioned appeal to their loyalty. "I am determined," he said, "to meet the English. Will you let your father go alone?" These words kindled the flame which till then had been only dormant. The Indians, rising with a shout, promised to follow him, and put on their war-paint. De Beaujeu, delighted, paid them the compliment of donning their costume, and prepared to march with them into the jungle.

Then, between 7 and 8 o'clock in the morning, the force marched out. It consisted of thirty-six French officers and cadets, seventy-two regular soldiers, a hundred and forty-six Canadians, and six hundred and thirty-seven Indians. With these, besides de Beaujeu, who commanded the party, marched Dumas and de Lignères. Contrecœur remained in the fort with the bulk of the white troops to await the result.

About 11 o'clock the same morning, Braddock quitted his encampment, and, marching in battle array, forded the Monongahela, made the detour which avoided the perilous defile, and, about 1 o'clock recrossed to the right bank. It has been erroneously stated by many English, and by some French writers, that "he had neglected all precautions of scouts or vedettes."* The English general had erred up to this point, erred—in the opinions of those most competent to judge—by an excess of caution; and if he did not show that failing on the 9th July, he certainly did not display its opposite. On the contrary, he took every possible precaution. After he had crossed, unopposed, the Monongahela for the second time, he halted his troops and allowed them to take their food. Then, when they had rested and were refreshed, he gave orders for the resumption of the march about 2 o'clock.

The distance which they still had to traverse was about eight miles. The path, a rough one, turned inland from the river, for a few yards, as far as a log hut known as Fraser's House, almost perpendicularly to it. From that point it made a curve till it reached a hollow way. Beyond this hollow way, flanking and commanding the path, and thus affording a splendid position for an enemy, was a wooded hill. Thence the way ran parallel with the river, along the base of steep hills, through a very dense forest,

* "Il méprisait trop les Indiens et les milices françaises pour prendre les mesures de précaution les plus ordinaires."—*Nouveau Dictionnaire Historique de sièges matiéubet lles morables.* That this is a mistake is proved by Mante's History, by the journals of officers engaged, and by Washington's testimony. *Vide* Parkman, Bancroft, and other American writers.

covered with brushwood and fallen trees. It was the very country for a surprise.

Recognising the possibility of such an attempt, Braddock had taken many precautions to guard against it. Well in advance of the column marched the guides, escorted by six Virginian light horsemen. A hundred paces behind these came the advanced guard, supported, at a distance of forty paces, by three hundred and fifty men, who again were closely followed by two hundred and fifty axemen to open the road, under Sir John Sinclair. Behind the axemen were two guns with tumbrils and tool-wagons, protected by a small rear-guard. The command of this division of the force, which was called the advance column, had been entrusted to Lieutenant-colonel Thomas Gage. The post of that officer was with the three hundred and fifty men who formed the supports to the advanced guard. At an inappreciable distance behind the advance-column marched Braddock with the main body, the artillery, wagons, and pack-horses moving along the path, whilst the infantry made their way through the partially cleared spaces on either side of it. The rear was brought up by a mixed body of royal troops and provincials.

The path was but twelve feet wide. Gage, then, had disposed his axemen on both sides of it to a distance of about a hundred yards on each flank, and, whilst the main body of infantry moved on slowly through the spaces thus provided for them, parties of skirmishers cautiously threaded their way through the trees and brushwood immediately beyond the cleared ground.

The rear-guard of the advance-column had just crossed the hollow way, and was moving under the shadow of the wooded hill, of which I have spoken, when the six Virginian horsemen suddenly fell back with the guides they were escorting; and the engineer, who was marking out the road, looking up, beheld a man, dressed like an Indian, bounding towards him. On seeing the engineer, this man, who was no other than de Beaujeu, turned round and waived his head-covering. Instantly the war-whoop of the Indians was heard, and a heavy musketry-fire from behind a semicircle of trees in front was opened upon the British column. Gage at once spread out his troops in a line of deployment as even as the ground would permit, and his men, returning the fire of their invisible enemy, caused amongst them so great an alarm that the Canadians and many of the regulars fled panic-stricken from the field. At the third volley the author of the expedition, de Beaujeu, fell dead. A few seconds later, the opening of a fire from the two guns, which, on the first alarm, had been moved to the front, produced great confusion among the Indians, but they did not abandon the field. Had the way in front been clear, Gage had virtually gained the day. Conscious of this, and acting as the brave and gallant soldier that he was, he pushed steadily forward, his men cheering lustily as they fired. For a moment it seemed as though they would make the success already achieved decisive.

Three causes prevented them : the capacity and presence of mind under imminent danger of the French officer in command, the density of the tree jungle, and the propinquity of the line of march to the hill on its right. On the fall of de Beaujeu and the flight of the Canadians, Dumas, who succeeded to the command, thought that the day was lost. But he was a brave man, a cool-headed and resolute man, and, far from allowing his thoughts to be divined, he excited the few white soldiers—they did not now number more than fifty*—who remained, by voice and gesture, and advanced at their head to stay the progress of the English, whilst he rapidly indicated to his Indian allies the feasibility of attacking their flanks from their impenetrable cover. These tactics produced the desired effect. The steady fire poured in by the French regular troops, encouraged by Dumas and de Lignères, checked Gage's column whilst it was advancing in the confidence of assured victory, and, before it could recover from its surprise, a withering musketry-fire from both flanks, but especially from the round hill on the right, threw it into confusion. It is possible that a resolute charge on the Frenchmen who barred the path, such a charge as was made

* At the close of the day he had only twenty.

under not very dissimilar circumstances by Sir Vincent Eyre at Arah in 1857, might even then have proved effective.* But such an idea, if it occurred to anyone, was not acted upon; the British soldiers broke their ranks and fell back, huddled together in small groups, abandoning the two guns which till then had rendered them so much service.

At his post with the main body, Braddock had heard the sound of fire. Leaving four hundred men under Sir Peter Halket to guard the baggage he hurried across the hollow way to the front. But before he could reach the scene of action the mischief had been done. The men of the advance column, stricken with panic and in utter confusion, endeavouring in a wild and senseless manner to escape the bullets which rained upon them from behind the trees, no sooner saw their comrades approaching, than, guided only by their instincts, they dashed wildly into their ranks, and communicated to them the panic by which their reasoning faculties had been overpowered. In thus acting they destroyed the one chance of safety which yet remained for the common mass. It was still possible that the main body, by a well-ordered charge, might have overwhelmed the French regulars who now held the gate of the defile, and that, having once overcome these and emerged from the vicinity of the fatal hill on which the Indians were now mostly congregated, they might have restored and even gained the day. But that day was irretrievably lost when the frightened soldiers of the advance column communicated their own panic to the soldiers behind them!

Then, indeed, it became a slaughter—a slaughter greater than it had been before. For in the same space more numerous groups

* This view seems to be borne out by the accounts of the eye-witnesses and writers of the time. Bancroft thus summarises the experience of these: "Had the regulars shown courage the issue would not have been doubtful; but, terrified by the yells of the Indians, and dispirited by the manner of fighting such as they had never imagined, they would not long obey the voice of their officers, but fired in platoons almost as fast as they could load, aiming among the trees or firing into the air. In the midst of this strange scene nothing was so sublime as the persevering gallantry of the officers. They used the utmost art to encourage the men to move upon the enemy," &c. &c.

were huddled, wildly firing, whilst exposed to the continuous fire of an enemy whom many of them never even saw.* The Virginians, indeed, formed an exception. These gallant colonists, most of them experienced woodmen, accustomed to trace the wild beast in his lair, and learned in all the secrets of jungle-fighting, had spread themselves out in the Indian fashion, and still maintained upon their savage foe the only fire which, from the commencement of the panic, had been at all effectual. It has been conjectured that had they been allowed to continue their irregular fire, they might have kept the enemy at bay till order should have been restored. But they were not allowed. Braddock had had no experience of bush-fighting, and an action which would not have been tolerated in a review in Hyde Park excited his indignation. With rising fury he ordered them back into line; with the same ignorant anger he discouraged every attempt made by the more self-possessed of the regular troops to fight behind trees. He would allow no movement which the regulations had not authorised. At length, when he had got his troops all huddled together, after some of them, in their panic, had fired upon a body of Virginians who, from a safe position behind the huge trunk of a fallen tree, were dealing deadly execution on the enemy, and had thus compelled them to fall back, Braddock directed an attack upon the fatal hill whence the most deadly firing proceeded. Colonel Burton, who was charged with this duty, persuaded with difficulty a hundred men to follow him; but he was soon struck down, and on his fall the men ran back, terror-stricken!

Meanwhile the terrible slaughter continued. The men were so dazed as to be useless. In vain did Braddock try to animate them. Four horses were shot under him, he then mounted a fifth, but he could effect nothing. His men had become targets, and they were as helpless to defend themselves as targets. On them the example of their general and their officers was quite thrown away.

* "Many of the officers, who were in the heat of the action the whole time, would not assert that they saw one." H. Sharpe, to Secretary Calvart, 11th August 1755, quoted by Bancroft.

Nothing could exceed the gallantry of these, and they suffered in proportion. Sir Peter Halket and his son were shot dead, Young Shirley was shot dead; Gates, who lived to be one of the heroes of the American fight for independence, Gage, of Boston celebrity, Orme, Morris, and Gladwin were wounded. Two horses were shot under Washington. Of eighty-six officers, sixty-three were killed or disabled, whilst of thirteen hundred and seventy-three non-commissioned officers and soldiers, only four hundred and fifty-nine came off unharmed.*

At length even Braddock recognised that a further halt on the spot would exhaust even the supply of targets to the enemy. Unwillingly, then, and with a heavy heart, he gave the order to retreat. Whilst striving to range his men so that this movement might be conducted in an orderly manner, a bullet pierced his lungs and he fell, mortally wounded, to the ground. Then, all was over; the bulk of soldiers, frenzied by the ordeal they had undergone, rushed madly for the ford, in the fond hope that the interposition of water between themselves and their foe would make them safe. In their wild and headlong flight they abandoned guns, wounded, baggage, the General's papers, the military chest, to the enemy. That enemy was the Indians! The French were far too weak and too tired to pursue.

The first ford was reached, but no halt was made there. Vainly did Washington and other officers attempt to induce them to stop. Overcome by a more than mortal fear, they could think only of rushing as fast as they could from the sound of the terrible war-whoop still ringing in their ears. They dashed on till they reached a point on the Monongahela singularly favourable for defence against Indians. Here Braddock, who had been so far carried by four men, and who, despite the pain of his wound, had had no thought but for the honour of the British arms, made a great effort to induce his soldiers to remain

* Parkman's *Montcalm and Wolfe*, pp. 219–20. The French had three officers killed and four wounded; only four of their soldiers were hit. Of the Canadians five were wounded; the loss of the Indians was not ascertained.

until the rear-guard under Dunbar should arrive. Had they listened to his entreaties all would have been well; for the Indians, having collected the plunder and the scalps, were resolving to return early the following morning to their homes, and the French commandant of Fort Duquesne was terribly fearful lest the strengthened English should return and attack him again. But as well might one have attempted, to use the graphic language of Washington, "to stop the wild bears of the mountains." Not more than a hundred men out of the whole force could be induced to remain; and within an hour these, too, followed their comrades. The disorderly flight continued during the night.

In the morning some sort of order was restored. Braddock was placed on a horse, but the pain from his wound being greater than he could bear, a litter was made, and four men were bribed by Captain Orme, by the promise of a guinea and a bottle of rum apiece, to carry him. Early on the night of the 11th the fugitives reached the farm of Gist, only six miles from Dunbar's camp, and here they met wagons and provisions sent by that officer. At last the scared soldiers could feel that they were safe!

The moment had now arrived when the coolness of the educated soldier should have asserted its sway over the exhausted passions of the mob. It is sad to record that, instead of this happening, the exhausted passions of the defeated soldiery found a fresh pasture-ground in the spirits of Dunbar and his followers. The force was at least safe from pursuit. In reality, it was strong enough to reassume the offensive; but admitting, for the sake of argument, that in the existing temper of the men such an advance would not have been prudent, it may be confidently affirmed that they might have halted where they were till reinforcements should have reached them. They held a position in which it would have been very difficult to assail them; supplies were abundant; their communications were safe. Their commander might have held the place safely until his force had either been reinforced or had become morally strong enough to retrieve the disaster. I cannot but think that Braddock, had he lived, would have adopted such a course. Several times during the retreat he had endeavoured to induce his men to halt and face the foe. But Braddock was dying.* The only order he had given—an order given on his arrival at Gist's farm, to succour the men who had fallen on the road—proved that he had not forgotten the duties of a general. But he was now too weak to command, and Dunbar, who succeeded him, displayed none of the qualities of a soldier. He at once resolved on flight, burned all the wagons—collected with so much difficulty—which were not required for the actual wants of the force,† and began his retreat to Fort Cumberland the morning of the 13th. At that moment Contrecœur was still trembling lest he should be attacked!

The panic-inspired retreat was a fitting climax to the disaster of the 9th. It filled to overflowing the cup of disgrace. If the reader will recall to mind that two at least of the colonists who were present throughout these fatal days became prominent leaders in the battle for independence which was inaugurated by another retreat,‡ just twenty years later, they will scarcely fail to draw the conclusion that those colonists must have been encouraged in their hopes of ultimate success by their recollection of the demoralisation which overtook a British force surprised on its march by an enemy, and by the panic which induced its leader to abandon ground which might have been successfully defended!

* He died on the 13th. His last words were, "We shall know better how to deal with them another time."

† He burned more than a hundred.

‡ The retreat from Lexington, 19th April 1775.

CHAPTER VII.

MAXEN.*

THE most noted ambush and surprise which occurred during the Seven Years' War was that of Domstädtl, on the 28–30th June 1758. On that occasion General Loudon, who commanded the Austrians, defeated a force fifteen thousand strong, destroyed the convoy it was escorting, and forced Frederic II. to raise the siege of Olmütz. The same distinguished general surprised and captured Fouquet and his army at Landshut in 1760; and surprised and took by assault the important fortress of Schweidnitz the year following. Of these three feats of arms the first fits itself with the greatest exactness into the plan of this work; for Domstädtl was an ambush as well as a surprise. I have, however, so recently written a detailed account of it,† as well as of the two other feats of arms I have mentioned, that I feel justified in passing them over in order to illustrate my subject by recounting the surprise of the army of General Fink at Maxen by Field Marshal Count Daun. This surprise is remarkable in its way, in so far that the general

* The authorities whence I have principally drawn the materials for this chapter are (1) *Geschichte des Siebenjährigen Krieges*, von J. W. von Archenholz; (2) *Memoirs of Daun*, by A. Henderson; (3) *Geschichte des Siebenjährigen Kriegs in Deutschland*, von G. F. von Tempelhoff; (4) *Geschichte des Siebenjährigen Kriegs*, herausgegeben vom Königlichen Preussischen Generalstabe: (5) *Zur Geschichte von Oesterreich und Preussen zwischen den Friedensschlüssen zu Aachen und Hubertusburg*, von Leopold von Ranke; (6) *History of Frederick II: of Prussia*, by Thomas Carlyle.

† *Loudon*: Chapman and Hall, 1884.

who carried it to a successful issue was the most cautious general who ever commanded in chief since the time of that most formidable opponent of Hannibal, Quintus Fabius Maximus.*

The earlier part of the campaign had not been favourable to the Prussian arms. On the 12th of August, Frederic had been completely defeated at Kunersdorf, and had been saved from destruction solely by the want of enterprise of the Russian commander, Count Soltikoff. The defeat, however, entailed upon him many serious consequences. It had left Saxony without defence, apparently at the mercy of two armies—one, the Reich's army, led, under the supervision of Count Daun, by General Maguire; the other, the main Austrian army, commanded by Daun himself. Before proceeding to describe the manner in which that commander used the opportunity thus offered to him, I propose to introduce him to the reader.

Leopold Joseph Maria, Count von Daun, was born at Vienna in 1705. He was brought up with the idea of entering the Church; but his father, his uncle, and his grandfather had been distinguished soldiers, and Daun had scarcely entered his teens when he resolved to adopt their profession. He served in the Imperial army with credit in Sicily (1718) and on the Rhine (1734–5), and when the Turkish war of 1736 occurred he had reached the grade of general of brigade. In that capacity he served with distinction in the campaign of 1739. Promoted, on its conclusion, to be a major-general, he took part in the war which broke out on the death of the Emperor Charles VI., as well as in the second Silesian wars, and was present at the battles of Hohenfriedberg (4th June 1745) and of Soor (30th September 1745), commanding the left wing in the latter. On the conclusion of the Peace of Dresden (25th December 1745) Daun married the Countess von Fuchs, the prime favourite of the Empress-Queen. This marriage, and the reputation he enjoyed as a capable officer, combined to obtain for him a command in the Netherlands against the French (1746–48). At the battle of Laffeldt, fought the 2nd of July 1747, and won

* "*Unus homo nobis cunctando restituit rem.*"— Quintus Ennius.

by the French against the combined English, Dutch, and Austrians—a battle which drew from Louis XV., who witnessed it, the significant remark, "The English fight for all and pay for all"—Daun displayed a steadfastness which won for him golden opinions. The year following, 1748, Daun was appointed permanently, with the rank of Field Marshal, to reorganise, in personal communication with Count Harrach, President of the War Administration, the Austrian army. During the eight years which followed he devoted all his energies to this important work. It was to him that Austria owed the very efficient state of her artillery at the outbreak of the Seven Years' War. Daun had observed that the victories of Gustavus Adolphus had been very much due to the excellence and superiority of his guns; that, in the first and second Silesian wars, Frederic II. of Prussia had enjoyed a marked advantage in that respect over his enemies. To deprive that sovereign of his superiority in the always impending war by the nurturing and developing of the Austrian artillery became then with him a passion, and he devoted all his energies to the work. Nor was he content with the success he achieved in that direction. He taught the generals under him how to choose and fortify camps; how, bearing in mind the lay of the ground, to place the guns advantageously; how to march lightly equipped—all points in which, in the wars immediately preceding, he had found the Austrian army deficient. Another matter on which he laid great stress was the improvement of the discipline, hitherto extremely lax. Reforms of this comprehensive character could only be carried out by a man of a singularly resolute nature; and it testifies much to the innate power of Daun that, in spite of tacit opposition on the part of many officers of high standing and great influence, notwithstanding the often very lukewarm support of the Council of War, he was able to accomplish as much as he did. More, a great deal more, ought to have been done, but that it was not done was not the fault of Daun. He was unable to triumph in very many matters over official self-complacency, and, when the war broke out, the infantry regiments were still on a peace footing, the cavalry regiments were mere skeletons, the commissariat department, the pontoon-trains, and the hospitals were in a condition such as to justify the sarcasm applied to their successors by William Pitt: "The statesmen of Vienna are always a year behind the rest of the world with an idea and with an army."

The shortcomings I have mentioned—shortcomings against which Daun had earnestly striven—only brought the name of the reformer into greater prominence when the sudden invasion of Saxony by Frederic brought matters between himself and his secret enemies to an issue; and, in the second year of the war, after Frederic had beaten Brown at Lobositz, had captured and enlisted in the ranks of his own army the Saxon soldiers at Pirna, had won the battle of Prague and was besieging that city, the Empress-Queen entrusted the command of a second army—the last hope of Austria—to the man whose services as an organiser were then universally admitted. With that army Daun fought and won the battle of Kollin (18th June 1757), the first battle in which Frederic had met defeat from the Austrians. In that engagement he displayed alike his merits and his defects as a general. The latter so evenly balanced the former as to make him, in the presence of an active, daring, aggressive adversary such as Frederic, almost a negative quantity. In fortifying his camp, in receiving an attack, in defending a position, he was admirable; but he was too cautious to attack, unless his superiority rendered success certain, and he never followed up a victory. The result was that he conquered only the ground on which he stood. To illustrate what I mean by a metaphor of the great historian of the Peninsular War,* the wall might go down before him in ruins, but, far from stepping over those ruins, he gave time to the skilful mason, who was his principal adversary, to repair them. It was so after Kollin. Subsequently he was defeated, when joined in command with Prince

* "In following up a victory the English general fell short of the French emperor. The battle of Wellington was the stroke of a battering-ram; down went the wall in ruins. The battle of Napoleon was the swell and dash of a mighty wave, before which the barrier yielded, and the roaring flood poured onwards, covering all."—Napier's *Peninsular War.*

Charles of Lorraine, by Frederic at Leuthen, 5th December 1757.

On the 31st October of the year following, Daun, prompted by his brilliant lieutenant, Loudon, took his revenge at Hochkirch, and gained a great though barren victory.* This was the last great battle he fought before the opportunity offered to effect the surprise which forms the subject of this chapter.

* It is to me surprising that a writer who professed such regard for veracity as Carlyle should have been so blinded by partiality for his hero as to repeat as authentic the story that after the battle of Hochkirch Pope Clement XIV. sent to Daun a consecrated hat and sword. The story is a pure invention, coined by Frederic in revenge for his defeat. Not only did the Court of Vienna declare at the time that the story had no foundation whatever, but it can be proved from the works of Frederic himself—works cited by Carlyle as his sole authority—that the idea emanated from his own brain; in a word, that he concocted it to throw ridicule on his adversary. I would invite any sceptic on this point to read the letters of the Marquis d'Argens to the King, published in the fifteenth volume of *Œuvres de Frédéric.* At page 70 d'Argens writes thus: "I never read a better joke than your Papal brief and your letter of Prince Soubise. Even your Majesty's enemies must confess that it is impossible to read anything more witty." At page 71: "I find the Papal brief so witty that *I shall translate it into Latin, two columns, one side French, the other Latin. This will give it more the appearance of authenticity, because all briefs to the Imperial Court and to its ministers are written in Latin.*" (The italics are mine.) Again, page 73: "I have the honour to send to your Majsty the Latin translation of the Papal brief. There is more true wit and spirit in this composition than in anything that has appeared in the course of the war." These extracts, to which notice was first called by Pezzl, in his *Life of Loudon,* published in 1791, prove beyond a doubt that the story was a satirical jest made by Frederic to avenge his defeat at Hochkirch by a man whose military talents he despised. Seventy-eight years later, 1869, the Herr von Janko published in his Life of Loudon (*Loudon's Leben,* page 77) the same extracts. The fable that the Pope did send a consecrated hat and sword has long since been abandoned by all the later French and German writers of repute. It speaks little for the impartiality of Carlyle that he, knowing, as he must have known, that the Court of Vienna had denied it at the time, that the letters of the Marquis d'Argens had disproved it—for he had Frederic's posthumous memoirs and Pezzl, both containing d'Argen's letters before him—should repeat it as a fact without attempting to explain, or even referring to, the contradiction. It is still more incomprehensible te me how so able, so conscientious, so painstaking, and generally so impartial a writer as Mr. F. W. Longman should (*Frederick the Great and the Seven Years' War,* Longmans, 1881) likewise aid in giving currency to the fable; and, having d'Argens' letters, Pezzl, and Janko, to refer to, should comment upon it as though it were an established fact which had never been contradicted. It affords a proof how hard it is to kill a well-conceived and oft-repeated lie. Amid this careless acceptation of fiction for fact by English writers, it is refreshing to find that the latest historian of Frederic, Colonel C. B. Brackenbury, R.A., has told the story of the battle of Hochkirch and its consequences without referring to the practical joke which Frederic played upon the intelligence of the world.

The earlier portion of the campaign of 1759 had not, I have said, been favourable to the Prussian arms. Frederic had attempted "to solve the Russian problem," and had failed. The defeat which he sustained at Kunersdorf, August 12th, ought to have decided the war. That it was not made fatal to him was due entirely to the want of enterprise of the Russian general who commanded in chief. Rarely, moreover, has a commander rallied so quickly from misfortune, or profited so much from the inactivity of his opponents as did Frederic on this occasion. In the first moment of defeat he had despaired, knowing that his enemies had but to follow him to complete his overthrow. But when, four days later, he saw that they still halted where they had fought, he recovered his confidence, and felt that he was still a match for such dawdlers, despite their numerical superiority.

The situation, however, was critical. Silesia and Saxony were alike threatened. Whilst Frederic, with his fast increasing army, covered Berlin against Soltikoff and Loudon, his brother Henry lay, with nearly forty thousand men, at Schmöttseifen, twenty-three miles southward of Liegnitz. Between the two lay the army of Daun, nearly seventy thousand strong, having its head-quarters at Triebel, in Saxony, its flank stretching as far as Hoyerswerda, in Silesia, eighty-four miles west by north-west from Liegnitz. At the same time, the Reich's army, nominally under the Duke of Zweibrücken, but really directed by the Irishman Maguire, encouraged by the turn events had taken, was besieging Dresden.

In the first moments of his despair Frederic had despatched a messenger to General Schmettau, who held Dresden for him, telling him practically that he must leave him to himself. A little later, however, recovering in the manner already described, he had directed General Wünsch to march with eight thousand men to the relief of that city. But on the 4th September Schmettau surrendered!

On the previous evening General Wünsch and his eight thousand men were at Grossenhain, twenty-two miles from Dresden. Wünsch had previously taken Wittenberg and Torgau, and now he was pushing on with eager haste from Grossenhain, hoping to be in time to force the raising of the siege. Had Schmettau delayed the surrender four-and-twenty hours, Dresden would have been saved. But, let the apologists of Frederic say what they may, the fact remains that Schmettau was fettered by the despairing letter of Frederic. Ignorant of the change in his master's fortunes, he only knew that he had lost a great battle, a battle which should have ruined him; while he himself was closely pressed by a vastly superior force. It is unjust, under those circumstances, to blame him very severely for his surrender.*

But that surrender was unfortunate. On the evening of the morrow of the entry of the Austrians into the city, Wünsch, still ignorant of the fall of Dresden, appeared before the suburb known as the Neustadt, and summoned Maguire, who occupied it, to capitulate. Maguire, anxious to detain him where he was, that he might overwhelm him in the morning, promised him an answer in two hours. Before that time had expired night had set in; Wünsch had discovered the surrender of the city, and had fallen back on Torgau, which he reached just in time to deliver it from falling into the hands of the Imperialists.

According to the programme of Daun the fall of Dresden was to have been the signal for decisive action on his part and on the part of the Russian general, Count Soltikoff. At a meeting which took place between those two commanders on the 22nd August, at Guben, in the lower Lausitz,† it had been arranged that the two armies should remain in their actual positions until Dresden should fall; that, on this occurring, they should act in concert to recover Silesia, beginning with the siege of the fortress of Neisse. They were now at liberty to carry this programme into effect.

Every military student must admit that it was a very poor programme, wanting in genius and audacity, a programme certain, in

* Even Carlyle admits that, "tried by the standard of common practice, Schmettau is clearly absolvable."

† Now forming a portion of the province of Brandenburg.

the presence of such an enemy as Frederic, to fail. Doubtless the recovery of Silesia lay at the heart of the Empress-Queen, and her generals knew that they could in no other way so well gratify their mistress. But the proper way to have recovered Silesia was to have crushed Frederic after Kunersdorf—a matter so easy that for three days after the battle he could see no way of escape: the certain way to lose Silesia for ever was to give time to Frederic to regain his strength, whilst the Austrians waited for an event which would never happen so long as Frederic was free to prevent it, but which would have instantly followed his capture. Never was an opportunity for finishing the war, and recovering for Austria all she had lost, so deliberately thrown away!

But Dresden had fallen, and now, as I have said, the two commanders were free to carry out the programme of Guben. But since that had been arranged many things, one main thing in particular, had happened. Frederic had recovered his vitality and his confidence, and Frederic had resolved to lose neither Silesia nor Saxony. He had, indeed, lost Dresden because his later messengers had not been able to reach Schmettau, but he had prepared a means of detaining Daun in Saxony until an opportunity should offer of smiting him hip and thigh.

On the 6th September Frederic, still ignorant of the fall of Dresden, and hoping to save it, had despatched an officer in whom he had great confidence, General Fink, with eight thousand men, to unite himself to Wünsch, to join touch with Prince Henry, and to attract to himself, as much as possible, the attention of Daun. The services which Fink had, up to this time, rendered his sovereign were of a nature to deserve the confidence with which he was treated. Born in Strelitz, in Mecklenburg, in 1718, he had been trained and attained the rank of major in the Russian army. In 1743 he transferred his services to the King of Prussia. Frederic took a great fancy to the Mecklenburger, and was so pleased at the decision and presence of mind he displayed that he gave him rapid promotion. The very year of which I am writing Fink had attained the rank of lieutenant-general, and Frederic had selected him as the one of all his officers the best fitted to carry out his plans with respect to Daun.

Fink entered with intelligence and alacrity into the King's views; reached Grossenhain the evening of the 9th, learnt there, for the first time, the surrender of Dresden, and pushed on to join Wünsch at Torgau; joined him there on the 11th, and, taking him with him, hastened on to, and recaptured, Leipzig (13th September)—the most effectual, and, to the slow mind of Daun, the most astounding reply to the loss of Dresden!

Up to this time the head-quarters of Daun had been at Triebel, in the extreme north-west corner of what was then Saxony. To facilitate the movements of his troops in his endeavour to maintain the severance between Frederic and Prince Henry he had established magazines at Guben, at Görlitz, at Bautzen, at Zittau, and at Friedland. He had, besides, a strong army corps at Lissa (on the Neisse). Prince Henry, on the other hand, anxious to gain touch with the King, had marched in the latter days of August, with the bulk of his troops, from Schmöttseifen to Sagan. Had he been opposed by Loudon instead of Daun it had been a diastrous march for him; for in his eagerness to effect his object he exposed his lieutenant, the renowned Ziethen, to the danger of being cut off. That officer had pushed on to Sorau, a short day's march from the main Prussian army at Sagan. There he sat, isolated, trusting, it may be presumed, to the known over-cautiousness of Daun. He was wrong in the main, though right by the bye. Daun had noticed the false position of his enemy, and had resolved to overwhelm him. He directed three strong columns, from the north, the south, and the west, to march on Sorau. Ziethen's destruction was certain; when, just as Daun was about to strike his still unconscious victim, he noticed a strong patrol of Prussian cavalry extending in the direction of Sagan. The sight of this patrol filled his mind with doubts. He halted his own column, satisfied himself that the calm about Sorau indicated that the entire Prussian army was either concentrated there and in communication with Sagan, and that he was being led into a trap—and fell back! Before the other Austrian columns could arrive, Ziethen had become aware of his danger, and had marched back to Sagan. Daun's conduct on this occasion was but a rehearsal of his sacrifice of Loudon at Liegnitz a year later!

Prince Henry and the King remained, then, separated the one from the other, unable to communicate except by spies. Their situation was still dangerous, though the unenterprising character of the two commanders-in-chief, Daun and Soltikoff, minimised the danger. The Russians and Austrians, almost in touch, occupied, on the 3rd August, a line of eighty miles; and whilst Soltikoff, at Lieberose, with fifty thousand, faced Frederick, who sat at Waldau with twenty-four thousand; Daun, at Trieben, with nearly seventy thousand, watched Prince Henry at Sagan with forty thousand. This long line of eighty miles occupied by the two armies thus separated the King from his brother. The central position of Daun was one in which a general like Napoleon would have rejoiced, for it gave him the chance of crushing the two armies opposed to him in detail.

Soltikoff remained till the 15th September in his camp at Lieberose. On that date he met Daun in council at Bautzen, and, urged by that commander, and encouraged by the promise of an additional corps of ten thousand Austrians, he marched towards Guben, with the intention of undertaking the siege of Glogau. Frederic, who had heard of the fall of Dresden on the 10th, no sooner noticed the movement of the Russian commander than, believing he intended to recross the Oder, he marched rapidly to surprise Daun; but when he learned, on the 19th, that Soltikoff, strengthened by the promised reinforcements, had reached Sommerfeld seriously bent on Glogau, he retraced his steps, and, determined to interpose between the Russian and that fortress, marched rapidly to Sorau. He reached Sorau on the 20th, and at once sent couriers to Prince Henry and to General Fouquet to request them to send him all the troops they could spare.

On that date the position, then, was as follows. Soltikoff at Sommerfeld (for he halted there a day) with sixty thousand men;

Frederic at Sorau with twenty-four thousand, and in communication with Prince Henry, now near Görlitz; Fonquet, with thirteen thousand men, at Landshut; Daun, now at Bautzen, with nearly sixty thousand.

But Frederic did not remain a day at Sorau. On the 22nd he reached Eckersdorf. There, satisfied that he had nothing to fear from the sluggishness of Soltikoff, he gave his troops a day's rest. Soltikoff had then a chance of anticipating the King in his design to cover Glogau by making a forced march to Suckau. But he would not depart from his ordinary pace, and the result was that Frederic reached Suckau before him, and thus interposed between him and his prey.

Meanwhile, Daun, having seen the King fully occupied with Soltikoff, had resolved to finish with Prince Henry. On the 22nd, then, he marched to Görlitz, reconnoitred the Prussian position on the Landskrone,* and, having taken stock of it to his satisfaction, issued the necessary orders for an assault upon it the following morning. Full of hope, he marched, early on the morning of the 23rd, to that assault. But hope suddenly yielded to astonishment. He found the Landskrone unoccupied. Prince Henry and his army had disappeared! Whither had he gone?

Yielding to one of those inspirations which are the choicest gifts of genius to her children, Prince Henry had resolved upon a *coup* which should confound his irresolute adversary. According to Archenholz, the Count of Kalckreuth, the Prince's aide-de-camp, then a young man who had seen only twenty-two summers, but who rose to great distinction in the Napoleonic days, strongly supported the plan. That plan had, at least, the merit of audacity. Prince Henry had resolved to give Daun the slip in the night, and, breaking through a weak point in his long line, to make a forced march of fifty miles, and, whilst Daun should be searching for him in the vicinity of Görlitz, to pounce upon the Austrian corps left under General Wehla at Hoyerswerda! When, then, on the evening

* A hill, four miles and a half from the town, 1,300 feet above the sea-level, but only 570 above Görlitz itself.

of the 22nd September he assured himself that Daun had returned to his camp he issued the necessary orders, quitted his position in light marching order, pushed down the valley of the Neisse as far as Rothenburg on that river, some twenty-three miles, bivouacked there for three hours; then, turning westward, marched eighteen miles to Klitten. Klitten is a little village nearly midway between the larger villages of Mücka and Uhyst, and lies exactly half-way between Rothenburg and Hoyerswerda. It is distant from Görlitz twenty-nine miles, though, to reach it by a

circuitous route, Prince Henry had marched forty-one. He was, in fact, between the Austrian main army at Görlitz and its strongest isolated corps at Hoyerswerda, and nearer to the latter by a distance of twenty-nine miles. At Klitten he halted again for three hours for rest and refreshment, and then resumed his march to Hoyerswerda.

At that place was Wehla with three thousand men. Occupying the westernmost point of the Austrian line, in communication with Daun, thirty-six miles distant at Görlitz, Wehla might well deem himself safe. Suddenly, however, on the morning of the 25th his scouts reported Prussians close at hand marching through the woods. In vain did he occupy the best available positions. Three thousand men were powerless against ten times their number. After a resistance which was from the first useless, which cost him six hundred men in killed, he agreed to lay down his arms and surrender!

The main result of this famous march was to bring all Saxony, Dresden alone excepted, under the power of the Prussians. For Soltikoff, renouncing his idea of attacking Glogau, re-crossed the Oder; and Daun, recovering from his astonishment, finally took up his post at Schildau, directly south of Torgau, on which place Prince Henry had marched. Whilst at Schildau he tried, but in vain, to draw the Prince to a general action. Then, discouraged by the defeat, at Pretoch, of an attempt made, under his orders, by the Duke of Ahremberg to cut off the Prince's supplies, and by the information which reached him that the Prussian army had received large reinforcements under General Hülsen, he quitted his camp, the 4th of November, and fell back into the Meissen district to cover Dresden. Prince Henry followed him, and took post three days later at Hirschtein some eight miles to the north-west of the Imperialist camp.

At Hirschtein Prince Henry was joined on the 13th of November by the King of Prussia. Frederic had emerged from the difficulties created by his defeat at Kunersdorf in the most marvellous manner. After having been for three months threatened, he had, thanks to the inactivity of his enemies and his own energy, become the threatener. At Hirschstein he found himself at the head of an army which practically held all Saxony except the Meissen country and the capital, strong enough to strike a blow at Daun. The fact that the King had joined the Prussian camp soon reached the ears of that cautious general, and impelled him to fall back hastily upon Plauen, three miles to the south-east of the capital. Here Daun intrenched himself. It is probable, however, that he would not have rested here, but would have fallen back into Bohemia, if Frederic, in his excessive haste to compel his retreat, and in his too great contempt for his enemy, had not exposed one of his corps in such a manner that even the hesitating mind of the Austrian commander resolved to stay to strike a retaliating blow.

To alarm Daun in his camp at Plauen regarding his communications with Bohemia, Frederic despatched General Fink with a corps of twelve thousand men to Maxen, whilst he directed Colonel Kleist with a smaller body to cross the Bohemian frontier. Such movements were above all others calculated to produce the desired effect; for, whilst the King threatened the Austrian camp from the north-west, an enemy at Maxen barred the direct line of communication with Bohemia.

Fink did take post at Maxen, and Kleist, crossing the Bohemian frontier, took many prisoners, burned many villages, and carried off much plunder. But Daun was not frightened. Carefully examining the situation he saw that Fink was, in a military sense, in the air. There was no link whatever between him and the main Prussian army. He recognised, then, that Frederic had repeated the blunder he had committed at Hochkirch—the blunder of despising the intelligence and under-rating the enterprise of the Austrian commander-in-chief. His wounded vanity roused him then to action, and he resolved to give his great adversary another lesson.

Fink had not been blind to the danger of his position. Before his corps had marched, he had represented to his sovereign how entirely he would be at the mercy of Daun if Daun should prove enterprising. But Frederic, recollecting only how the Austrian commander had wasted the opportunities of the three months immediately preceding, had replied: "You know I can't stand the making of difficulties; you must manage it." Upon this Fink marched, but anxious, like the good general he was, to secure one way of retreat in case he should be attacked, left General Lindstadt with three thousand men to occupy the pass of Dippoldiswalde, on the Weisseritz, eleven miles south of Dresden. With the remainder he marched on and occupied Maxen. Fink reported his proceedings to Frederic, who was then at Wildsruff, nine miles west of Dresden, and within twelve miles of the Austrian head-quarters. In reply, Frederic directed him to call to himself the detachment he had left in the pass of Dippoldiswalde. Daun, he argued, was quite capable of overwhelming so small a detachment, whereas the concentrated troops of Fink would be able to give him a warm reception.* Fink obeyed, leaving only a few light horse at Dippoldiswalde.

Possibly, had nothing else happened, Frederic's calculations regarding Daun might have proved correct. But Daun had many detachments in the neighbouring villages; his light cavalry regiments were constantly scouring the country, and he was enabled, by their means, to intercept some of the later letters from Fink to the King. These letters revealed to him the exact situation of the Prussian general, and the confidence of the King in his own inactivity. Concluding from this that Fink was left to his fate by his master, he resolved, I repeat, to read that master a lesson.

The position of his several corps enabled him to do this without exciting extraordinary suspicion. He had with him at Plauen some fifty thousand men; there was a strong garrison in Dresden; Brentano, with about five thousand men, was at Döhlen, four and a half miles to the south-west of Dresden; whilst seven thousand men of the Reich's army, under the Duke of Zweibrücken, were marching down the Elbe towards Dohna to take there a position to the north-east of Maxen. He made, then, the following dispositions. Leaving nearly half his army in the camp at Plauen, he resolved to march in person with twenty-seven thousand men to attack Fink's position from the south-east; to direct Brentano to assail it from the north; the Reich's army from the north-east and east. General Sincere, meanwhile, with three thousand cavalry, was to issue from Dresden, and, by an attack on some provision-carts expected by Fink, to draw off the enemy's attention.

The position held by Fink at Maxen was an irregular plateau, interspersed with rocky elevations, and surrounded on three sides

* "It will be better that you concentrate your corps; because you will then be in a position to receive the enemy with more effect. The few battalions at Dippoldiswalde could be easily overwhelmed, because, if the enemy should undertake anything he will come with a strong force."—*Archenholz*, 8th edition (1864), p. 286.

by valleys, reached only by very steep descents. On the fourth side, that approached from Dippoldiswalde, it was less difficult, though still far from easy. Dippoldiswalde lies nearly seven miles to the south-west by west from Maxen. Between it and Maxen, is the pass of Reinhartsgrimma—a narrow gullet overhung by lofty heights—a place easily defensible, but which Frederic had not allowed Fink to defend; thence the road to a point near the plateau was comparatively easy. At this point is a height second only in altitude and in importance to the high plateau of Maxen—a height called Hausdorf. The occupation of this would secure a good position for an army attacking Maxen.

The range of hills which covered Maxen on the west extend in a semicircle to the north, where, very difficult of access, they form a strong support on the left to the village of Trohnitz, then a walled village. Nearly midway between this line and Maxen is the village of Schmörsdorf, forming a sort of advanced post to another range of hills called the Schmörsdorf hills; between these and the plateau of Maxen are rugged ravines.

A lower range of a less pronounced character covered the approaches from the east and north-east. This range, not less difficult, abuts, on its north-eastern angle, in the village of Dohna, a very strong post, nearly four miles from Maxen.

Fink was expecting a convoy of provisions to reach him by way of Dippoldiswalde on the 19th of November. Always suspicious of danger, he climbed, on the 18th, to the highest point on his position to take a survey of the country, to behold, not his convoy, but a corps of the Reich's troops marching down the Elbe in the direction of Dohna. It did not follow that because this corps was marching on Dohna it had any intentions on Maxen; rather, indeed, the contrary, for, to reach Dohna, it had to expose its flank to Maxen. But, nevertheless, Fink, always vigilant, sent an order to General Wünsch, commanding on his left front, to look to the safety of Dohna. The view did not in any way disquiet him.

Far otherwise was it the next afternoon. Again ascending to the highest peak, and casting his eyes in the direction of Dippoldiswalde, he beheld, indeed, in the far distance, approaching that place, carts which seemed to be his provision-carts; but he beheld likewise an enemy—an enemy well provided with cavalry—attacking them. The same evening he received a few lines from Frederic, enclosing a report received by General Ziethen from a deserter to the effect that Sincere was marching on Dippoldiswalde and Brentano on Döhlen, whither also a corps of the Reich's army was proceeding.

There was not very much to disquiet Fink about his position in this report. He already knew about the Reich's corps, and had provided for them at Dohna. Probably it was Sincere who was attacking his provision-carts; but, he held, Sincere could not be very strong, for he occupied no high position in the enemy's army. Evidently, he thought, it was a dash upon his provision-carts, and nothing more, the movement towards Dohna on the previous day having been concerted to draw his attention north-eastwards. He drew additional confidence from the fact that Frederic, who still lay at Wildsruff, at no great distance from him, knew as well as he did of the movements of Sincere, of Brentano, and of the Reich's corps, and that he was in no way disquieted by them. Ziethen, the active, intelligent Ziethen, had been able to discover no more for his master. Had he not the right to conclude that the reason was because there was no more to be discovered?*

The outlook, then, on the evening of the 19th, formidable as far as the provision-carts were concerned, was not apparently threatening to Fink himself. Had he been of a different opinion there was still time for him, supposing, as he did suppose, that he had only the light horsemen of Sincere to deal with, to withdraw, even by way of Dippoldiswalde, for the enemy were still on the further side of that place. But his position, he judged, was not threatened. He confined himself, therefore, to calling in his outposts and keeping his troops on the alert. He discovered his mistake—a mistake

* This was certainly Frederic's opinion. On the margin of Ziethen's despatch he had written with his own hand: "Er wird entweder mit den Reichern oder mit Sinceren einen Gang haben" (You will have a brush either with the Reich's folk or with Sincere).

shared to the full by his master, who, with the same information, had arrived at the same conclusion—the following morning, the 20th, when, on glancing from the highest point, he beheld the main Austrian army firmly planted at Dippoldiswalde, Brentano marching on Trohnitz, and a third enemy—probably the Reich's corps—attacking Dohna! It was a veritable and a very unpleasant surprise!

Daun, in fact, had outwitted Frederic just as much as he had outwitted Fink. He had somewhat ostentatiously despatched Sincere with three thousand men from Dresden to make a display, and then to join Brentano. Hoping that this movement would be reported to Frederic, and that he would not be disquieted by it, he had sent his own cavalry to beat up the convoy, and marched (19th) with his infantry to Dippoldiswalde, planted them there for the night, returned himself to his camp at Plauen to watch Frederic, and, rendered confident by his quiescence, had transmitted renewed orders to Brentano, Sincere, and the Reich's corps, and had then ridden back at daylight to Dippoldiswalde, had reached it at 7 o'clock, and had at once ordered his attack. Considering the cautious nature of the man, it was a wonderfully bold and daring exploit, for Frederic was with an army larger than his own detached corps not very far behind him at Wildsruff,* whilst Fink, in a strong position, was in front of him!

The mind of the Austrian commander-in-chief was intensely relieved when he discovered that the formidable pass of Reinhartsgrimma had been left unguarded. Had Fink only marched thither the previous evening and occupied it in force Daun would, perhaps, have hesitated long ere, with Frederic within twelve miles of him, he had ventured to attack it. But now, to march with all haste on the unguarded position was his one joyous thought. Still, unguarded though the position was, Daun found it no easy matter to thrust his army through it. The soil was slimy with rain; the

* Wildsruff is about nine miles from Dresden, which lies directly east of it, and twelve miles from Dippoldiswalde, to the south of it. Plauen, where Daun was, is about three miles nearly due south of Dresden; seven miles due south of Plauen is Dippoldiswalde; and seven miles north-east by east from Dippoldiswalde is Maxen.

pass was narrow; the way was uneven; the guns stuck so fast that much precious time was lost; and it is said that, but for the excellent spirit and resolution of his officers and men, Daun, who had always Frederic within twelve miles of him on the brain, would have renounced the enterprise. His men, however, intelligent as the Austrian soldier always is—intelligent beyond the intelligence, often, of their commanders—worked with a will and got him through. Relieved now from all thought of Frederic—for the pass of Reinhartsgrimma was between them—Daun pushed on with all haste to the hill of Hausdorf. Whilst he is thus hurrying forward I propose to return to Fink.

The sight which greeted that gallant soldier as he gazed from the highest point of the Maxen plateau on the morning of the 20th November was such as to cause him the gravest anxiety. The movement which had been possible the previous evening was now impossible. The three thousand light horsemen of Sincere attacking his provision-carts beyond Dippoldiswalde had developed into an Austrian army, twenty-seven thousand strong, not merely occupying that place, but in full march to assail him. Did he cast his eyes to the north or the north-east no relieving vision met his gaze. There, too, on Trohnitz and on Dohna, enemies were marching against him. With his twelve thousand men he had to defend his position against more than three times their number!

But he never flinched. Under those trying circumstances, serving a King who never forgave disaster, though he might be the direct cause of that disaster, Fink did all that a capable commander could accomplish. Wünsch, he knew, would hold Dohna as long as possible; at Trohnitz he posted what infantry he could spare to hold that place, whilst he massed his cavalry at Schmörsdorf, to charge the enemy in the less hilly ground between that village and Trohnitz, should they force Trohnitz: under his own immediate orders he concentrated his remaining troops on the plateau of Maxen, abandoning Hausdorf; he was not strong enough to defend both.

In consequence of this concentration, unavoidable under the

circumstances, Daun was able to gain a second strong position, the height of Hausdorf, without firing a shot. He reached its summit about 11 o'clock, halted there his infantry whilst he ranged his batteries, and then opened fire.

The cannonade—for the Prussian guns, previously placed in a suitable position, promptly returned the Austrian fire—lasted for about an hour and a half. It effected little damage, but, under cover of the smoke, Daun was able to form his infantry for the rush which, he intended, should prove decisive. Cautious in all things, he deemed it more prudent, when the attack-formation was complete, to begin by a strong musketry fire. Under this fire, his troops advanced and stormed position after position. At each the resistance was fierce, for Fink, a real soldier, had used every advantage offered by the ground, and the Prussians, under Fink's leadership, died hard. On the other hand, the Austrians, greatly inspirited by the success of their chief's plan, and confident in their numbers, gave their enemy no time to rally; pushed them from post to post; until at length, shouting victory, they expelled, at 5 o'clock, the last Prussian from the village of Maxen! Fink, still stubborn, and full of resistance, then evacuated the plateau and fell back with the remnants of his force on the village of Schmörsdorf.

Whilst Daun had been thus assailing Maxen from the south-west, Brentano and Sincere had been attacking Trohnitz from the north, and the Duke of Zweibrücken Dohna from the north-west. Brentano's task was difficult, for the enemy's position was naturally strong; it was extremely well defended, and he had too many horsemen for mountain warfare. However, pushing on step by step, he gained Trohnitz about the same time that Daun obtained possession of Maxen. Before he could range his troops for a decisive advance on Schmörsdorf, Fink had reached the latter place. Learning that Wünsch at Dohna was still successfully resisting the attack of the Duke of Zweibrücken, Fink hoped for a moment that, by overwhelming Brentano before the cautious Daun should follow him to Schmörsdorf, he might still save his little army. Animated by this hope, he ordered his cavalry to the front. But, meanwhile, Brentano had ranged his troops, and was moving forward, the guns in front. On noticing the advance of the Prussian horse he opened out an effective fire—a fire the more effective because the ground, though less unfavourable for cavalry movements than any other of the battle-field, was still unfavourable. Fink's cavalry, then, were swept away before they could charge home; then Brentano, sending his cavalry and infantry to the front, drove Fink from the village of Schmörsdorf, and forced him, as the shades of evening fell, to take refuge on a hill of the same name—a hill to the east of the village descending by a steep slope to a rivulet called Rothwasser, on the other side of which was a detachment of the Reich's army.

The attack of the Duke of Zweibrücken on Dohna had failed. It had not been made, indeed, with any very serious intent, for the instructions given to the Duke included the holding of the line of the Rothwasser from a point nearly opposite Maxen as far as

Dohna. Neither from that village nor from any point along that line was Wünsch to be allowed to issue. The attack on Dohna, then, though technically it had failed, had so far answered the purpose of the Austrian commander-in-chief in that it had kept Wünsch within the trap.

Such, then, were the relative positions of the several divisions of

the two armies when night fell. Daun with the main body of the Austrian army in full possession of the village and plateau of Maxen; within three-quarters of a mile of the village of Maxen, and in full communication with him, Brentano and Sincere occupying Schmörsdorf, and, a mile further on, Trohnitz; the Duke of Zweibrücken guarding the line of the Rothwasser as far as the point opposite Dohna; Fink on the hill of Schmörsdorf, and Wünsch within Dohna. Between them the two last-named could muster still over eleven thousand men.

Night is always fruitful of opportunities to a man who is bold and venturesome. Fink was bold and venturesome, and he was capable of executing the thoughts which burned within him. He hailed, then, the approaching darkness as likely to afford him the means of extricating himself from his terrible position. He knew that Wünsch held Dohna; he felt tolerably certain that the cautious commander-in-chief opposed to him would venture nothing till the morning. The distance between the position he held and Dohna was about three miles. On two hills between the two points—the hills of Falkenhain and Bloschwitz—he had detachments. Here, then, was his chance. He would concentrate during the night on Dohna, and break his way through the Reich's army. Once escaped out of the trap of which Dohna was a salient angle, it was hard indeed if, though on the wrong side of Dresden, he could not find means of communicating with the King!

Full of this intent Fink sent orders, as soon as night had fallen, to the commanders of the posts at Falkenhain and Bloschwitz to move as expeditiously as possible on Dohna. He then tried to do the same himself; but he found it impossible. Daun and Brentano had caused every outlet to be occupied. The night was dark, the ways were difficult, and on all sides he was hemmed in. Perceiving the impossibility of escape for himself, he then ordered Wünsch, whose troops were mostly cavalry, to thread his way backwards between Trohnitz and Schmörsdorf and endeavour to break through an unguarded part of Daun's position. Wünsch attempted to obey, and, after a great deal of floundering, reached a point whence, had night been prolonged and his men fresh, he might possibly have emerged. But his troops were wearied by their flounderings, day was dawning, and he was still within the trap. On every side hope had given way to despair.

With the coming morn, then, there remained but one act to accomplish—the painful, the harrowing act of surrender. For with the light of day came the sound of the booming of the Austrian guns from the plateau of Maxen, playing on that isolated hill, from which all escape had been cut off. There was positively no other chance for Fink unless he would see eleven thousand men mown down. Painful as the idea of surrender was to the gallant Mecklenburger, it was yet, according to Archenholz, accompanied by one small compensating reflection. "He wanted to cut his way through," writes that historian; "and, assembling his generals, proposed to them to do so. But the absolute impossibility of forcing their way through the strongly occupied passes left him no choice but the entire sacrifice of his troops or surrender. Fink believed he would render the King no service by the first course, because, there being so many Austrian prisoners in Prussian hands, it would be easy to redeem his army by simple exchange. He therefore listened to the pleadings of humanity." He sent, then, a flag of truce to Daun with an offer to capitulate on terms.

Daun would grant no terms. He insisted, and rightly insisted, upon absolute surrender. With the generosity, however, which has almost always characterised the proceedings of the Austrians, he proposed to respect the private baggage of the Prussians. Fink strove hard, Fink pleaded hard, for more favourable conditions; urged that as Wünsch commanded an independent corps he might not be included. His pleadings were to no purpose: Daun was inexorable. On the morning of the 21st November nine generals, five hundred and twenty-nine officers, and eleven thousand men laid down their arms and were despatched as prisoners to Dresden.*

* In computing the numbers of the prisoners Carlyle is, as usual, unjust to the Austrians. He admits nine generals and five hundred and twenty-nine officers, but he whittles down the rank and file to 2,836. Fancy 529 officers to 2,836 men! It is almost a proportion of one officer to five men. Noting this incongruity, he

The baggage of the officers was respected, but seventy-one guns, twenty-four standards, and ninety-six flags fell into the hands of the victors. A few Hussars who managed to escape brought the sad news to Frederic.

Such was the surrender of Maxen. For an extremely cautious general it was a noteworthy feat of arms. Who was to blame for it? Frederic blamed Fink, brought him to a court-martial which sentenced him to a year's imprisonment, and would never employ him again. But the real culprit was Frederic himself. Why did he, separated but a few miles at Wildsruff from Daun at Plauen, why did he allow Daun to steal between him and his subordinate? It was Frederic's business to see that that subordinate—whom he had deliberately placed in the air to frighten Daun, for he threatened his line of communication—was not overwhelmed by Daun. But, beyond sending to Fink a deluding message to the effect

endeavours to account for it by adding: "Besides the killed, wounded, and already captured, many had glided clean off." He crowns his incongruity by admitting in the very next line: "It is judged that Frederick lost, by all these causes, about 12,000 men." I would simply remark that there was no possibility that any large portion of the Prussian force could "glide clean off." If we admit that a thousand men were killed and wounded—a large proportion—we have still 11,000 to account for. Carlyle accounts for 2,836; and, to diminish the credit of the Austrians, asserts that 8,000 men had been already captured or had deserted. Let us grant again that 2,000—a large proportion—had been already captured, there remain 6,000 who "had deserted." When could they have deserted? Not a man had deserted before the 19th, and after that date they were hemmed in. If 6,000 could have deserted between the 19th and the 21st, why not 8,836? It will thus be seen that Carlyle's contention is absurd—as absurd as the story which he endorses and constantly refers to of the consecrated hat and sword. Archenholz, a native of Dantzig, who certainly was not inclined to err by showing predilection for the Austrians, writes thus: "The total remnant of sixteen battalions and thirty squadrons, nine generals and eleven thousand men, laid down their arms; only a few Hussars escaped and brought to the King the terrible news." Pezzl, an Austrian writer, puts the total number of Prussian prisoners at 14,000. The *Conversations Lexikon* (Brockhaus) writes, under the head "Maxen," thus: "The place is historically noteworthy for the reason that here, on the 21st November 1759, the corps of the Prussian general, Fink, which *still counted* 12,000 combatants with seven generals and 550 officers, surrendered in open field to Daun as prisoners!" When I find Archenholz concurring in the main with the specially selected contributors to the *Conversations Lexikon*, and both with the Austrian historian, Dr. Precechtel, who, describing how Daun attacked Fink on both flanks, thus concludes, "He forced him to surrender with 12,000 men and all his artillery, and with the generals Wünsch, Gersdorf, Bredor, Platen, Rebretisch," I can arrive at no other conclusion than that which I have stated in the text.

that he was threatened by Sincere and by the Reich's troops, and that he might expect to have a bout with one or the other, he did nothing. With plenty of light horsemen, with so splendid an officer as Ziethen to command them, he kept no sufficient watch on the camp at Plauen. The fact is Frederic repeated the mistake he had committed at Hochkirch. His contempt for Daun's enterprise was so great that he coolly placed temptation in his way, confident that he would not seize it. The only error that may be charged against Fink is that he did not quit Maxen and concentrate at Dippoldiswalde when he saw on the 18th the attack on his provision-carts. But it is by no means certain that such a move would have saved him. He had, besides, Frederic's positive order to remain concentrated at Maxen. No discretion was allowed him. It was thus doubly Frederic's duty to see that in the dangerous position he had allotted to his subordinate that subordinate was not overwhelmed by the enemy whom he himself was watching! Frederic, then, was the culprit; Fink was made the scapegoat!

The victory of Maxen gave Daun a splendid opportunity. He seemed at first inclined to use it. A Prussian corps, three thousand strong, commanded by General Dierecke, was on the banks of the Elbe not far from Meissen. Frederic on learning of the disaster at Maxen had called it in; but, just at the very same moment, Daun had directed one of his best officers, General Beck, to march with five thousand men against it. Unfortunately for Dierecke the bridge at Meissen was broken, and the Elbe was so full of ice as to be navigable only with difficulty; he had to sustain, then, on the 3rd of December the full attack of Beck. During the night he managed to transport one half of his force to the opposite bank, but the morning of the 4th found him with fifteen hundred to oppose five thousand. After resisting as long as was possible he had no course but to surrender.

Encouraged by this, Daun now made up his mind to attack Frederic himself. He could mass sixty-five thousand men against Frederic's forty thousand. But when the time for action arrived and he noticed that Frederic was ready to receive him, his moral courage gave way. Far differently would Loudon have behaved. Under such a commander, a commander as brilliant, as decided, and as enterprising as Frederic himself, the Austrians might, that winter of 1759–60, have brought the war to a conclusion. But Daun was only enterprising when success was absolutely certain. Confident, after Maxen, that he could drive Frederic from Saxony, he finally resolved to content himself with the humbler task of maintaining only its capital.

This, then, was the result of the surprise of Maxen. Frederic, in his too great eagerness to force Daun to retreat into Bohemia and to leave Dresden at his mercy, had placed Fink in a false position. Daun, not terrified to the extent Frederic hoped he would be, had fallen upon Fink and destroyed him. As the consequence he wintered in Dresden instead of in Bohemia; and Dresden remained till the close of the war in the possession of the Austrians.

CHAPTER VIII.

ST. GOTHARD.*

THE treaty of Campo-Formio, made between the Emperor and the French Republic the 17th October 1797, secured peace for continental Europe. It was the work of General Bonaparte, the fruit of his immortal victories of the year preceding, followed up with wonderful effect in the spring of 1797. By it he gained for France the Low Countries and the Ionian Islands, whilst in Italy he constituted the Cisalpine, and proclaimed the Ligurian, republics. In return, and as a poor compensation, the Emperor obtained the dominions of the extinguished republic of Venice.

The treaty concluded, General Bonaparte returned (5th December) to Paris. But his active genius could not rest quiescent, and the Directory, dreading the continuance of his presence in the capital, aided him to carry out the plans he was forming for an enterprise which should startle the world. It resulted from their combined action that on the 19th May of the year following General Bonaparte sailed with a fleet and army to conquer Egypt.

Even before his departure the Republic had continued the work

* The chief authorities upon which I have drawn for this article are: (1) *Précis des événements militaires sur les campagnes de* 1799 *à* 1814, par le Comte Mathieu Dumas; (2) *Mémoires de Masséna*, par le Général Koch; (3) *Geschichte des Feldzugs von* 1799 *in Deutschland und der Schweiz*, von Karl, Erzherzog von Oesterreich; (4) *Histoire des campagnes de Souvaroff;* (5) *Notice historique sur le General Lecourbe;* (6) *Histoire critique et militaire des campagnes de la révolution de* 1792 *à* 1801, par le Baron Jomini.

he had begun. In Italy, under its influence a Roman republic had been proclaimed, 20th March 1798; in Switzerland, at the same time, the Helvetian republic had been constituted. To traverse the intrigues which had thus, in a time of peace, and whilst a congress for the settlement of its affairs was sitting at Rastadt,* changed the constitution of Europe, and threatened to change it still further, the Powers of Europe, strongly pressed by Great Britain, began to form a new coalition. To the plan of this coalition, known in history as the Second Coalition, there acceded, in December 1798—the Congress of Rastadt, be it recollected, still sitting—in addition to the prime author, Russia, Turkey, and Naples: a few days later, the Emperor, the southern States of Germany, and Portugal. A stimulant to the action of the allies was furnished immediately afterwards by the expulsion by the French Republic of the reigning family from Naples, and the constitution of the Neapolitan territories as the Parthenopean republic, 14th January 1799.† The tension between the secretly hostile States was stretched by this act to the point of bursting. The negotiations at Rastadt did not march. Towards the end of the month of February the diplomatic arsenal was exhausted by both parties, and thenceforward it became the intention of each to choose the most favourable moment for the renewal of the war. The French Republic made the first move. No sooner did the Directory learn that Russia was about to send an army into Italy whilst the Emperor should act on the Rhine, than it transmitted orders to its generals to act as though war had been declared.

As opposed to Germany, not inclusive of Russia, the contending nations were not unequally matched in numbers. The French had 162,000 men under arms: Germany 169,000. In the first week of the month of March 1799 they were ranged as follows:

On the part of France, the army of observation on the Rhine from Düsseldorf to Mannheim, including the garrisons, numbered

* The Congress of Rastadt met the 9th December 1797, and sat all through 1798 and until the 28th April 1799.

† Signed the 22nd June following.

only 25,000 men. It was commanded by General Bernadotte. This force was to advance and occupy the valley of the Necker.

The army of the Danube, numbering 42,000, and commanded by General Jourdan, stretched between Schaffhausen and Mösskirch, and from the latter point towards the Danube at Sigmaringen. At Schaffhausen it gave touch to the army of Switzerland.

The army of Switzerland, 45,000 strong, commanded by General Masséna, communicated with the eastern shore of Lake Constance by Bregenz, and stretched by way of Zurich to Chur in the Grisons.

The principal army of Italy, commanded by General Scherer, and 50,000 strong, occupied a position behind the strong places of Peschiera and Mantua.

On the Imperial side, the army of observation, 24,000 strong, was cantoned in and about Würzburg in the valley of the Main.

The army of Swabia, 66,000 strong, and commanded by the Archduke Charles, occupied a position parallel to that of General Jourdan. The Archduke had his left at Kempten, his centre at Memmingen, whilst his right extended as far as Ulm. That right was covered by another army, 18,000 strong, commanded by General Sztarrai, which, posted on the left bank of the Danube, watched the movements of General Bernadotte. A fourth army, equally 18,000 strong, to be promptly increased to 30,000, commanded by Generals Bellegarde and Loudon,* occupied the passes of Tirol. In the Grisons, Generals Auffenberg and Hotze commanded a corps of 7,000 men, shortly afterwards augmented to 18,000, to maintain in the strong position of Feldkirch the communications with the Archduke. In Italy, 36,000 men, under the orders of General Kray, occupied a strong position on the lower Adige, covered by the fortress of Verona and the strong point of Porto-Legnago.

Such was the general position when the war broke out in the first week of March 1799.

To write a detailed account of the proceedings of this most

* Nephew of the famous Generalissimo of that name.

interesting campaign would take many pages. I must ask the reader to be content with a general sketch until I come to the event which illustrates this chapter.

The storm burst in Switzerland. There, we have seen, General Masséna—"the spoiled child of victory," as he was subsequently called by Napoleon—commanded. With a view to sever the communication between the Archduke and his lieutenants to the south-east of Lake Constance, Masséna had disposed his forces in the following manner. The Austrian position stretched from Feldkirch twenty-two miles and a half south of Bregenz, to Chur, which formed its centre, some twenty miles in the same direction; thence to the castle of Reichenau,* six miles west by south-west from Chur, at the junction of the Vorder and Hinter Rhein. Masséna directed, then, Oudinot to make a false attack on Feldkirch, whilst he himself should pierce the enemy's centre by crossing the Vorder Rhein opposite Luciensteg, and General Dumont, supported by Loison by way of Dissentis, should cross the river at a point higher up, and obtain a flanking position on Chur. At the same time he directed Lecourbe, who lay with a weak division at Bellinzona, to penetrate—supported by Dessolles, who was to march into the valley of the upper Adige—by way of the Via Mala into the Engadine, so that he might thus communicate with the army of Italy under Scherer.

These movements, executed with skill, vigour, and celerity, were nearly all successful. With the centre divisions of his army, Masséna crossed the Rhine, 6th March, and stormed the fort of Luciensteg: Dumont crossed the river, the same moment, at the point indicated, forced the pass of Kunkels and the bridge of Reichenau, and compelled the Austrian detachment which was resisting Loison at Dissentis to surrender. In the course of two days Masséna had become master of the Rhine from its source to its

* In this castle was the famous educational establishment founded by the Burgomaster Tscharner of Chur at the end of the eighteenth century, and the part-proprietor of which at this time was the famous Heinrich Zschokke: it was here that Louis Philippe of Orleans, afterwards King of the French, gave lessons in the French language and literature.

entry into Lake Constance, and had taken five thousand prisoners and fifteen guns. The intrepid Lecourbe, meanwhile, had crossed the upper Rhine and pushed on to Tusis in the valley of the Albula. From this point he moved on as rapidly as possible into the valley of the Inn, supposing that he was supported by Dessolles. But unavoidable circumstances had hindered the march of that general, and Lecourbe thus found himself, with his weak division, exposed to the attack of all the Austrian forces stationed in Tirol.

There are few of the generals of the first French Republic of whom so little is known as Lecourbe. Few have up to 1801 a worthier record. But his stern republicanism, his intimacy with and friendship for Moreau, made that record from that year to 1814 an absolute blank. Lecourbe was born at Lons-le-Saulnier in 1760. The son of an officer, he had entered the army, as the army was constituted under the old *régime*, at an early age, but, obtaining no promotion after a service of eight years, had retired. When at the commencement of the Revolution the National Guards were organised, Lecourbe was made commandant of the National Guard of Lons-le-Saulnier. Soon afterwards he joined the army of the Rhine at the head of a battalion of the Jura. His skill and courage procured for him there a rapid advancement. At Hondschoote (August 1793), and in storming the lines of Wattignies (October 1793), he particularly attracted the notice of the generals commanding. Promoted, shortly afterwards, to be a general of brigade, he sustained, at the battle of Fleurus, 26th June 1794, at the head of three battalions, for seven hours, the assault of ten thousand Austrians. Employed, after that battle, successively with the armies of the Sambre and Meuse, of the Rhine and Moselle, of the Danube and of Switzerland, he continued to give proofs of firmness and capacity. During the retreat from the intrenched camp of Mayence, at the end of 1795, he, commanding the rear-guard, and repulsed the pursuing enemy for twenty-four hours: he might then have joined the main body, but, having received no orders to fall back, he remained in position, though surrounded, and finally, after inflicting great losses on the enemy, cut his way through them and successfully rejoined his leader. Promoted to be a general of division the following year, he held a command at the terrible battles of Radstadt, the 6th and 9th July 1796, and by his conduct contributed greatly to the brilliant victory obtained on the second day.

With such a record Lecourbe might well be trusted, even without the support of Dessolles, to make head against the Austrian forces in Tirol, greatly outnumbering him as they did. Soon was he to feel their presence in his vicinity. Whilst, in fact, he was marching along the Inn valley upon Martinsbruck,* Loudon, with fourteen thousand men, fell upon his rear. Notwithstanding his inferior numbers, and the information which reached him simultaneously that the Austrian Bellegarde had blocked the passes in front of him, Lecourbe halted, turned, defeated Loudon with loss, and then resuming his forward march, assailed Martinsbruck.

Lecourbe was still before that place when Dessolles joined him. The arrival of this brilliant officer and his troops so strengthened Lecourbe that, though he was not yet the equal of the enemy in numbers, he resolved to assume an active offensive. He attacked, then, Martinsbruck in front with his own division, whilst he dispatched Dessolles across the Stelvio, the highest pass in Europe, then covered with snow, into the Münsterthal, better known to mountaineers, perhaps, as Val Mustair; and Loison to assail Nauders. The three attacks succeeded admirably. Dessolles, to whom the most difficult part had been assigned, climbed the Stelvio, surprised the Austrians who should have guarded its defensible points, descended the mountain by the right bank of the Rammbach and fell upon and surprised Tauffers, a loftily-situated village at the entrance of the Val Avigna, carefully fortified and occupied in strength, taking four thousand prisoners and many guns. Tauffers is but a mile and a half from Mustair, which

* A post on the road from Chiavenna to Nauders, in a gorge, the continuation of the Finstermünz pass, three miles and three quarters on the Chiavenna side of Nauders.

gives its name to the valley. Loison, meanwhile, had seized Nauders, and Lecourbe had firmly planted himself within Martinsbruck.

The situation of General Loudon was now well-nigh desperate. Every avenue of retreat but one was occupied. Dessolles was at Tauffers, Loison at Nauders, Lecourbe at Martinsbruck. The one weak point of the force which hemmed him in was the chain above Tauffers which led to the Venosta valley. That chain was but lightly occupied. Loudon, massing the few men who still remained to him, succeeded in piercing it and in joining Bellegarde who was marching to his relief. The two Austrian generals fell back to cover Bozen. Lecourbe had reason to be satisfied with his success. He had gained command of the two most important valleys of Tirol and had opened the communication between the French armies in Switzerland and in Italy.

Whilst Lecourbe had been thus engaged, matters had progressed less favourably for the French arms in other parts. Notwithstanding the early success which I have recorded, Masséna had been repulsed in the attacks which he had made upon Feldkirch. The last of these attacks, made on the 23rd March, and conducted by himself in person, had cost him so severely in men and in officers—the *élite* of his army—that its failure had had all the effect of a defeat in the field. Masséna had, in consequence, recrossed the Rhine, and had brought Oudinot's corps to Rheineck, the point where the Rhine flows into Lake Constance.

Nor had Jourdan been more successful in Germany. That general, posted with forty-two thousand men between Schaffhausen and Mösskirch and between the latter point and the Danube at Sigmaringen, had received the most positive orders to attack the army of the Archduke without delay. Attack implied advancing; and advancing whilst maintaining communication with the army of Switzerland implied undue extension of a line already sufficiently attenuated. But the orders of the Directory were imperative, and Jourdan obeyed them. He advanced first to Mengen on the one side and to Markdorf on the other. Learning then that the army which was being organised under Bernadotte to support him would not be in position before the 30th, he hesitated to advance further lest he should be turned by the valley of the Neckar. But the news of Masséna's first success, and renewed orders from the Directory, overcame this hesitation, and he moved forward to take up a strong position, covered in front by the Ostrach and the Aach, two mountain torrents, which, issuing from nearly the same point, course with great rapidity, the one towards the Danube, the other into Lake Constance. The front cover they offered to an army posted between the river and the lake was thus complete. In other respects the position was well chosen. The left, commanded by St. Cyr, occupied the village of Mengen; the centre, led by Souham, was at Pfullendorf; the right, under Férino, at Barendorf; the advance division, under Lefebvre, at Ostrach.

The position occupied at the outbreak of hostilities by the army of the Archduke Charles has been indicated in a previous page. That illustrious prince had been fully alive from the first to the advisability of striking a blow at Jourdan before Bernadotte should come to support him. Strongly reinforced since the war had broken out, he had the superiority in numbers, his army was well in hand, and he was prepared to take the first opportunity which Jourdan should offer him. The position the French general had taken between the Ostrach and the Aach seemed to offer such an opportunity. Accordingly, on the 22nd March, the Archduke sent two columns, sixteen thousand strong, to threaten the right and left of the French position, whilst with fifty thousand he led a serious attack on the centre. The attack, in spite of a very strong resistance, succeeded. The French centre was pierced; Jourdan, however, fell back in good order and took up a new position between Singen and Tuttlingen.

Notwithstanding his retreat, Jourdan was still bent on renewing a forward movement. He could not leave Masséna, who was now operating on the right bank of the Rhine, uncovered. He fixed, then, upon Stockach as a point to occupy, because that place commanded the roads alike to Switzerland and to Swabia, and would restore to him the communications of which his retreat had deprived him.

It was on the little river which gives its name to the town of Stockach that the Archduke had taken position. His left was on some heights between Nenzingen and Wahlwies, behind one of the many windings of the Stockach; his centre was beyond the river on the plateau of Nellemberg; his right upon the prolongation of the same plateau, occupying the entire length of the chaussée between Stockach and Liptingen, also beyond the river. It was a weak position, and the necessity of maintaining Stockach was the only excuse for occupying it. The Archduke, moreover, scarcely expected Jourdan to attack him, and he had consequently neglected to throw up works which would have streugthened it where it was weakest.

The very day on which Jourdan had fixed to assault this position, the 25th March, the Archduke had made a reconnaissance in force to discover, if it were possible, the intentions of the French general. There resulted, then, a clash, which neither leader, less than the other the Austrian, had anticipated. The Archduke fell back rapidly on the positions I have described. He was, however, so closely pressed by the French that his right was completely beaten, and the French general had it in his power to gain a great victory. In his eagerness, however, to make that victory absolutely decisive, Jourdan despatched St. Cyr to cut off his enemy's retreat, weakening himself, in consequence, too much at the decisive point. The Austrians felt at once the relaxation in the attack; they rallied, regained their lost positions, and remained masters of the field of battle.

To Jourdan the result was fatal. Not only was he compelled to abandon all idea of a forward movement, but it was more than probable that he would be forced to recross the Rhine. Dispassionate consideration advised a retreat upon Switzerland and a junction with Masséna; but such a movement would have placed Jourdan under the orders of that general. To this he could not reconcile himself. Falling back, then, to the entrance to the

20

defiles of the Black Forest, he posted his army there in a position he deemed strong, and, leaving it under the command of the chief of the staff, hastened to Paris to endeavour to obtain there the means of strengthening and of providing it with necessaries. Fortunately for that army, the Archduke was fettered by orders from Vienna which forbade him to march towards the Rhine till the French should have evacuated Switzerland.

In Italy hostilities had commenced somewhat later. On that country the Directory had concentrated all its efforts, and, as a consequence, the army there had been raised by the third week of March to a strength of nearly one hundred and sixteen thousand. Of these thirty thousand occupied Rome and Naples; another thirty thousand were distributed as garrisons; there remained fifty-six thousand for active operations. The command of this army had been desired by Moreau, but, he not being in favour with the Directory, it had been refused to him. After having been vainly offered to Joubert and Bernadotte, it had been accepted by Scherer, at the time Minister of War. Under Scherer, Moreau served nominally as a volunteer, actually as second in command.

Of the fifty-six thousand men thus disposable, five thousand under Dessolles were acting with Lecourbe in the Tirol, and a like number had been despatched under Gauthier to occupy Tuscany. The remaining forty-six thousand were composed of six divisions, commanded respectively by Serrurier, Delmas, Grenier, Hatry, Victor, and Montrichard.

Opposed to this French army was the Austrian, commanded for the moment by Baron Kray. Although the total Austrian force in Upper Italy consisted of at least seventy thousand, Kray had thirty-six thousand only available for active operations. Of these, twenty thousand were at the central point of Verona: ten thousand were at Porto Legnago, and six thousand were distributed on the heights of Pastrengo, Cyse, and Calmasino, which had been fortified with great care. The right wing extended to the lake of Garda; the left was posted on the Adige, over which had been thrown two bridges to maintain communication, and, if necessary, to facilitate retreat.

Both generals, the French and the Austrian, had attacking orders:—the Frenchman was the first to act upon them. On the 26th March, the morrow of the day on which Jourdan had been beaten at Stockach, Scherer despatched the divisions Serrurier, Delmas, and Grenier, to cross the Adige and attack the Austrian left; the divisions Hatry and Victor, under the superior command of Moreau, to threaten Verona; the division Montrichard to make a demonstration against Legnago.

The battle raged with varying fortunes throughout the day. In front of Verona Moreau was successful, in so far that he occupied the several points outside the city and forced Kray to retire behind its walls. But the three divisions on the French right were repulsed, and Montrichard, finding himself in the presence of greatly superior forces, had judiciously fallen back on Moreau. The success obtained by this general had been at least counterbalanced by the repulse of the divisions under Scherer.

Three days later Scherer renewed the attack. His design was to force the Adige between Verona and Legnago, and, then, masking Verona, to cut the Austrian communications with Bozen. With this object he despatched, 30th March, Serrurier's division to cross the Adige at Polo, whilst with the bulk of his army he forced that river between Verona and Legnago.

This was a very dangerous operation in the face of a competent general occupying a central position at Verona. Kray showed his thorough appreciation of it by overwhelming Serrurier and forcing him to retreat with heavy loss, the prisoners alone counting fifteen hundred. Scherer renounced the further execution of the plan.

Six days later Kray took the offensive. Debouching from Verona he marched with a design to take a flanking position, and to pin the French army between the lower Adige and the sea. An intercepted despatch warned Scherer of this design, and he took measures to assail the Austrians on their march. The two armies met, the morning of the 5th April, between Butta-Preda and

Magnano. The advantage was at first all with the French, whose left under Moreau gained a position which cut off the Austrians from Verona. The victory seemed gained, when, at the supreme moment, Kray, massing his forces, fell upon the somewhat prolonged French right, drove it from the field, and took such a position with respect to the left, that Moreau was forced to retreat. That night the French army fell back on the Molinella and the next day on the Mincio. It had lost at the battle of Magnano seven thousand men, of whom four thousand had been taken prisoners. Scherer continued the retrograde movement, first to the Oglio, and thence, 12th April, to the Adda.

His army was reduced by this time to twenty-eight thousand men. But Macdonald was marching with all speed from Rome to reinforce him, and he had only to keep his force well in hand to be ready, if not to take the offensive, at all events to render the task of the enemy difficult and dangerous.

But Scherer offered another example of the danger of placing a man with no self-reliance in a difficult position. Instead of concentrating his little army he dispersed it over a distance of about seventy miles. He posted two divisions under Serrurier at Lecco at the point where the Adda issues from the lake of that name; the division Grenier at Cassano; that of Victor at Lodi, whilst he dispersed that of Montrichard in positions looking towards the Modenese and the mountains of Genoa to give a hand to Macdonald.

The French army was thus dispersed when, on the 27th April, it was assailed by an Austro-Russian army, under Field-Marshal Souvoroff and Count Mélas, but under the supreme direction of the former. The Russians composed but one-third of the army, which amounted in numbers to ninety thousand. But it was the character of their commander which inspired confidence.*

* Alexander Vassilievitch, Count Souvoroff, was in his sixty-first year. He was descended from a Swedish family named Souvor, which had emigrated to Russia in 1622. The son of an officer who had reached under Catherine I. the rank of General-in-chief and the dignity of Senator, the young Alexander had been enrolled in the army at the age of thirteen. A lieutenant at the age of twenty-five, he was

On the approach of the Russo-Austrian army Scherer had endeavoured to concentrate about Cassano; but his army was not yet in its assigned position, when, on the night of the 26th, the Russian general, Vukassovitch, crossed the river on a flying bridge and took a position near Brivio. Early the following morning, an Austrian column, under General Ott, passed the Adda near Trezzo, and rolled back Grenier from Cassano. At the same time Mélas attacked that place in front, whilst another Austrian corps assailed the division Serrurier at Lecco and forced it to surrender. Before night had set in the line of the Adda had been forced. That same night Scherer resigned his command.

That command devolved upon Moreau. Rarely has a general been called upon to assume responsibility under circumstances so discouraging. The army was reduced to twenty thousand men; it had just received a severe defeat; and it was in the presence of ninety thousand men, animated by victory and full of confidence in their general. But the occasion was one well calculated to call forth the solid qualities of the new commander. Moreau was always great in difficulty and in danger. He possessed that

a lieutenant-colonel at twenty-eight, and served in that rank two years afterwards in the hard-fought battle of Kunersdorf. A general of brigade in 1768, he defeated the Poles under Pulowski and took Cracow by assault. In 1773 he served against the Turks, helped to beat them in three successive battles when serving under Marshal Rioumiantsof, and gained, in conjunction with Kamenski, a decisive victory at Kasladji. After the peace of Kainardji he was employed in suppressing risings in the interior of Russia, and, notably, the formidable insurrection of Pugatschef. In 1783 he brought the Tartars of the Koudan and of the Boudjak under the yoke of Russia. At the battle of Kinburn against these people all his attacks on their intrenchments had been repulsed, and his men were thoroughly disheartened. Souvoroff, however, succeeded by his personal example in rallying them. At the siege of Ochakoff in 1788, serving under Prince Potemkin, he again distinguished himself by his reckless valour. In the combined war of Austria and Russia against the Turks he had greatly added to his reputation as an independent commander. His capture of the town of Ismael was one of the most brilliant feats of arms of that war. He had crowned his services to his sovereign by the conquest of Poland and the storming of Praga and Warsaw. He was then promoted to the rank of Marshal, and entrusted with important civil functions. For some reason the Emperor Paul deprived him of his grades and honours in 1798. These, however, were—on the intervention of the Emperor of Germany, who had requested as a personal favour that Souvoroff might command the army which was to co-operate with his own troops against the French—restored to him, and, with a very great reputation, he set out for Verona in the winter of the same year.

coolness and calmness under all possible circumstances which were the special attributes of Hannibal, of Cæsar, of Turenne, of Marlborough, of Clive, of Frederic, of Loudon, of Masséna, and of Wel lington. His mind never lost its balance. He possessed, moreover, one advantage which went far to counterbalance the many difficulties of which I have spoken: he possessed the love and confidence of his soldiers. If he could not perform impossibilities, he would always take the fullest advantage of the resources at his disposal.

In the trying position to which he had now succeeded Moreau was true to himself. The line of the Adda forced, he could not hope to save Milan. At least, however, he could prevent the enemy from reaping some of the military advantages which the prompt occupation of that city would secure to them. With his twenty thousand men, then, he covered Milan, whilst, under his orders, the military parks, ammunition and stores evacuated the city. Two days were spent in accomplishing this result. Moreau then fell back on the Po, designing, whilst maintaining his communications with France, to keep touch with Tuscany in order to hold a hand to Macdonald. With this end he marched in two columns: the one, escorting the parks and heavy baggage, proceeding by the regular road from Milan to Turin, the other taking the road to Alexandria.

Arrived at Turin, Moreau despatched the impedimenta with which he could dispense to France, armed the citadel, roused, or attempted to rouse, the enthusiasm of the citizens, and then rejoined the column he had despatched to Alexandria. Then he placed his army in a position splendidly adapted for defence. Behind the point where the Tanaro joins the Po, he had first-rate river defence, whilst he guarded all the routes to Genoa. Occupying Casale, Valenza, and Alessandria, he had a chain of posts on the two rivers, and he was able to concentrate his troops with great celerity at any point which might be attacked. There, then, he waited with imperturbability the movements of Souvoroff.

It must be admitted that Moreau was aided not a little by the slowness of Souvoroff. This general, in the hope of completely crushing the French, had waited the permission of the Aulic Council to utilise the corps of General Bellegarde, which we left covering Bozen. The permission having been granted, Souvoroff—who had detached corps simultaneously to besiege Peschiera, Mantua and Pizzighettone; to attack the castle of Milan; to blockade Alessandria, Tortona, Ferrara, and Bologna; to occupy the passes of Susa, Pignerol, and the Col d' Assiete—marched with between forty and fifty thousand men to crush Moreau.

At the end of the first week of May the Russo-Austrian army occupied the ground opposite the position so skilfully chosen by the French general. Souvoroff fixed his head-quarters at Tortona. The next few days were spent by him in reconnoitring. Then, having decided upon his plan, he, on the night of the 11th May, despatched between two and three thousand men to a point above the confluence of the two rivers, opposite the village of Mugarone, and transported them across the main arm of the Po to a wooded island in one of the branches of that river. From this island the men of the detachment waded to the right bank, and set to work to establish themselves there. At the same time he massed the bulk of his troops on the Tanaro, and made as though he would force the passage. But Moreau was not deceived. Informed in good time of the passage of the Po by the detachment, he left his main army to confront Souvoroff, whilst he moved with his reserves on Mugarone and fell upon and completely destroyed the enemy he found there.

Thus baffled in a front attack, the Russian general, who had just received information that an insurrectionary movement had been organised to act on the rear of the French position, resolved to march on and take Turin and assist as much as possible that movement. He had set out with this object, when Moreau, divining his projects, fearing for Macdonald, and dreading lest the ostensible movement of the Russo-Austrians should conceal a plan for taking him in the rear, moved with the bulk of his forces on Alessandria, with the view to make thence a reconnaissance. Should he find that fortress blockaded in force, it was his intention to fall back as quickly as possible on the Riviera; but should there be only a small detachment in observation he would fall upon it, and then retire leisurely by the main road of the Bocchetta to the mountains of Genoa and hold out thence his hand to Macdonald.

He carried out this plan with skill and energy. Debouching from Alexandria, he found the enemy in observation in too great force to be swept away. Feeling, then, the extremely critical nature of his own position, threatened in his rear by a country already in insurrection and soon to be occupied by an enemy largely superior to himself in numbers, and yet greatly dreading lest Macdonald should be overwhelmed, he came to a resolve which stamps him not only as a general of the first class, but as a general on whom his brethren in arms could depend for loyal support; a quality in which, during the times that were to follow, the generals who were made marshals were to show themselves singularly deficient.

About Moreau there was a simplicity of character, an absence of selfish aims, a regard for the real weal of his country, which mark him for special admiration. Envisaging the actual position, he saw that whilst by a hurried retreat he could save himself, such a retreat would sacrifice Macdonald; that he could not stay where he was, for the insurrectionists had already occupied Ceva, which commanded the only road practicable for guns; that he could not join Macdonald himself without sacrificing the communication with France. He resolved, then, to despatch Victor with twelve thousand infantry to occupy the mountains of Genoa, to hold thence the hand to Macdonald, whilst he, himself, with the cavalry, the guns, and a few infantry, should retreat by a bye-road leading to the Riviera behind Ceva.

Victor marched accordingly by Aqui, Spigno, and Dego, and occupied without opposition the crest of the Appenines. The task of Moreau was much harder. He had to spend four days in cutting across the mountain a road practicable for his guns. Fortunately he was not hindered by the enemy, and he succeeded, after incredible toil, in reaching a point on the Riviera, not too far from the post occupied by Victor to forbid a common action. Of the twenty thousand men he had led from the Adda he had left two thousand in the fortress of Alessandria; three divisions, those of Victor, Lapoype, and Montrichard, were in the mountains above the valley of the Upper Trebbia; he, himself, with the rest, was at Novi. The communication with France and with Macdonald seemed equally assured.

Souvoroff, meanwhile, had marched upon and taken Turin, and had captured there an enormous quantity of munitions of war. Kray was still besieging Mantua; Bellegarde was keeping watch in the mountains; other corps were still before other strong places. Though the total Russo-Austrian army numbered at least a hundred thousand, in no part was there a single force counting more than forty thousand; and that force was with Souvoroff at Turin.

Whilst Moreau was thus greatly acting in the north, Macdonald was advancing from the south with thirty thousand men. The junction of Macdonald would give the French general an army fifty thousand strong, superior at any single point to the Russo-Austrian army. This consideration was never absent from the mind of Moreau. The military student can judge, then, how his hopes were raised when the information reached him that Macdonald had opened communications with the division Montrichard.

Macdonald, in fact, had reached Florence the 25th May. Had he pressed on at once, before the alarm had been given to Souvoroff, the opportunity so longed for by Moreau would have offered. But Macdonald was suffering from a wound; his men had endured great privations, the organisation of his army was not as perfect as he would have wished. He committed, then, the fatal mistake of halting a fortnight at the enticing capital of Tuscany.

Fatal mistake indeed! No sooner had Souvoroff heard of the successful arrival of the French general at Florence than he broke up from Turin, and transmitting orders to his several generals—some under any circumstances, some only if attacked—to fall back upon him, he hastened towards Piacenza. Before he arrived there, Macdonald, who had called to himself the divisions Victor and

Montrichard, had marched on that place, driven back the Austrian general, General Ott, and had taken there a firm position. The division Lapoype, now placed at Bobbio, maintained his communication with Moreau. This general had determined that whilst Souvoroff attacked Macdonald in front, he should assail the flank and rear of the Russian general. But this plan was practicable solely on the condition that Macdonald should not engage the enemy until he was sure that Moreau was near enough to act. This condition was not fulfilled.

I regret that the design of this chapter will not allow me space to describe in detail the three days' battle of the Trebbia. It must suffice to state that on the 17th June, Macdonald, who had repulsed Ott and defeated an Austrian corps stationed on the lower Po, encountered, at a village six miles beyond Piacenza, Souvoroff and the Russo-Austrian army. Macdonald had about thirty-six thousand men under him, but at the moment of collision he could dispose of little more than a third of the number, whereas Souvoroff had his whole force, numbering forty thousand men, well in hand. The result of the day's encounter was, as might have been anticipated, unfortunate for the French. Macdonald, though overpowered, retreated in good order behind the Tidone, and thence to the Trebbia, resolved to cross that river on the morrow, and to await behind it the junction of his other divisions which might be expected on the 19th.

But Souvoroff was too good a general to accord his enemy the delay requisite for his purpose. Early on the morning of the 18th he crossed the Tidone with his whole force, and attacked the French drawn up along the banks of the Trebbia and with that river behind them. In the early part of the day the French left was turned, a Polish regiment was nearly destroyed, and the republican army was driven across the river. Undaunted, however, Macdonald recrossed it with ten thousand men, became in his turn the assailant, and fought so well that if he did not gain the victory he rendered it impossible for the enemy to claim it. The same night he recrossed the Trebbia prepared to defend its left bank, and full of hope that his army might be concentrated before he should be attacked. The Russians remained impassive the next morning waiting for the French attack. At 2 o'clock Macdonald, notwithstanding that his reinforcements had not arrived, attempted to cross the river. During the whole afternoon he combated against greatly superior numbers, unable, however, to gain a solid footing. When night fell he gained, unpursued, the left bank, and fell back in two columns on Piacenza, having lost twelve hundred killed and nearly five thousand wounded, and leaving seven hundred prisoners, several pieces of artillery, and three stands of colours, in the hands of the victors!

Moreau had not appeared on the field of action. Hoping that Macdonald would delay at Piacenza till he was able to co-operate with him, he had marched as soon as his preparations were completed—the 18th June—to Alessandria; had fallen upon the force blockading that place, and defeated it with great loss, 20th June, taking three thousand prisoners. This occurred, be it remembered, the day after the third day's battle on the Trebbia. It had the effect of causing Souvoroff to countermand his intended pursuit of Macdonald in order to turn upon Moreau. Moreau, however, on hearing the result of the battle on the Trebbia, fell back on Novi. He was joined there by Macdonald. The following month he received information that he had been superseded by Joubert; he was at the same time forbidden to undertake any new operation pending the arrival of that general.

Souvoroff, meanwhile, obeying the orders of the Court of Vienna, which chained him to Italy till its strong places should be reduced, had been pressing the siege of the several fortresses. The orders of the Directory, just referred to, prevented Moreau from attempting to render his task difficult. One fortress after another fell, and when on the 21st July he obtained possession of Alessandria, and a week later of Mantua, he might well hope that the time had arrived when he might devote himself to the ulterior object contemplated by his master, the expulsion of the French from Switzerland.

The one remaining obstacle was the presence of the French army at Novi. That army, consisting of forty thousand experienced troops, was now in splendid order and well provided. Joubert had joined it just before the fall of Mantua had increased Souvoroff's army by twenty-five thousand men. Had it been possible to delay the surrender of that strong fortress it had been Joubert's intention to act upon the advice of Moreau, who remained to serve under him, and to assume the offensive, by attempting the relief of Tortona. The fall of Mantua, however, rendered an offensive movement impossible, and Joubert, at a council of war summoned when he heard of that event, decided to fall back on to the crests of the Appenines.

At this moment the French army was formed in a half circle, on the slopes of Mount Rotondo, dominating the entire plain of Novi. Its left, formed of the divisions Grouchy and Lemoine, was extended in a half circle in front of the ravine of the Riasco, just beyond the village of Pasturana. Its centre, composed of the division Laboissière, occupied the heights on the right and left of the village of Novi. To the right of it the division Watrin guarded the approaches to Mont Rotondo on the side of the road leading to Tortona. The division Dembrowsky blockaded the fortress of Sera-Valle, then occupied by the Austrians. The reserve of cavalry, commanded by General Richepanse, was formed up behind the left wing. That wing was commanded by General Perignon; the centre and left were directed by Saint-Cyr.

At the head of sixty-eight thousand men Souvoroff attacked this position at 5 o'clock on the morning of the 15th August. His first assault was led by General Kray against the French left. The regiments forming that wing had not yet taken their assigned positions when the Austrians were upon them, and they began soon to give ground. At this crisis Joubert, who had been watching the Tortona road, galloped up, accompanied by Moreau, rallied his men, and was leading them forward when he was pierced in the breast by a ball, and fell dying to the ground. "Forward, my lads, forward, fight for the Republic," were the last words of the dying hero. Moreau, who was at his side, at once assumed the command, rallied his men, and drove the Austrians down the mountain-slope. Though the attack was renewed against the left on several points, it was, in the end, successfully repulsed.

Not more fortunate in its results was the attack on the French centre. As this did not take place until an hour after the assault on the left, Saint-Cyr had had time to range his men in perfect order, and the attacks against it, directed by the Russian general Bagration, were repulsed with loss.

In spite, then, of the death of their commander, Fortune, up to midday, smiled on the French. Moreau and Saint-Cyr had displayed a capacity which more than compensated for numbers. But at 12 o'clock fresh reinforcements reached Souvoroff. He ordered then a general attack along the line. Kray was again sent against the left, Bagration, reinforced by the new arrivals under Derfelden, against the centre, whilst a despatch was sent to Mélas to urge him to press on to attack the right. During four hours the French supported heroically the weight of the Russian right and centre. Vain were the efforts of Souvoroff himself. Kray was again repulsed; Bagration and Derfelden were rolled back and even partially followed; the attack had failed; when at 4 o'clock the fresh divisions of Mélas assailed the right not only in front but on their rear. This attack decided the day. The French right fell back, Souvoroff renewed the assault on the centre at the same time: the hills covering Novi were carried: the communication with Gavi was threatened. With true instinct Moreau ordered then a retreat on that place. This was accomplished with great difficulty and at an enormous sacrifice of men and officers. As the shades of evening fell, however, he succeeded in rallying rather less than one half of the beaten army in front of Gavi. Ten thousand men lay on the field killed or wounded. The French had lost their commander, four generals of division, thirty-seven guns, and many prisoners. The defeat delivered Italy to Souvoroff.

That general was now in a position to execute the designs

of his Court relative to the expulsion of the French from Switzerland. To comprehend the position of affairs in that country at this precise time I must return to the point where I had left the armies of Jourdan and Masséna.

Defeated at Stockach, Jourdan had, as already related, left his army under the command of his chief of the staff, Ernould, at the entrance to the defiles of the Black Forest, and had proceeded to Paris to obtain from the Directory the succours he deemed necessary to enable him to re-establish his position. But Jourdan had but just quitted the army, when Ernould, frightened by some demonstrations on his flanks, fell back (3rd April) in disorder on the Rhine. A few days later the Directory transferred (9th April) the command of his army to Masséna, who was to hold it conjointly with the command of the army of Switzerland.

This act of the French Government saved France. Events were about to occur which would require the firm, self-reliant and tenacious character of "the spoiled child of victory" to meet and to overcome.

For Masséna occupied a position requiring all the care and all the prudence of a general of the first rank. Though Lecourbe had, by his masterly movements, opened communications with Italy, he himself had been badly repulsed at Feldkirch. Acting in a country which projects, which thrusts itself in as a wedge, between Germany and Italy, it was indispensable that he should hold his troops well in hand. Calling to himself, then, Lecourbe on the one side, and the army of the Danube on the other, he fell back and distributed his army on the line of the Limmat, from the slopes of the Alps to the confluence of the Aar and the Rhine.

His entire strength did not at this moment exceed fifty thousand men. In accordance with the plan stated in the preceding paragraph, he posted these in the following manner: the right wing, composed of the divisions Lecourbe, Ménard, and Lorge, under the command of General Férino, extended from the summit of the St. Gothard to the lake of Zurich; the centre, consisting of the divisions Oudinot, Vaudamme, Thureau, and Soult, was on the Limmat; the

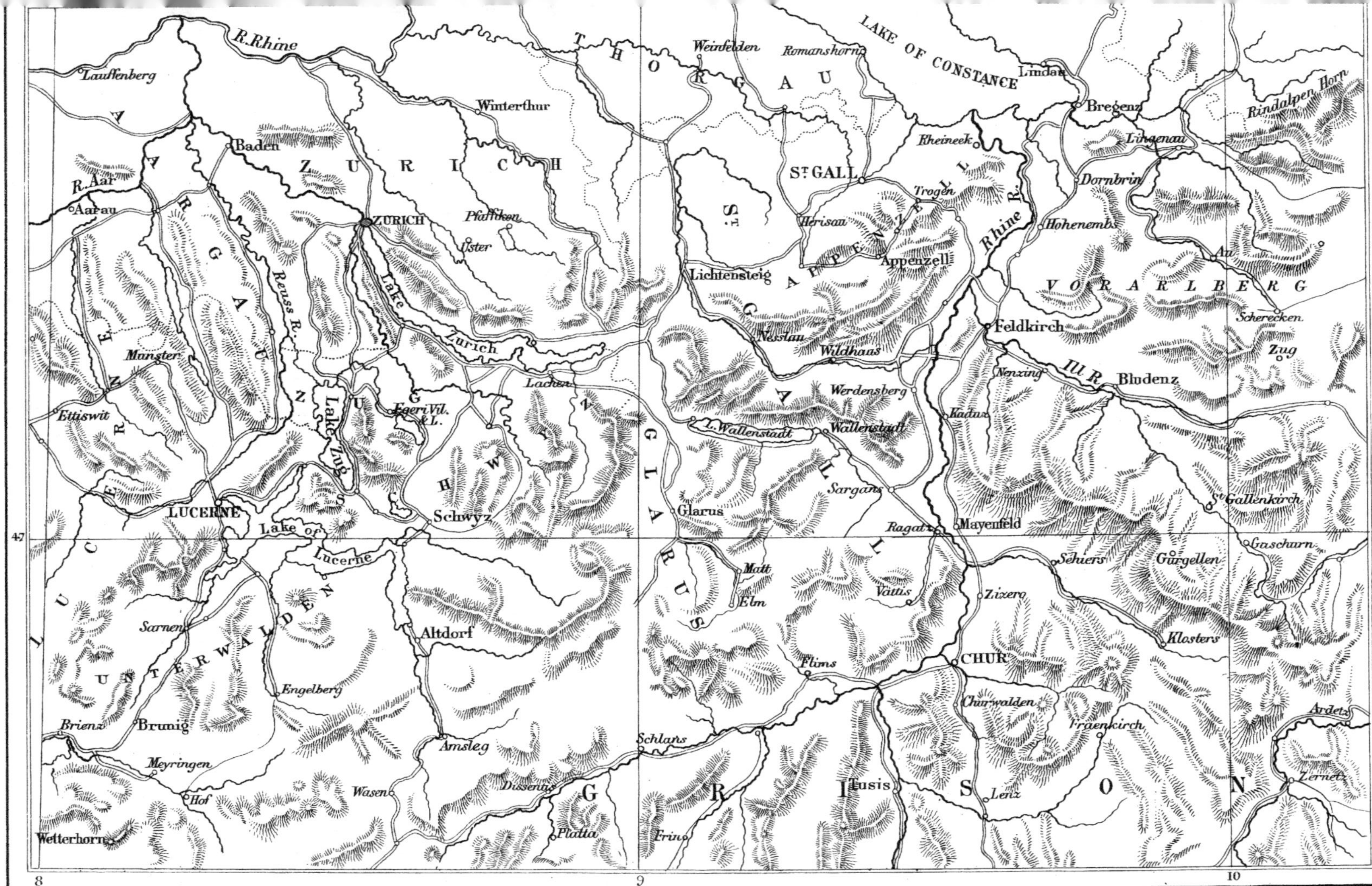

SCENE OF MASSÉNA'S OPERATIONS IN 1799.

left guarded the Rhine between Basel and Strasburg. Masséna made his head-quarters at Basel.

Opposed to him were three armies, all, fortunately for him, acting, at the moment, independently of each other. In Tirol was the force of Bellegarde, thirty thousand strong, and which the retreat of Lecourbe and Dessolles, not effected without some desperate fighting, had left there supreme; in Voralberg, twenty-eight thousand men under Hotze, the general who had so well defended Feldkirch; and between Lake Constance and the Danube was the Archduke, forty thousand strong, flushed with victory over Jourdan. Directed by a firm and skilful hand, such a force was strong enough to crush even Masséna. But, we have seen, the orders of the Aulic Council had paralysed the Archduke after Stockach, and neither Bellegarde nor Hotze was at the moment under his orders.

The concentration of the French Army on the line of the Limmat had only been completely accomplished on the 19th May. On that day Lecourbe, after having sustained some terrible combats against Bellegarde, established himself securely in the valley of the Reuss. It was just time: for the Court of Vienna, rallying from its stupor, had placed Hotze under the command of the Archduke; and that prince, encouraged by the news which reached him of the arrival in Gallicia of a strong corps of Russian troops under General Korsakoff, was preparing to act with vigour.

Naturally it was an object with Masséna to prevent, if possible, the junction of the two Austrian armies, and he made, on the 24th and 25th of May, an attack with two strong divisions on the advanced troops of the Archduke at the village of Andelfingen, whilst, with the remainder of his force, he assailed General Hotze between Frauenfeld and Winterthur. Matters promised well at the outset. Masséna drove General Nauendorf from Andelfingen and seized the bridge-head. But, under the severe fire from the opposite bank, he could not retain it;—and he was forced back. The attack on Hotze was more serious and sustained. But by 5 o'clock in the afternoon that general had repulsed all the assailants.

The following day, the 25th, the advance of the Archduke rendered it impossible to contest seriously the junction. Masséna, however, fell back fighting, and only yielding under pressure, to the position he had chosen on the Limmat.

But even this position was not now tenable. Whilst the Arch duke, united now with Hotze, occupied Winterthur, Bellegarde had assured himself of the passage over the St. Gothard, had occupied Glaritz, and threatened Lucerne. Lecourbe held, indeed, a strong position at Wasen, in the Urserenthal, but the line of the Limmat was not the less turned.

Masséna, sensible of this, fell back into a new position on the Glatt; but, finding his left flank menaced there, he retreated behind intrenchments he had caused, in prevision of such an occurrence, to be thrown up at Zurich, leaving Lecourbe on the St. Gothard at Wasen, constituting the support of the right of the new line of defence of which the camp of Zurich was, in the centre, the salient angle.

The French army was in this position when, on the 29th of May, a fresh Austrian division made a tremendous effort to cut off Lecourbe. At the instance of Souvoroff, Bellegarde, with his force of thirty thousand men, had been, as already stated, ordered to Lombardy, whilst, to replace him, the Russian general had despatched a corps, ten thousand strong, under Haddick, an officer of whom he had formed a high opinion, to overwhelm Lecourbe; or, at all events, to drive him from the slopes of the St. Gothard.

I have said that Lecourbe was at Wasen, a place well known to Alpine travellers, about seventeen miles beyond Altdorf, to protect there the passage of the guns and heavy baggage across the St. Gothard. On their arrival he had moved to Altdorf, on the lake of Lucerne. He still kept, however, a strong detachment at Airolo, beyond the summit of the pass, the chief place in the upper valley of the Ticino, about twenty-four miles beyond Wasen,* under

* The stages between the two places are—Wasen to Andermatt, 6¾ miles; Andermatt to Hospenthal, 2¼; Hospenthal to the Hospice, 7½; the Hospice to Airolo, 7½. The Devil's Bridge is midway between Wasen and Andermatt.

the command of General Loison. Loison had, in fact, covered Lecourbe's retrograde movement when that General, recalled by his chief, was attacked and harassed by Bellegarde, and he had been stationed at Airolo to assure the retreat of the guns and heavy baggage to Altdorf. He remained there, an outlying division, to guard the right flank of the French army. He had with him about three thousand men. Here, at Airolo, he was suddenly attacked, the 29th May, by Haddick, with the ten thousand men detached by Souvoroff.

Loison had placed his troops to the best advantage. About one mile from Airolo, the path—as it was in those days—leading under the village of Madrano on its left, enters the defile of Stalvedro, crosses then to the right bank of the Ticino, and traverses the hamlets of Piotta and Ambir. Here the valley widens, and corn-fields give evidence of a milder climate. The reader will infer from this description that the key to Airolo was the pass of Stalvedro. No one knew this better than Loison, and he had done his utmost, with the means at his disposal, to render it impregnable.

He did not, indeed, succeed. But the fight for the pass of Stalvedro was one of the most hotly contested of the whole war. Loison animated his troops by his presence and his example; but he had, at last, to give way to numbers. There was no halting place either at Airolo, at the Hospice, or at Hospenthal. Haddick pursued with unrelenting vigour. At the Devil's Bridge Loison's retreat was seriously compromised by the narrowness of the way; and he could only emerge by sacrificing six hundred prisoners. He was still followed to Andermatt, to Wasen, and even beyond Amsteg, ten miles and a half further down, and but nine from Altdorf.

There are two kinds of pursuit: the one, careful and well-considered, which never loses sight of or imperils the plan of the campaign; the other, a pursuit reckless and neglectful of all precautions. Had Haddick stopped his pursuit of the French at Wasen, he would have achieved a marked success and have seriously alarmed the French general for his right. For the fleeing troops of Loison would have certainly exaggerated the number of their enemies, and Masséna might well have imagined that Lecourbe was left to struggle against thirty instead of only ten thousand. The reckless continuance of the pursuit beyond Amsteg greatly neutralised the effect of the victory at Airolo. Lecourbe, always on the watch, had, on the first intimation of the defeat of Loison, marched with his whole force towards Amsteg. Three miles from that place he came upon the pursuers, tired and in disorder. His own men were fresh and eager for revenge. He attacked the Austrians with great fury, drove them in disorder through the gorge of the Schöllinen, and across the Devil's Bridge. It was only by cutting an arch of that bridge that Haddick saved his army from utter defeat.

The termination of the attack of the 29th of May left Masséna secure, for the moment, respecting his right. Haddick had been taught caution. But in front of Masséna there was still an army double in numbers to his own, and the general of that army meant mischief. On the 4th of June desultory attacks and skirmishes presaged the coming storm. On the following day it burst.

At daybreak of the morning of the 5th June, in fact, the Archduke attacked the French intrenchment at all points. The defence was as vigorous as the attack. "Few actions," wrote the impartial chronicler of the wars of the Revolution and the Empire, Count Mathieu Dumas, "have cost so much blood." Four Austrian and two French generals were wounded. When night fell Masséna still remained master of the intrenchment. Never had he more distinguished himself. Where danger was, there, with the calm, cool, self-reliant mien which never deserted him on the battle-field,* was the spoiled child of victory. He repulsed the enemy, but his losses had been enormous. A fresh attack would ruin him; and that the Archduke contemplated such an attack was certain. Masséna felt strongly that the safety of the country required him to preserve to France the one army available to cover her eastern frontier. That night, then, be defiled over the bridges of Zurich

* "Dull in conversation, but in danger acquiring clearness and force of thought." —Napoleon.

and Wettingen, and took post on the Albis range, between Zurich and Zug, a position so strong as to be impregnable. His retreat was conducted without opposition. He had to abandon, however, the great arsenal of Zurich.

In that secure position Masséna resolved to await the future movements of the Archduke. Vainly did the Directory, terrified by the news that Korsakoff was advancing with thirty thousand men to co-operate with the Austrians, urge Masséna to assume the offensive, promising even to strengthen him with twenty thousand men, so that he might overwhelm the Archduke before Korsakoff should arrive! Masséna felt that a defeat where he was would be fatal; that the safety of France depended upon his guarding her eastern frontier. One movement, however, and that a very important movement, he did order. He directed Lecourbe to regain the St. Gothard, by recovering Airolo and the Stalvedro defile, and to re-occupy the Grisons. Lecourbe executed these orders with skill and success. Early on the 14th August he embarked with his own corps on board a flotilla he had collected to seize Brunnen and Schwytz, on the eastern shore of the lake, thence to sustain three other columns which he had despatched, at the same time, in well-chosen directions. Acting upon his orders, Gudin, with five battalions, forced the ridge of the Grimsel, and, joining Thureau in the Valais, drove the Austrians from the source of the Rhone and the Furka; similarly, a second column, led by the intrepid Loison, traversed the Sisten passes and descended on the enemy at Wasen; a third marched from Engelberg upon Erstfelden; a fourth upon Altdorf. All these attacks proved successful. After a fierce contest Loison gained Wasen. Altdorf having been also successfully occupied, the Austrians, who had been expelled thence by Lecourbe, were assailed in front and rear, and were compelled to seek a hasty flight by the Maderaner Thal, in the direction of Tavetsch. Meanwhile Thureau had attacked the Austrians near Brieg, and had forced them to retire by the gorges of the Simplon to Duomo d'Ossola; this defeat had enticed from his position on the Grimsel and the Furka,

Colonel Strauch, who, leaving but fifteen hundred men to guard those passes, had rushed with the remainder of his corps to the aid of his countrymen in the Valais. This action, though it checked the French advance in the upper Valais, lost for Strauch the important passes which it had been his special duty to protect; for Gudin, debouching from the Aar valley, forced the higher ranges of the Grimsel and drove the slender Austrian defenders down the other side. Strauch himself, placed between two fires, was glad to escape by the Nufenen pass, eight thousand nine feet above the sea-level—a pass which has suffered more, almost, than any other pass in the Alps from avalanches—to Faido, on the Ticino, whence he rejoined the scattered remnants of his force who had made their way to the Italian side by paths only known in those days to chamois hunters.

Such was the result of Lecourbe's splendid action on the first day. Disposing of nearly thirty thousand men on five different points, he had been victorious, though the distances were too great for him to be certain of the full nature of his success the same day, on all. But Lecourbe was a master of that self-reliance which is the greatest strength of a general. He was prudent because he knew when to be bold; because he knew that a blow not followed up is only a blow with a padded glove. Though still uncertain of the amount of success achieved by Gudin and Thureau, he determined to push that gained by Loison and the two other corps he now had united under himself to the utmost. At day-break on the 15th, then, he pushed forward from Wasen, and drove the enemy before him through the rugged walls of the Schöllinen to the Devil's Bridge. Here the Austrians made a desperate stand. In the midst of the conflict some powder-bags which had been attached to the arch, exploded. The effect can scarcely be described. The combatants on the bridge suddenly disappeared in the torrent below, whilst a chasm, thirty feet wide, separated their surviving comrades from one another! There was a pause—horrible for the moment—but it did not last long. The determination to win drowned every other feeling.

The blowing up of the bridge checked Lecourbe for the remainder of the day. Meanwhile, however, information reached him of the complete success of Gudin in the Valais, and that he was marching from the Furka by the hamlet of Realp to take the enemy in rear. That night, then, Lecourbe planked the chasm, and sent his men across it with the early dawn. The Austrians, meanwhile, had become aware of Gudin's movements, and had fallen back to the Kreuzli pass, an almost impregnable position, seven thousand seven hundred and ten feet above the sea-level, flanked by the Crispalt, over ten thousand feet high, to the west, and by the Oberalpstock, close upon eleven thousand feet high, to the east. Assailed there by Lecourbe, they maintained themselves till the evening; but the next day they gave way before that general aided by Gudin, and, having lost many killed and wounded, a thousand prisoners, and three guns, they retired, broken, to Ilanz on the Vorder Rhein.

Meanwhile, the fifth column, with which Lecourbe had embarked on board the flotilla, but the command of which he had made over to Chabran, to enable him to direct the movements on the St. Gothard, had been not less successful. Chabran had driven the Austrians from Schwytz into the Muttenthal, had defeated them again at Einsiedlen; and had forced them, the day following, to take refuge in the canton of Glarus. This operation had completely cleared Lecourbe's left.

As the result of the three day's fighting, Lecourbe re-occupied the Hospice, and pushed forward a strong detachment to Airolo, and another to secure the defile of Stalvedro. Not only had he killed and wounded two thousand of the enemy, and taken four thousand prisoners and ten guns, but he had wrested from them the most important post on the St. Gothard, with all its approaches and lateral valleys.

On the third day of Lecourbe's splendid operation, Masséna, weakened by the absence of the divisions he had lent to that general, was attacked by the Archduke, strengthened by the arrival at Schaffhausen of twenty-three thousand Russians under Korsakoff. Able thus to dispose of sixty-one thousand, the Archduke hoped to be able to crush Masséna, who, at the moment, had but thirty-seven thousand.* Leaving Hotze to occupy Zurich with eight thousand, the Archduke, with fifty-three thousand, reached Gross Dettingen, just below the junction of the Reuss and the Aar, before his movements had been discovered by the French. The river here was swift, and its bed was rocky. To cross it the pontoons were ready. When, however, the Austrian engineers began the operation of fastening them, they found they had left behind the grappling irons. Their inventive genius was not sufficiently alive to meet the difficulty, and the Archduke, for want of those irons, was forced to fall back, *re infectâ*!

It was not the desire of the statesmen of Vienna that the Archduke should commit himself to a decisive action against Masséna until Souvoroff should be ready to crush Lecourbe. Had Masséna been a lay-figure, their plans would have been faultless. But the most brilliant soldier of Republican France was not the man to allow himself to be played with. He had long since penetrated the Austrian designs, and he had resolved to foil them. When, then, the Archduke, baffled in the manner stated at Gross-Dettingen, marched to the Upper Rhine, leaving Hotze to support Korsakoff in the canton of Glarus, Masséna fell upon the former at Näfels, 19th of August, defeated him, and forced his allied enemies to occupy a defensive line forty miles in length, extending from Zurich to Chur.

To those who have had no experience of the incredible folly of men who call themselves statesmen, it will seem strange that at such a time the Archduke, with the great bulk of the Austrian forces, should have been sent by a positive order from Vienna to the Upper Rhine, there to conquer places which would be wrested

* Soult, with ten thousand men, occupied the Linth to its junction with the Lake of Zurich; Masséna himself, with the divisions of Mortier, Klein, Lorge, and Mesnard, thirty-seven thousand strong, occupied the ridge between Zurich and Zug. Lecourbe, with the divisions Molitor and Gudin, twelve thousand strong, guarded the St. Gothard and the valleys of the Reuss and the Upper Linth: Thureau, with eight thousand men, guarded the Valais, and Chabran, with eight thousand, the environs of Basel.

from him as soon as Masséna should achieve success at the decisive point of the scene of action. For, undoubtedly, at this crisis, the decisive point was at Zurich; and at that point, thanks to the enforced absence of the Archduke, Masséna could dispose of a greater number of men than his enemies.

The month of September had come; Souvoroff had begun his march up the valley of the Ticino; and the Archduke was still on the Upper Rhine. The opportunity offered which a commander possessing genius could use to advantage. Masséna seized it, and employed it in a manner which gained for him the admiration of his contemporaries—an everlasting niche in the temple of Fame. The allied commanders had agreed that whilst Hotze and Korsakoff should hold Masséna at Zurich, Souvoroff should overwhelm Lecourbe, form a junction with Hotze and Korsakoff, and assail Masséna on his front and flank. Again did they make their plans as though the French general were a lay-figure. Masséna surprised them by proving he possessed vigour and energy beyond the vigour and energy of other men. Finding that the allies were concentrating the greater part of their forces between the ramparts of Zurich and the banks of the Sihl—a little river which rises in the Alps of Schwytz, and, flowing parallel with the Lake of Zurich for eighteen or twenty miles, after approaching within a mile of its shores, joins the Limmat at the north-east end of the town—he massed thirty-nine thousand men on the Limmat, and, making a feigned attack on the town of Zurich, crossed the river at Kloster-Fahr, with but slight opposition, early on the morning of the 25th September, seized the road to Winterthur—the only road communicating with Germany—and, before nightfall, summoned Korsakoff, now enclosed in Zurich, to surrender!

The Russians were, in fact, hemmed in as in a cage. They were shut up in a space so confined that their very numbers hampered them. The arrival during the night of the Austrian right wing, detained on the Limmat by the false attacks which Masséna had caused to be directed against them, did not improve matters. The position of the allies was well-nigh desperate. They had the choice of surrender or of forcing a passage. They chose the second alternative.

At daybreak on the morning of the 26th, Korsakoff made a most desperate effort to force the Winterthur road. It was an almost impossible feat, for whilst he was dashing his whole army, or as much of it as he could display, on that road, strongly defended by two French divisions, other French divisions were forcing their way into the town, prepared, then, to harass his flanks and his rear. The struggle which followed was one of the most determined of the whole war. Its results were decisive. Korsakoff, indeed, and with him rather less than one half his force, succeeded in forcing the road. But he lost all his guns, his baggage, his ammunition, and his stores; eight thousand of his men were killed and wounded, and five thousand were taken prisoners. Nor were the Austrians less unfortunate. They were driven, with the loss of many men, including their general, who was killed, of three thousand prisoners, twenty guns, of all their baggage, and of the flotilla they had constructed on the lake of Wallenstadt, across the Rhine!

Before this event, so disastrous to the allies, had happened, Souvoroff had begun the march which was to bring him upon Lecourbe. That capable commander, whilst occupying the Hospice, had, I have stated, posted an advanced corps at Airolo and the defile of Stalvedro. There stood Gudin, with a brigade numbering nearly four thousand men, covering alike the direct road by the Hospice and the path leading to the Furka. Gudin, the representative of the ancient family, de la Sablonnière, the nephew of a distinguished officer who had served since 1752, was himself one of the ablest officers in an army which was to make "the tour of Europe." His influence over his men was enormous, first, because they believed in him, and secondly, because his firm discipline, based upon a character as strong as it was irreproachable, made them a machine in his hands. On this occasion Gudin was not wanting either to himself or to the reputation he had already acquired.

Suddenly attacked at Airolo and Stalvedro on the morning of the 23rd September, he disposed his soldiers so well that Souvoroff, who led his troops in person to a direct attack, whilst he despatched six thousand men, under Rosemberg, by the Val Blegno, to turn the St. Gothard by Dissentis and the Crispalt, was again and again repulsed. The Russian soldiers, exposed in the ascent, fell in hundreds under the fire of the French marksmen, securely resting in positions, carefully selected, where they were invisible to their enemy.

Vainly did Souvoroff urge forward his men, and, when they fell back, declare his resolution to stay and die on the spot where, for the first time in his career, he had seen them retreat! Gudin's brigade stood firm, uninjured, barring to him the passage. Convinced at last that it would be impossible to force his way by a front attack, Souvoroff had recourse to the manœuvre he should have tried in the beginning, and sent strong detachments to the right and left to turn his enemy. Recognising that his position at Airolo was no longer tenable, Gudin fell back slowly to the Hospice. But Lecourbe had likewise, for the reason to be described, fallen back; the Hospice was not defensible; and Gudin, according to a pre-arranged programme, marched rapidly that night by the Furka, and took post on the summit of the Grimsel. His splendid defence had, meanwhile, given Lecourbe time to draw to himself the six thousand men of whom he could dispose. His own position, he knew, was full of danger. He could not hope to stay the burst of Souvoroff at Hospenthal; and, whilst he expected every moment to see Rosemberg debouch upon his rear, he was aware that his retreat might be cut off by a detachment which, he had learned, had been despatched through the Maderaner Thal to Amsteg.

That which Lecourbe had foreseen was at the moment actually occurring. Rosemberg was driving the French detachment on the Crispalt before him down the valley of the Urseren, whilst Auffenberg was marching with all speed to cut off Lecourbe's retreat by the gorge of Schöllinen. For Lecourbe to stay where he was would be madness. Recognising this, the French commander, with heroic wisdom, massed his six thousand men, threw his guns and heavy material into the Reuss, crossed the Devil's Bridge and broke it down, and, sending a strong detachment to secure the Schöllinen gorge, prepared in this strong defensive position to keep the enemy at bay as long as it might be possible.

The following morning, the 24th September, Souvoroff, who had been joined in the night by Rosemberg, pressed forward from the Hospice by way of Hospenthal and Andermatt to the Devil's Bridge. On the opposite side the soldiers of Lecourbe, strongly posted, met his advancing men with a continuous musketry fire. The slaughter of the Russians was fearful. The men of their advanced guard were literally annihilated. The columns behind, hearing the firing in front, pressed on at the double quick, and by their weight alone forced the leading companies headlong into the chasm below. The butchery lasted long. At length Souvoroff had recourse once more to a turning movement. Then Lecourbe, having held his enemy as long as possible, and knowing that he was still some few hours in advance of Auffenberg, led his men to Altdorf, and thence across paths, which he had previously surveyed, to take up a position whence he could watch and harass. His coolness never deserted him. Before he left the shores of the lake of Lucerne he took the precaution to destroy everything in the shape of a boat which could float upon its surface.

This is the operation which is known in history as the surprise of the St. Gothard. In the true sense of the term, it was not a surprise; for Lecourbe had been posted on the St. Gothard to ward off an attack which was expected. The date of the attack was, naturally, uncertain. If, in that sense, the sudden attack of the Russian commander-in-chief may be regarded as a surprise, it must be admitted that it found Lecourbe on the alert. Being on the alert, not only did he delay his enemy as long as possible, but with very slight loss to himself in men, with the loss of his guns alone, he inflicted on him enormous damage, and all but ensured his absolute destruction.

For, when on the morning of the 26th, Souvoroff reached Altdorf, he found himself in a valley, shut in by a lake on the one side and the mountains on the other, a valley exhausted of food, and whence there was but one outlet. That very day Masséna was crushing Korsakoff at Zurich. The absence of any news from his colleague caused Souvoroff the greatest embarrassment. He could not stay at Altdorf, and he could only gain a safe position by thrusting his army through the Schächenthal, the valley in which stands the village said to be the birthplace of William Tell. He had, indeed, no other resource. He threw himself then with all his force into that pass, and thence across the Kinzig Kulm. His experience was a very rough one. He had to abandon all his artillery and baggage; his men had in many places to move in single file, whilst Lecourbe, hanging upon his rear, destroyed them by hundreds. From the Kinzig Kulm he descended upon Muotta, whence he hoped to force his way to Schwytz, and join Korsakoff on the lake of Zurich. From Muotta Souvoroff reached, still harassed and distressed, the Müttenthal. Thence, under ordinary circumstances, he could debouch by Schwytz, or ascend the valley and throw himself on the south of the Bragel. But it seemed too late even for that last resource. On the side of Schwytz was the division Mortier, on that of the Bragel the division Molitor, occupying the defile of Klonthal. After resting his troops for two days, Souvoroff had no choice but to attempt the Bragel. On the 30th December he set out with this purpose. He marched but a short distance, however, before Masséna attacked his rear, whilst Molitor held him in the defile of Klonthal.

After some very fierce fighting, the Russian general forced his way to Glarus. There no other resource was available to him but to ascend the valley of Engi, and to cross thence into that of the Rhine. This route was infinitely more beset with hardships than that which he had already traversed. He attempted it, however, and after four days of terrible suffering, and the loss of nearly half of the force till then remaining to him, he reached Chur, and, from Chur, Ilantz on the Vorder Rhein. He had then but ten thousand men under his orders. In fifteen days twenty thousand Russians and from five to six thousand Austrians had succumbed. What was of not less importance, the defeat of the Russians dissolved the alliance. At Zurich and on the St. Gothard Masséna and his heroic comrades saved France!

One word as to the fate of Lecourbe. This cool and capable general commanded the following year the right detached corps of the army, which, under the chief command of Moreau, won the battle of Hohenlinden. When, four years later, Moreau was placed on his trial for conspiracy against the First Consul, Lecourbe made great exertions in the defence of his old general. He used to accompany Madame Moreau to the trial, was present there all day and every day, and repeatedly indicated his sentiments by violent gestures. The First Consul never forgave him. He struck his name from the list of officers, and ordered him to reside, first at Lons-le-Saulnier, afterwards at Bourges. There Lecourbe lived during the continuance of the Empire. In 1814 Louis XVIII. restored him to his rank in the army and conferred upon him the title of Count. During the Hundred Days, though pressed by Ney and Bourmont, he refused at first to recognise Napoleon. But the feeling that the country was in danger triumphed in the end over his personal sentiments; and towards the end of May Lecourbe accepted the command of the corps of observation of the Jura. Opposed here to the Archduke Ferdinand, he held him at bay till the catastrophe at Waterloo completely changed the situation. He died the following October. As an officer he was clear-headed, calm, cool, and always collected on the field of battle; very self-reliant, daring up to the extreme limit of prudence, and particularly versed in mountain-warfare. Until Souvoroff met Lecourbe, Souvoroff had been always victorious. Lecourbe gave the Russian general a lesson which hurried him, disgraced by his master, to his grave; for Souvoroff survived the disaster I have recounted little more than seven months.

CHAPTER IX.

INKERMAN.*

When the Czar Peter, commonly known as Peter the Great, ascended, in the year 1689, the throne of Russia, the only commercial seaport that country possessed was Archangel. His father, Alexis Mikhaïlovitch, had imbibed the vague idea of entering into commercial relations with Persia by means of the Caspian, and had caused a ship to be built for that purpose by the Dutch. This ship had descended the Oka and the Volga as far as Astrakhan, but it was burnt there by the Cossacks of the Don. Of the crew but two returned to Moscow, and one of these, Karsten Brand by name, became eventually the first constructor of the Russian navy. Following in the footsteps of his father, Peter, four years after his accession, made the sea voyage to Archangel, and pushed even as far as Ponoi, on the coast of Lapland. The year following he renewed the voyage with several vessels, and, full of hope for the future, nominated Prince Feodor Jourievitch Romonadofski admiral of the fleet the building of which he only contemplated. Turning round to look for a sea upon which he could develop his maritime aspirations; barred as yet from the Baltic; rejecting the White Sea because it lay too far north, and the Caspian because it was

* My principal authorities for this article are (1) Rüstow's *Der Krieg gegen Russland*, 1855–6; (2) Bogdanowitsch *Der Orient Krieg*, 1853–6; (3) Hamley's *Story of the Campaign of Sebastopol*; (4) Kinglake's *Invasion of the Crimea*, vol. v.; (5) W. H. Russell's *British Expedition to the Crimea*; (6) *The Progress and Present Position of Russia in the East*, 4th edition: John Murray.

not sufficiently important, Peter decided on the acquisition of the Black Sea. To carry out this object he laid siege to the fort of Azof, twenty-five miles to the east of Taganrog. This siege involved him in his first war with Turkey. So stern was the defence, and so poor were the means of attack, that Peter turned the siege into a blockade. The blockade lasted more than a year; but, during that period, the Czar had, with marvellous energy, completed the construction of a fleet consisting of twenty-three galleys, two galleasses, and four fire-ships. This fleet entered the Sea of Azof in May 1696, and defeated that of the Turks within sight of the fortress. On the 29th July following Azof surrendered to the Czar.

Having thus obtained possession of a place which he regarded as the key of the Black Sea, Peter set to work to increase and to improve his fleet. For that purpose, and for the purpose of witnessing, in order to introduce into his own country, the civilisation of which he had heard so much, he traversed Revel, Livonia, Brandenburg, Hanover, and Westphalia to Amsterdam. There he adopted the trade of a carpenter. At Zaandam he remained seven weeks, working as a shipwright. On his return to Amsterdam he superintended the construction of a ship-of-war carrying sixty guns, destined for Archangel. Three months were then spent by him in England, in acquiring knowledge of the sailor's craft. He returned to Russia in time to punish, with terrible severity, the Strelitz, who had revolted, and to replace them by twenty-seven regiments of infantry and two of dragoons, trained on the system he had learned during his travels.

The year following there broke out war with the most renowned warrior of the age, Charles XII. of Sweden. A series of repeated defeats was necessary to teach the Russians how to conquer. At the battle of Narva Charles, with eight thousand men, defeated an army of thirty-eight thousand Russians. This blow, severe as it was, did not discourage the unconquerable soul of Peter. "I know," he said, "that we shall have to suffer many defeats at the outset; they are necessary to force Russia from her apathy." He raised new troops, founded new cannon, infused his own spirit into his nobles, and, whilst Charles, in his self-confidence, disdained to interrupt him, he trained them to a high state of discipline. In the years immediately following Charles played into his hands. His long stay in Poland and in Saxony, and then the ill-considered march to the Ukraine, gave Peter an opportunity of which he availed himself at Pultawa (8th July 1709) to dispose for ever of Swedish rivalry. After the triumph in Moscow which celebrated this great victory, Peter began the campaign which was to unite to Russia the valuable provinces of Livonia and Carelia—the north coast of the Gulf of Finland. Before, however, he had secured as fully as he would have liked his territories in the north, circumstances compelled Peter to return to his first idea—the expansion of his empire in the direction of the Black Sea. At the instigation of Charles XII., whose provinces he was threatening, Turkey had declared war against him (1711). On the whole, Peter accepted the challenge without reluctance. He believed that Turkey was tottering, that her Christian subjects were ripe for revolt, and that it would be easy for him to found a solid rule in the provinces from which he would expel the Osmanli.

But here, again, Russia had to learn by dire experience. Her long career shows that she is a very apt learner. Having concluded an alliance with the Hospodar of Moldavia, Peter traversed that province with his army, and crossed the Pruth. Here, however, he found himself, in July 1711, surrounded by a vastly superior force led by the Grand Vizier of the Ottoman Empire, Muhammad Ali. This force, whilst abstaining from a general action, cut off his supplies and effectually prevented his retreat. Peter and his army were gradually being starved into surrender. It had been, then, in the power of the Grand Vizier to nip for a long time the newly-awakened aspirations of Russia; for those aspirations centred in the Czar, and with the premature death of the Czar they might have expired. In this crisis, and when Peter himself had lost all hope, the Russian army was saved by a woman, Catherine Alexeievna, for long mistress of the Czar, but who had just been made his lawful though unavowed wife.* Peter, writes the Sieur de Villebois, in his secret memoirs of the Court of Russia during the reigns of Peter the Great and Catherine I., believing himself utterly lost, retired into his tent, where, confused, cast down, and utterly without hope, he gave way to complete despair, refusing to be seen by, or to speak to, anyone. He was in this condition when Catherine forced the sentry, entered the tent, and impressing upon him the necessity of displaying some firmness, declared to him that she had an expedient which yet would deliver them. She knew well, she said, from the picture which Count Tolstoy had drawn of them in his despatches, the characters of the Kaimakan and the Grand Vizier; she was confident that they were both open to bribery, and that by loading them with presents it would be possible to obtain a peace far less onerous than that which, in their circumstances, they had a right to expect; she was, moreover, she added, acquainted with a man who was thoroughly well fitted to carry out such a negotiation, adding that not a moment was to be lost in sending him to the Kaimakan. Without waiting even for Peter's answer she quitted the tent; and, returning in a few moments with the soldier to whom she had alluded, gave him his orders in the presence of Peter, and despatched him on his errand. During this time Peter had not uttered a word, but the soldier had scarcely left the tent before he exclaimed, regarding his wife with admiration: "Catherine, the expedient is marvellous; but where shall we get the money to pay these two rascals, for they

* Catherine was born at Germunared, in Sweden, in 1682. Her father, John Rabe, was a quartermaster in a Swedish regiment. Her real name was Martha. She had married, in 1701, a dragoon of the garrison of Marienburg, in Livonia, but the capture of that fortress by Russia, in 1702, caused her to be transferred to Moscow, where she entered the household of Prince Menschikoff. There she was seen, in 1704, by Peter, who, captivated by her beauty, took her to live with him as his mistress. Then it was that she assumed the name of Catherine Alexeievna. She bore Peter, in 1706, a daughter, named Catherine, who lived only two years; another, in 1708, named Anne, afterwards Duchess of Holstein Gottorp; a third, in 1709, Elisabeth, subsequently Czarina. Peter married her in a church in the environs of Warsaw, the 29th May 1711, just before entering upon the campaign on the Pruth described in the text. He made the marriage public the 19th February of the year following.

will not, you may be sure, be satisfied with promises?" "The necessary means are here," replied Catherine; "I have my own jewels, and in a few moments I shall have all that are in the camp. All I ask of you is that you will not be cast down; that you will re-animate the poor soldiers by your presence. Go, show yourself to your troops. Leave the rest to me: I will answer that I shall be in a position to satisfy the greed of the Turkish Ministers, were it even greater than I imagine." Whilst Peter followed her counsel Catherine went out amongst the officers and men, and induced them all by appeals to their patriotism to deposit with her whatever treasures they had. When the messenger returned she was able to satisfy the demands of the two Turkish chiefs.

The peace which Catherine thus purchased (23rd July 1711), to save the Russian army, cost Peter the fortresses of Azof and Taganrog, and annihilated his projects on the sea the possession of which was necessary to the carrying out of his projects against the Black Sea. But the check was not less necessary for him than the defeat of Narva. Whilst it hardened his projects, it taught him the diplomatic means by which, in conjunction with increased military sources, they could be carried into effect. Not for an instant did he lose sight of his original aim—the expansion of Russia to the south, the south-west, and the south-east. To the story of the carrying out of that aim by Peter and his successors this chapter will be devoted, leaving with but little notice the conquests in Northern and Central Europe.

Foiled in his attempt on the Black Sea, Peter turned his attention to the Caspian and the countries bordering that sea. In 1717 he sent an embassy to the Khan of Khiva. The misconduct of the members of that embassy brought about their destruction. In 1718 he opened negotiations with Persia. He discovered that the country was in danger of being attacked by the Afghans of Kandahar, and was too weak to resist that attack. The slaughter of some of his subjects by the mountaineers of the Lower Caucasus eight years previously gave him, then, a pretext for sharing in the spoil. Preluding his invasion by a manifesto full of expressions of personal regard for the Sháh, in whose interest he was acting, and concluding with an appeal to Heaven, he invaded Persia at the head of an army fifty thousand strong (July 1722), and forced her to yield the towns of Derbend and Baku, the provinces of Ghilan, Mazanderan, and Astrabad. This was his last eastern expedition. He died two years and a half later, the 8th February 1725. He had lived less than fifty-three years: but he had founded the Russian Empire such as we know it; he had formulated plans for its indefinite extension of which those of his successors who have not been dethroned have not dared to lose sight.

I pass over the uneventful reign, of two years' duration, of Catherine, and of that which followed, lasting but three years, of Peter II. The reign of Anne—1730–40—was not altogether glorious. Nadir Shah recovered for Persia the towns and the provinces which Peter the Great had conquered. Under her reign, on the other hand, there began that interference of Russia in the internal affairs of Poland which was destined to be so unfortunate for that country. Influenced by something akin to dread at the increasing power of a country she already recognised as a rival, the Porte stirred up the Khan of the Crimea to oppose forcibly the stealthy efforts which Russia made just about this time to possess herself of the Sea of Azof. The war which followed was not altogether favourable to the northern Power. Russia captured, indeed, Fort Azof, and Fort Otshakof. Marshal Münnich defeated the Turks at Choczim, and overran Moldavia; but the Russian troops, badly fed, were decimated by famine and disease. So great were these losses that Anne was glad to sanction the signature, at Belgrade, 13th September 1739, of a treaty by which she restored to the Porte all her conquests, the fort of Azof, which was to remain dismantled, excepted.

On the death of Anne, her grand-nephew, Ivan, then two years old, held the imperial dignity for a year. But Elisabeth, a daughter of Peter the Great, possessing the resolute will of Peter, soon disturbed this arrangement. On the 6th December 1741 she sent the youthful Czar a prisoner to the fortress of Schüsselberg, and packed off his parents and supporters, including amongst them Marshal

Münnich and Count Ostermann, to Siberia. A war with Sweden gained for Russia (17th August 1743) a boundary which assured the safety of St. Petersburg, till then dangerously exposed, whilst the participation of Elisabeth at a later period in the Seven Years' War increased the European influence of the country. In her reign, too, first began that trafficking with the Valis of Georgia which led eventually to the absorption of that important province into the Russian Empire.

Peter III. and Catherine II. succeeded Elisabeth in January 1762; but the murder of the former in the following July left Catherine sole ruler. Under the rule of this princess Russia made enormous strides. Having first undermined Poland, the Czarina employed all her efforts, when the throne of that country became vacant in 1763, to secure it for her early lover, Stanilaus Poniatowski. Soon after the election of that weak nobleman, Catherine began those intrigues with Austria and Prussia which ended, in 1772, in the first partition of unhappy Poland. Alarmed at the threatened absorption of a kingdom which, though her enemy, had more than once hindered the prosecution of Russian designs against herself, the Porte declared war in 1769.

For Turkey that war was a mistake. Russia made efforts commensurate with her long-cherished views. Her navy, officered to a very great extent by English officers, destroyed the Turkish fleet, kindled insurrection in Greece, and spread disaffection in Syria and Egypt. Turkey, exhausted and bleeding at every pore, was glad to secure peace, in 1774, by making many sacrifices. By the Treaty of Koutschouc-Kainardji, signed the 21st July of that year, Russia acquired the free navigation of the Black Sea and of all the Ottoman seas—her right, however, being limited to the maintenance of but one ship of war thereupon—the possession of the Sea of Azof, with Taganrog and Kertch, suzerainty over the two Kabardas,* the independence of the Tartars of the Crimea. To maintain her new frontier, she erected, during the two years

* The districts commanding the northern slopes of the Caucasus.

immediately following, thirty fortresses between the Black Sea and the Caspian.

To gain absolute possession of the last-named sea, Russia directed her next efforts. To this end a Russian fleet was constantly maintained on that sea, with orders to sow discord between the Khans on the coast; to support, with regard to justice, the weak against the strong; to burn every Persian vessel it might happen to encounter.

Nor was her policy in the Crimea less directed to the long-cherished end. Having made the Khans of that peninsula independent of the Porte, she proceeded to make them dependent upon herself. Profiting by her experience in Poland, she revived an ancient law, in virtue of which the sovereignty was declared elective. Thus exciting the ambition of the leading families, she proceeded to formulate a plan for their submission to herself. First, she formed an alliance against Turkey with the Emperor Joseph II. When the allied armies were ranged in a position for effective action, she kindled in the Crimea the revolution which she had been carefully preparing. The alarmed Khan of that peninsula fled to Russia for protection: but as his late subjects proceeded, in virtue of the law sanctioned by Russia in prevision of such a case, quietly to elect a new Khan, who at once placed himself under Russian guidance, no pretext for Russian interference arose. Not to be thus baffled, Russia made one. She persuaded the new Khan to insult the Turkish Governor of the little island of Taman. The fierce Osmanli strangled the insulting messenger, whereupon the Khan was induced to ask for a Russian army to expel the offender. A Russian army consequently entered the Crimea, never again to leave it. When the veil was dropped, and the real purpose of that army became manifest to the Tartar population, they refused to acknowledge the new master. The refusal sealed their doom. Thirty thousand of them, men, women, and children, were slaughtered in cold blood. With the Russians a reign of terror always follows the defeat of an Asiatic race which has resisted their arms. In our own days we saw it after Khiva, we saw it after Geok Tépé. In the Crimea it was certainly effective. After the slaughter of the thirty thousand, there were no more protests against the rule of Russia. She had fraudulently acquired the Crimea in a time of profound peace. The Porte was in no condition to resent the act. Her conquest was therefore ratified by a Convention signed at Constantinople in 1784, which gave her likewise the island of Taman and a great part of Kuban.

The year following, Georgia became a dependency of Russia, and a military causeway was constructed from the Russian frontier across the mountains to Tiflis.

The position of Russia with reference to the nations on her several frontiers at this period has been thus tersely and accurately described* :—

> It is impossible to regard without astonishment the extent of the views Russia had developed with her growing strength and the unbounded ambition they displayed. While engaged in partitioning Poland with her allies, she was dismembering Turkey for her own individual aggrandizement, and even then avowed her design to have a third capital on the Bosphorus. While subjugating the tribes of the Caucasus, she was acquiring kingdoms beyond them, and seeking contests on the further shores of the Caspian. She had added to her dominions an immense increase of territory, and a million and a half of subjects in Poland—the whole of Little Tartary and the Crimea—the isle of Taman and country of Kuban, containing a population equally numerous—the principalities of Georgia, Imeretia, Mingrelia, and the passes of the Caucasus which were now included in her territories; and she had obtained the undisputed dominion of the Euxine Sea and the passage of the Dardanelles. The utmost cravings of ambition might have been satiated, if ambition had been capable of satiety. But Poland had still some provinces to be divided; Courland was not yet a Russian government; Sweden retained Finland; Turkey had territory to cede, and a spirit of independence to be humbled; and Persia had not yet contributed her full share to the triumphs and conquests of Russia.

Catherine II. had inherited the spirit of the race whom, by a fiction, she represented. She had won over the Emperor Joseph II., and had persuaded him (1787) to enter into her plans for the dismemberment of Turkey. In the war which followed, though the allied armies met with success, it was not the success which they had anticipated. The resistance of the Osmanli was stubborn.

* *The Progress and Present Position of Russia in the East.* 4th Edition. John Murray. 1854.

The other Powers of Europe meanwhile took the alarm. In 1789 Prussia sent an army into Poland, England despatched a fleet to the Baltic, Sweden declared war against Russia. Russia has always recoiled before the resolute foreign policy of a strong enemy capable of injuring her. She did so on this occasion. Deserted by the empire no longer ruled over by Joseph II., she concluded at Jassy, 9th January 1792, a treaty whereby her frontier was advanced to the Dniester, thus opening the Black Sea to the provinces she had taken from Poland, and she was guaranteed the possession of Georgia. Notwithstanding these advantages, the result was regarded in St. Petersburg as a check.

The following year was completed the second partition of Poland —a partition profitable mainly to Russia. Certainly, one of the parties to the shameful transaction, Austria, would gladly yield her share of the plunder, if her co-partners would agree to do likewise. By this partition, completed at Grodno the 17th August 1793, Poland lost a great part of Lithuania, Volhynia, and the rest of Podolia and the Ukraine. When, to recover these territories, the Poles, inspired by the renowned Kosciuszko, rose in 1794, the three robbers united to complete the spoliation. On the 28th March 1795, the duchy of Courland was incorporated with the Russian Empire. When Catherine died, the 17th November 1796, it was reckoned that she had added, during her reign of thirty-four years, about three million seven hundred thousand square miles of territory to the Russian Empire.

Paul I. reigned only five years. He fell, the victim of a conspiracy, the 24th March 1801. The part taken by his son and successor, Alexander I., in the great contest with Napoleon is well known; not so well known, however, are the endeavours he was making throughout that period to extend his territories by the spoliation of Persia and Turkey. By the Treaty of Gulistan, concluded with the former Power in 1814, Persia renounced all claims on, and agreed to the transfer to Russia of Georgia, Imeretia, Mingrelia, Daghestan, Shirwan, Ganjeh, Karabagh, Baku, Derbend, and a part of Talish. The proceedings of Alexander with respect to Sweden and Turkey require, here, a more detailed record.

Protected by the strict alliance with Napoleon, concluded at Tilsit—one of the great mistakes made by the French Emperor—Alexander despoiled Sweden of Finland, East Bothnia, and the Islands of Aland (17th September 1809). The war with Turkey had broken out in 1807. In this war the Osmanli showed themselves not unworthy of their former renown by defeating the Russians at Silistria, 26th September 1809; but their administration was corrupt, the Janissaries had become virtual masters of the situation, and there was no central power at Constantinople to direct the energies of the Empire on a given point. On the part of Russia the war, as the year 1812 approached, languished. The threatening armaments of Napoleon made Alexander anxious to concentrate his resources. He desired peace with the Porte. On the other side, Napoleon, domineering and self-confident, despised the aid which Turkey, inspired by French officers, might have rendered him. The consequence was the conclusion (28th May 1812) of the peace of Bucharest—an agreement by which the river Pruth was constituted as the boundary between Turkey and her northern enemy.

The peace of Bucharest lasted till 1821. In the interval, however, Russia had neglected none of the means in which she was a proficient to undermine the fidelity of the Christian populations of the border-provinces of the Ottoman Empire. Her efforts were successful. On the 6th April Prince Alexander Ypsilantis raised the standard of the Cross against the Crescent at Jassy, and then began that war which terminated, the 25th April 1830, in the acknowledgment by the Porte of the independence of the land of Themistocles.

The part which Russia took in this war was eminently characteristic. She had fomented it, and she had appealed to the rulers who had joined with her in striking down the despotism of Napoleon to unite with her in insuring the freedom of a land so dear to the classic mind. It is due to those rulers and their ministers to record that on this occasion they were not deluded.

Not only did they refuse to unite with Russia; they forced Russia, at the Congress of Verona, August 1822, to abandon the intention she had announced to act independently of them. Before that congress met, however, Russia had, 9th August 1821, withdrawn her Ambassador from Constantinople. Withheld by the counsels of Prince Metternich—who, since the fall of Napoleon, had exercised an ever increasing influence over the mind of the Czar—from open hostilities against the Sultan, Alexander I. did not the less heap upon Sultan Mahmond II., one of the noblest of reformers, the last extremities of arrogance and insolence. But Mahmoud did not fall into the snare, and when Alexander I. died at Taganrog, the 1st December 1825, the most powerful of monarchs and the most disappointed of men, the peace between Russia and Turkey had not been broken.

The accession of Nicholas to the throne of the Czars was inaugurated by conspiracy and bloodshed, 26th December 1825. The firmness of the new sovereign carried him victoriously through the dangerous crisis. It cannot be doubted but that this event determined the character of his rule. Thenceforth repression and severity, the furtherance, at all costs, of the mission of Holy Russia, were to be the guiding maxims of the Czar. An event which occurred soon after his accession furnished him with the opportunity he desired to break with Turkey. On the 20th October 1827, an accident brought about, near Navarino, a collision between the Turko-Egyptian fleet on the one side, and the combined fleets of England, France, and Russia on the other. This collision, which, it has been stated, was due to a misreading of a despatch from England by the English admiral, Admiral Codrington, and which was characterised by the Duke of Wellington as "an untoward event," destroyed the Turko-Egyptian fleet. It was natural that the destruction of his fleet should impress Mahmoud with the conviction that he would shortly experience the open hostility of the neighbour whose ships had contributed to that calamity; and as natural that he should make preparations to meet the impending invasion. Nicholas eagerly caught at the fact of these preparations for defence to declare war against the Sultan, 26th April 1828. The 7th May following a Russian army crossed the Pruth.

The war, a most disastrous war for Turkey, lasted only eighteen months. It was terminated, the 14th September 1829, by the treaty of Adrianople. By this treaty Russia, which by the mouth of her Czar had publicly disclaimed all intention of aggrandisement, acquired Anapa, Poti, Achalzik, and Achalkalaki, and with them the eastern shore of the Black Sea; the virtual possession of the islands formed by the mouths of the Danube; obtained for the Danubian principalities virtual independence; removed many thousand Armenian families to her own territories; established for her own subjects in Turkey an exemption from all responsibility to the national authorities; burdened the Porte with an immense debt under the name of indemnity for the expenses of the war and for commercial losses; and retained the Danubian provinces and Silistria in pledge for the payment of a sum which Turkey could not hope for many years to liquidate.

Such were the onerous conditions imposed by a Czar who had publicly disclaimed all intention to aggrandise his dominions! Possibly Nicholas was influenced by the conviction that he had escaped an enormous disaster by the merest accident. When Marshal Diebitsch entered Adrianople at the head of the Russian army, on the 20th August 1829, that army was in the last stage of exhaustion. One great effort on the part of the Osmanli, and it must have succumbed. But the Ottomans had no general of their own, and, this time, no Baker Pasha to animate and to lead them. The brave Sultan Mahmoud himself heard only of the rapid march of the enemy; not of their losses and their exhaustion. Surrounded by ambassadors and eunuchs he was persuaded into signing a peace which was doubly disastrous, inasmuch as it yielded territory when a little firmness would have produced a result directly opposite!

The Revolution of July 1830 brought about a change in the foreign policy of Russia. From 1815 up to the time of the occurrence of that popular outbreak the Czar had cultivated the most friendly relations with the elder branch of the Bourbons. With the Orleans family, which supplanted it, the Czar Nicholas had no sympathy. From that time to the year 1850 he drew closer the bonds between himself and despotic Austria and Prussia. It was with the secret support of those powers that, in 1832, he crushed the independent organisation of Poland, making of that conquered country a Russian province. He pursued at the same time, with equal vigour and foresight, his plans of aggrandisement in the East. The revolt of the famous Pacha of Egypt, Muhammad Ali—vulgarly but incorrectly styled Mehemet Ali—had brought danger to the gates of Constantinople. Unwilling to see the rule of the doomed house of Othman supplanted by a more vigorous branch of the same race, Nicholas offered to Sultan Mahmoud the assistance of Russia. Mahmoud's last army had just been defeated at Konieh; the victor, Ibrahim Pacha, son of Muhammad Ali, was traversing Syria; so great was the prestige he had acquired that the important city of Smyrna opened to him her gates without firing a shot, before even she had been summoned. It was under circumstances such as these that Mahmoud accepted the offer of his powerful rival. Then was witnessed the spectacle of a Russian fleet entering the Bosphorus and landing with extraordinary rapidity twenty-five thousand Russian troops at Scutari, 3rd April 1833. What wonder that, terrified lest the Czar should take advantage of his position, England and France should interfere, and force upon Turkey and Egypt the convention of Kutayah, 4th May 1833! Nicholas was equal to the occasion. He could not, indeed, in the face of the united maritime Powers, seize Constantinople, but, as the price of his intervention, he forced upon Turkey (8th July 1833) the treaty of Unkiar-Skelessi, by which the latter bound herself to allow no foreign vessel to enter the Dardanelles!

In the direction of the Caucasus and the Caspian Fortune did not favour the arms of the Czar. He took advantage of the death of Fatteh Ali Shah to invite his successor to send an army against

Herat. How that expedition failed, mainly through the gallantry of an Englishman, Eldred Pottinger, who happened to be in the beleaguered city, is a matter of history. The attempts of Nicholas to force the barrier of the Caucasus, though carried on by Prince Woronzoff from 1845 to 1854 with unrelenting vigour, were baffled by the skill and valour of the Circassians, inspired by a love of liberty and led by a hero worthy to rank with Washington and Kosciusko, the illustrious Schamyl. Meanwhile, Nicholas had suppressed, or taken the strictest measures to suppress, all national feeling in Poland; and he had issued in 1839 a ukase, most cunningly inspired, by virtue of which the Christian subjects of the Porte were placed under the spiritual rule of the orthodox Greek Church!

The revolution of 1848 apparently affected the foreign policy of Russia in that it struck for the moment to the ground the authority of the two despots upon whom, till then, she had leaned for support, the Emperor of Austria and the King of Prussia. The Czar, however, took advantage of a rising in Wallachia (1848) to occupy, in complete understanding with the Porte, the two Danubian Principalities; and, as payment for that service rendered, he exacted from the Sultan the treaty of Balta-Liman (1st May 1849), in virtue of which the right of Russia to intervene in case of any disturbance in the provinces under Turkish rule, during the seven years that were to follow, was admitted. In another case, however, the imperious will of Nicholas was baffled. The House of Habsburg-Lorraine, defeated by the Magyars of Hungary, had appealed to the Czar for assistance. Nicholas, who loved to pose as the arbiter of Europe, had granted that assistance willingly. Prince Paskievitch had entered Hungary, and, after a short struggle, had forced the last National army to lay down its arms at Világos (13th August 1849). Many leaders of the Nationalists, amongst them the Polish general Bem, had fled to Constantinople. Russia and Austria demanded their surrender. The Porte, countenanced by an England guided by the firm hand of Lord Palmerston, refused. Nicholas vented his indignation by at once breaking off intercourse with Constantinople (12th November). The very next day the British fleet anchored, under orders from Lord Palmerston, in Besika Bay. The Czar speedily recognised that he had made a false move. To enable him to draw back he required, however, some salve to his vanity. This the Porte granted him by assigning to the refugees the city of Konyeh (Iconium) as a place of residence. Russia then (31st December 1849) resumed diplomatic relations with her neighbour.

I have now brought the narrative to a period when there began those political movements which led to the war of which the SURPRISE which it is the object of this chapter to describe was a striking incident. It is necessary to relate the circumstances which led to that war, but I shall relate them as concisely as I can.

The revolution of 1848, whilst it had greatly diminished the power of the other monarchies of the Continent, had enormously increased the influence of Russia. Alone of all the Continental countries Russia had been unmoved by the shock which had revolutionised France, forced the Emperor of Austria to abdicate, compelled the King of Prussia to grant a constitution, and driven his brother, the present Emperor of Germany, to seek refuge in England. From the day when, as it were by a wave of his hand, he had suppressed Hungary, the Czar posed as the arbiter of Europe. This position flattered the vanity and exalted the pride of Nicholas. In all that he had attempted he had succeeded. He had raised the other Continental Sovereigns from the dust of the Revolution, and he had treated with scorn the advances which the heir of the Bonapartes had made him when that prince assumed the imperial dignity. There can be but little doubt that in 1852 he deemed himself strong enough to make a decisive effort towards the attainment of the eternal aim of Russian policy—the replacement of the Osmanli by his own people at Constantinople. Fortunately, as he deemed it, the opportunity soon offered.

Turkey had, I have stated, successfully resisted (1849) the demands of Russia for the yielding of the Hungarian fugitives. In 1851 she had demanded and obtained the withdrawal of the Russian troops from the Danubian Principalities. When, then, in January and February 1852 the French ambassador at Constantinople, Count Lavalette, required the Porte to make some concessions to the Latin Christians respecting the Holy Places at Jerusalem, the Porte, oblivious of, or disregarding, the fact that concession to the Latin Christians meant a slight to those of the Greek Church, granted the demand (8th February 1852). About the same time, Austria, under the resolute guidance of Prince Schwarzenburg, had made certain demands upon the Porte which, in spite of his death in April of that year, the Porte granted. These concessions to other Powers so annoyed the Czar that he could no longer hold his hand. Believing himself sure of the support of Austria and Prussia, and desirous to isolate France, he made, in the winter of 1853, proposals to the British ambassador at St. Petersburg, Sir Hamilton Seymour, which had for their object the partition of the Ottoman Empire. These proposals, which began in January 1853, were reported to the British Government, and were in April of the same year finally refused. Meanwhile the Czar had directed all his efforts to the furtherance of his aim. In February 1853 he despatched to Constantinople an embassy extraordinary at the head of which figured Prince Menschikoff, the same prince whom he had sent on a similar mission to Persia immediately after his accession. The mission of Menschikoff meant war—to be brought about by the best means which might offer, diplomatic or undiplomatic. Menschikoff, determiued to succeed, plunged at once into the second of these alternative courses. On the 16th March he handed to the Sultan a note, couched in the most peremptory language, in which he complained of the action of the Porte with respect to the Holy Places, and demanded guarantees for the maintenance of the rights of the Greek Church. The Sultan, anxious by all the means in his power to avert the storm, replied by issuing two firmans, which had for their object the removal of all subjects for strife between the Latin and Greek Churches. Menschikoff refused to be satisfied with this concession, and insisted that the Sultan should consent to a formal treaty guaranteeing the rights of the Greek Church, and giving to the Czar the authority to assert and maintain those rights. The Sultan, strong in the support of England and France, refused to consent to a measure which, by bestowing upon Russia the right to interfere in the internal affairs of his country, would have been equivalent to an abdication. He broke off, therefore, negotiations with Menschikoff, and quitted Constantinople (21st May). As an indication of the justice of his intentions he issued, however, on the 6th June, a firman addressed to the spiritual heads of his subjects of the Christian and Jewish persuasions, in which he formally confirmed all the rights and privileges which, from time to time, had been granted to them. The very day that the Sultan issued this firman a note from the Russian Minister was handed to him. This note contained the declaration of the Czar that he regarded the refusal of the Porte to agree to Prince Menschikoff's proposal as a personal insult; and that unless, in the space of eight days, that proposal were accepted, he would order the occupation of the Danubian Principalities. When, on the expiration of that term, the Sultan still refused, Nicholas issued a manifesto, dated 26th June, in which he declared that he had directed his armies to march into Moldavia and Wallachia, in order to hold in his hands a security for the re-establishment of the rights of Russia and of the Greek Church. The Czar was as good as his word. On the 2nd July Prince Michael Gortschakoff entered the Principalities at the head of a Russian army!

The two great Western Powers, meanwhile, had watched the action of Russia with the most excited attention. England was, so to speak, behind the scenes. The indiscreet communications of the Czar to Sir Hamilton Seymour had left the Ministry without a doubt as to the ultimate objects of the Menschikoff mission; and, although many members of that Ministry were weak-kneed and vacillating, it yet contained two men, Lord Palmerston and Lord John Russell, in whose eyes the maintenance of the greatness of the Empire was ever a paramount consideration. Averse to war on principle, those two statesmen were resolved to have recourse to that expedient rather than to allow the dismemberment of the empire of England's ancient ally. Motives of a different character led the policy of France to a similar conclusion. Napoleon III. had been but just proclaimed Emperor of the French. The accession to the imperial throne of a member of the family which all Europe had combined to overwhelm in the second decade of the century, was received favourably only by the Power which had the most contributed to the overthrow of 1815. But whilst Austria and Prussia had contented themselves with a bare acknowledgment, the Czar had replied to Napoleon's notification of his accession by a studied insult. In the action of Russia towards Turkey the new Emperor saw an opportunity of accomplishing something more than merely vengeance for that insult. A firm alliance with England, and a successful war waged hand-in-hand with that country against Russia in the cause of law and order, would give him the place in the councils of Continental Europe which Nicholas had occupied since 1830, and more especially since 1848, whilst it would go far, he believed, to establish his popularity in France. Napoleon III., then, entered heartily into the views which were held by Lord Palmerston and Lord John Russell.*

Already, in June, the two Western Powers, acting in concert, had despatched their fleets to Besika Bay. By this act England and France hoped to convince the Czar that persistence in his designs against the Porte would bring to the aid of the Sultan two powerful allies. When, to balance this intervention, the Czar turned to the two Powers whom he had so long controlled, he found that they, too, had slipped from his grasp. Vainly did he meet the Emperor Francis Joseph at Olmütz (24th September), and King Frederic William IV. at Warsaw (2nd October). Both Sovereigns positively refused him the support he required; and although a Conference of the Powers met in Vienna on the 21st

* The Ministry was composed as follows: Lord Aberdeen, Mr. Gladstone, the Duke of Newcastle, Mr. Sidney Herbert, Mr. Cardwell, Sir James Graham, the Duke of Argyll, Lord Cranworth, Sir W. Molesworth, Lord Granville, Sir C. Wood, Lord Palmerston, and Lord John Russell. The first six of these were Peelites: the entire "grit" of the Ministry was concentrated in the two last.

July 1853, and continued to sit till April of the following year, its deliberations led to no practical result. Looking back from the standpoint of accomplished events, it is easy to see now that the Czar did not believe there would be war; that he was convinced that a policy of brag and bluster, displayed by the occupation of the Principalities, would intimidate Turkey; that an offensive alliance against him of France and England was impossible. Nicholas had some reason for the last belief. Lord Aberdeen, Mr. Gladstone, and the Peelite party as a body, were pro-Russian in their sympathies; there was, as recent events have proved, no "grit" whatever in Lord Granville; but the Czar had not taken sufficient count of the English resolution of Lord Palmerston and Lord John Russell.

On the 9th October, Prince Gortschakoff refused the summons of Omar Pasha to withdraw the Russian army across the Pruth. On the 23rd of the same month Omar Pasha sent a corps across the Danube at Kalafat, but almost at once withdrew it. On the 25th, the allied British and French fleet entered the Bosphorus. But all hopes of preserving peace had not been abandoned. The Western Powers were endeavouring to persuade the two principals to accept modifications of the Menschikoff proposals. A sudden act of wanton aggression—an act which on the part of Nicholas was a gigantic blunder—an act which finds its parallel in the act of wanton aggression lately perpetrated by General Komaroff on the Afghan frontier—brought to an end the hopes of the peacemakers. Towards the end of November a Turkish squadron in the Black Sea, which had been engaged in covering the transport of munitions of war to the Circassians, anchored in the roadstead of Sinope. On the 30th November a Russian fleet, greatly superior in numbers, which had quitted the harbour of Sebastopol the previous day, attacked and destroyed it, killing four thousand Osmanli. As soon as information of this transaction transpired, a cry of indignation rose all over Europe. The Western Powers ordered their fleets to enter the Black Sea, and transmitted to the Czar an ultimatum, the non-acceptance of which would be regarded as a declaration of war. The Czar disdained even an answer to the ultimatum, and on the 27th and 28th March 1854, England and France formally declared war against Russia.

The Western Powers had wisely resolved that the declaration of war should be not a menace only. They sent, in the spring of 1854, fleets to the Baltic and armies to the Bosphorus. The actual destination of the allied armies was at first undetermined. They disembarked in the first instance at Varna (29th May 1854). But events marched very quickly. Omar Pasha proved a very capable commander: Austria forced Russia to evacuate the Principalities (August–September): Schamyl defeated the Russians in the Caucasus. It appeared then to the commanders of the allied forces, Lord Raglan and Marshal de St. Arnaud, very desirable to attack a vulnerable point of their enemy's empire. After some consultation it was decided to direct a blow at the Crimea, and especially at the fortress of Sebastopol in that peninsula.

In pursuance of this resolve, nine army divisions—four English, four French, and one Turkish—sailed from Varna the 1st to 3rd September, and arrived at Old Fort, near Eupatoria, on the 14th of the same month. The same day forty thousand men, a large number of horses, and a powerful artillery, disembarked. On the 18th, a troop of English Hussars—the 11th—reconnoitring in front, were pursued by seven troops of Cossacks. On the 19th the allied armies commenced their march, had a skirmish with the Cossacks the same day on the Bulganak, on the banks of which they bivouacked for the night, and on the day following, the 20th September, fought and won the battle of the Alma!

I regret I cannot enter upon details of that famous battle. It must suffice to state that Prince Menschikoff, who commanded the Russian army, occupied "high, steep knolls terminating in plateaus, behind which rose another and higher range of heights," * covered by a village, by flat meadows and vineyards, and by the river Alma. His army numbered forty thousand infantry, six thousand cavalry, and he had one hundred and eighty guns. The attack-

* Hamley's *Story of the Campaign of Sebastopol.*

ing force consisted of twenty-six thousand English with fifty-four guns, twenty-four thousand French troops with seventy guns, and four thousand five hundred Turks. The battle began about one o'clock, and was won by four o'clock. The Russians, who had lost about five thousand men, of whom nine hundred, mostly wounded, were made prisoners, fled in disorder, throwing away their arms and knapsacks. The loss of the British was twenty-six officers and three hundred and twenty-seven men killed, seventy-three officers and fifteen hundred and thirty-nine men wounded; that of the French three officers and two hundred and thirty-three men killed, fifty-four officers and one thousand and thirty-three men wounded. The loss of the Turks was small. Two guns only were captured.*

The allies remained till the morning of the fourth day, 23rd September, on the Alma, and then pushed forward to Sebastopol. Menschikoff, on his side, after collecting as many of his troops as he could, made for Bachtchi-sarai, with a view of maintaining his communications with Perekop. The two armies were thus marching within a short distance of each other, both in perfect ignorance of the fact! In the course of the march Marshal St. Arnaud was compelled by sickness† to transfer his command to General Canrobert. On the 26th the allied army arrived before the north face of Sebastopol. It has since been asserted, but, I believe, without sufficient reason, that if they had at once assailed it, the place must have fallen. Such was not the opinion of the two commanders-in-chief. They did not think that a sudden attack on that face would bring the full measure of success which they needed, for, after due consideration, they resolved to transfer, by a flank march, their forces to the ground fronting the southern face, where they could rest safely upon the harbour of Balaklava, which would thus become their base. Marching the same day with that object, the English suddenly and

* "The principal trophy was Prince Menschikoff's carriage with his papers. In one despatch the General assures the Czar that the position selected on the Alma must detain the allies at least three weeks, and that he confidently hoped it would be found altogether impregnable. It was taken in three hours."—HAMLEY.

† He died on the 29th September.

unexpectedly fell in with a Russian force, estimated from ten to fifteen thousand strong, as much surprised as were our countrymen at the rencontre. The skirmish which followed—known, from the large white house near which it took place, as the skirmish of Mackenzie's Farm—was favourable to the English inasmuch as the Russians took to flight without fighting, leaving some carriages with baggage and ammunition on the plain. Without further sight of an enemy the allies continued their march, and on the 28th took up their position in front of the southern face of the fortress, and began the ever memorable siege.

Before proceeding to give a brief summary of the operationt which followed the arrival of the allies, it seems desirable that I should describe (1) Sebastopol itself, and (2) the position of the besiegers.

The fortress of Sebastopol is built on the southern shore of a bay which penetrates from west to east into the south-western coast of the Crimea, and forms one of the roomiest and safest harbours in the world. For that reason the Russian Government selected it to be the permanent resort of the Black Sea fleet. The building of the present fortress was begun by Prince Potemkin, the 5th May 1784, in the reign of Catherine II., on the ruins of the Tartar village of Achtiar; yet until the year 1805 it was itself only a village, neglected as being of but little account, and almost forgotten. To Alexander I., and to his successor Nicholas, especially to the latter, belongs the credit of making the place worthy of the position. Alexander made of the village a town; Nicholas made of the town a fortress of the first rank. From Cape Kherson, the extreme south-western point of the Crimea, the coast stretching eastward is cut into by a number of creeks penetrating southward, many of them very deep. The principal of these are the creek of Fanary, with the bay variously known as the Cossack, Kamiesch, or Reedy creek; the Pestschanaya, or Sandy creek; the Strelitz, or Sluice creek; and the Quarantine creek. To the north-east of the latter project the Cape and Fort Alexander, and opposite these, to the north, the Cape and Fort Constantine. The space between these forms an entrance about two-thirds of a mile wide into the roadstead or bay of Sebastopol. This bay is five and a half miles long, and has an average width, not calculating the side basins, of twelve hundred yards. The depth throughout varies from fifteen to eighteen metres; it has an excellent bottom, and ends in the east at the mouth of the Tchernaya, close to Inkerman. On the south coast the bay has likewise several creeks; the Artillery creek; the southern, or Line-of-battle-ship creek; the Arsenal or Dockyard creek; and the Keel creek; all natural harbours sheltered against storms by the chalk cliffs rising high behind them. All these harbours were provided with fortifications, magazines, and workshops.

On the southern front of the fortress which dominated this magnificent roadstead the allied forces took up, I have said, their position on the 28th September; the English having their head-quarters at Balaklava, resting on their fleet in the harbour of that name; the French posted on the Kherson peninsula, their fleet anchored in the Bay of Kamiesch. As the allies were not in sufficient number entirely to blockade the place there was a gap to the north and to the east in their communications, and Prince Menschikoff had it in his power to introduce at will, by this gap, reinforcements. The position is thus graphically described by a gallant soldier who took part in the siege*: "Looking at a map of the Crimea, the reader will see that a valley extends from the inner end of the harbour of Sebastopol, where the Tchernaya runs into it, to that of Balaklava. From the former harbour to the ruins of Inkerman the valley is from twelve to fifteen hundred yards wide; then the heights on either side separate, till, at the point where the road to Mackenzie's farm crosses the Tchernaya, they are nearly four miles asunder. Here a rounded cluster of gentle eminences divides the valley into two defiles; these, sweeping round from south-east to south-west, unite in one plain, which, traversed by small hills, spreads to the gorge of the valley of Balaklava, and up to the heights right and left. Thus this valley, extending from one harbour to the other, forms a wide neck to a small peninsula, of which Cape Kherson is the extremity, and on which the allied troops took their position.

* Hamley's story of the campaign of Sebastopol.

This peninsula, having steep cliffs at the sea-shore, consists of a high undulating plain, or range of hills, cleft by deep gullies that descend gradually to the basin in which lies Sebastopol. From a point opposite the basins of Inkerman to that where the high road from Sebastopol descends to Balaklava, the range of heights bounding the valley is unbroken, except at a point easily defensible where the Woronzoff road crosses it. But to the left of the point opposite the ruins of Inkerman the ground south of the Tchernaya slopes upward so gradually as to oppose no serious obstacle to the advance of troops to the heights, while the English division posted there was not on the ridge looking into the valley, but on another ridge in rear of it. Thus the space between the allied batteries of attack and the heights opposite Inkerman was, while unintrenched, the weak point of the position."

Again: "The harbour of Balaklava lies, as has been said, in a cleft between high and steep mountains. Beyond the inner extremity of the harbour this cleft continues itself for about half a mile in a small cultivated valley. A row of low isolated hills extends across the entrance of the valley and up the heights on each side, to the plains of the peninsula on the one hand, and to the cliffs above the sea on the other, thus forming a natural line of defensive posts. At about three thousand yards in front of these, on the plain, sweeping from the valley of the Tchernaya, is another range of isolated hills, the left of which is within cannon-shot of the heights held by the allies, and the right one near the village of Kamara, which lies on the mountains forming the southern boundary of the plain. This last range of hills, crowned with small intrenched works armed with artillery, and garrisoned by the Turks, formed the outpost of the allies in front of Balaklava. Thus, the position extended from the sea-shore in front of Sebastopol round the heights of the peninsula to the Woronzoff road, and thence across to the last hill on the plain near Kamara; while an inner line of posts extended across the entrance of Balaklava valley up to the heights of the peninsula on the left and round to the sea-cliffs on the right, enclosing valley, town, and harbour."

The allied batteries opened fire the 17th October. The same day the fleets of France and England stood in and engaged the forts at the mouth of the harbour, Fort Alexander and Fort Constantine. The result of the naval attack went to prove that wooden ships armed with guns of the old pattern were no match for the heavily-armed stone fortresses to which they were opposed. Nor was the experience on the land side much more encouraging. It is true that the effects of the English fire were visible on the Malakhoff, and that a shell from their batteries fired a magazine in the Redan. But, on the other hand, a magazine in a French battery blew up at half-past eight in the morning; and another explosion followed a little later. These explosions caused the entire cessation of the French attack. The English, too, soon discovered that the apparent advantages produced no permanent results. "The Russian works," wrote Hamley from the spot, "being of earth, like our own, were repaired with equal facility, and the disabled guns were replaced by fresh ones from the arsenal."

From the land side, however, the fire on both sides was continued with the same fury every day, neither being able to boast of the advantage. On the 20th the operations developed a new feature. A Russian force made a demonstration on the cluster of low heights which divide the valley of the Tchernaya into two defiles, whilst a body of their cavalry, with some guns, posted itself on the Bachtchi-sarai road, near the bridge which there crosses the Tchernaya. It soon became evident that on both these points the Russians were in considerable force, and that their numbers were being constantly increased. On the 23rd the Russian guns on the eminences and in the valley opened fire on the outposts held by the Turks. A troop of horse artillery, and a field battery, supported by the Scotch Greys, promptly despatched to Balaklava, found themselves opposed there to the fire of several batteries and guns in position, and, completely overmatched, retired as soon as its ammunition was exhausted. The Russians then pushed on and carried the works on the hills nearest Kamara. Two other strong Russian columns advanced then over the slopes of the other hills, nearest to the position which the English division had taken up. Then the Scotch Greys and the Enniskillens, supported by the 5th Dragoon Guards, made that splendid charge which not only checked this advance, but drove the Russians over the slopes. This charge of the heavy cavalry was magnificent. Another part of the enemy's columns, pushing forward from the right, was about the same time checked by the 93rd, drawn up in line, and firm and solid as though they were rooted to the earth.

The Russians, however, were still on the two hills nearest Kamara, whence they had driven the Turks. The English troops posted at Balaklava occupied two points of the ridge nearest their main position, whilst the 1st Division and part of the Light, despatched at 9 o'clock in the morning to support them, were descending into the plain. An artillery fire at a very long range from the advanced guns on both sides was proceeding.

Then it was that the Light Cavalry Brigade made the famous charge which has immortalised them. "*C'est magnifique mais ce n'est pas la guerre*!" exclaimed General Bosquet to Mr. Layard, who was present as a spectator. Never was a charge made alike so gallant and so useless. The Cavalry charged into a trap. When they reached the guns which were their goal they found the whole Russian force before them, whilst a body of Russian cavalry interposed to cut off their retreat. "Assailed on every side by every arm, their ranks utterly broken, they were compelled to fight their way through, and to regain our position under the same artillery fire that had crashed into their advance. Singly, and in twos and threes, these gallant horsemen returned, some on foot, some wounded, some supporting a wounded comrade. The same fire which had shattered their ranks had reached the heavy cavalry on the slope behind, who also suffered severely. Our loss would have been greater but for the timely charge of a body of French cavalry, which, descending from the plateau, advanced up the heights in the centre of the valley, where they silenced a destructive battery."*

The result of the day's fighting was that the Russians were left in

* Hamley.

possession of the two hills nearest Kamara. The advantage was, however, more nominal than real, for the allies had till then been holding a too extended front.

The loss of prestige, as some considered it, consequent upon the occupation of those hills, was more than repaired the next day by the repulse, with great loss to the assailants, eight thousand strong, of an attack led by General Federoff, from the fortress itself, upon the second division, commanded by Sir de Lacy Evans. The service rendered on this day was as brilliant as it was useful: brilliant, because an English force fifteen hundred strong repulsed a Russian force of more than five times its strength; useful, because the repulse prevented the Russians from seriously occupying an intrenched position on a ridge from which a pitched battle would have been required to dislodge them.

The advance made by the Russians in the direction of Balaklava had demonstrated to the English general the necessity, unless he should choose to abandon Balaklava altogether, of strengthening his position opposite Inkerman. The ruins which take their name from the locality "stand on the edge of a cliff-like precipice on the Russian side of the valley, about a mile from the head of the harbour of Sebastopol. They consist of a broken line of grey walls, battlemented in part, with round towers. The yellow cliff they stand on is honeycombed with caverns; in the valley close beneath runs the Tchernaya, fringed with trees. Beneath them the ground slopes upwards to plains covered with coppice, and on two high points stand lighthouses to guide ships entering the harbour. Masses of greystone protrude abruptly through the soil about the ruins, of such quaint, sharp-cut forms, that in the distance they might be taken for the remains of some very ancient city."*

Between these ruins and a cluster of heights across the Tchernaya was the valley of that river, and it was known that those heights were occupied by a Russian force of twenty-four thousand men, and seventy-eight guns, under General Liprandi. To oppose that army

* Hamley.

the force in front of Sebastopol would necessarily be much reduced. On the other hand the Commissary-General declared that without Balaklava he could not feed the army. Lord Raglan decided, then, to make all the sacrifices that were necessary to hold that place. He caused the position opposite to Inkerman to be covered by a continuous intrenchment extending from the plateau across the entrance of the valley, up the hills, round to a mountain path communicating with the Woronzoff road near the sea, and garrisoned by about eight thousand men of the three allied nations. The 2nd Division, commanded, in the absence from sickness of Sir de Lacy Evans, by General Pennefather, still occupied the position which it had maintained against Federoff on the 26th—a position which, from its having been assailed before, was the most liable to be attacked again. About three quarters of a mile in rear of the 2nd Division was the brigade of Guards, watching the approaches of the Kherson from the east, whilst they formed a support to the division in their front. About a mile and a quarter from the Guards began the position of the French army of observation commanded by General Bosquet. The duty imposed upon him was the defence of the Kherson against attacks from the east. He had under him three brigades, those of Bourbaki, d'Antemarre, and Espinasse, and occupied a space of about two and a half miles along the edge of the Kherson, from the Woronzoff road to the Col. Next to Bosquet, on the southern side of the Col, were the Turks; on the lower ground, fronting the road across the Col, was the brigade of General Vinoy; on the extreme right lay the brigade of Sir Colin Campbell guarding Balaklava. The two last-named brigades were so placed that they could fall on the flank of any hostile force attempting to gain the Kherson by forcing the pass at the Col.

The necessity which had imposed itself upon Lord Raglan to face the outside Russian army of Liprandi, the numbers of which were daily increasing, at the same time that he continued the siege of Sebastopol, caused him very grave anxiety. The same cause filled the Russian commander-in-chief, Prince Menschikoff, with rejoicing. The lengthened line of the British intrenchments, occupied by troops who had no supports, assured him of an opportunity which, well used, could not fail to destroy the enemy who had had the audacity to profane the soil of Holy Russia. On the 31st October he transmitted to the Czar a despatch in which he declared that if the weather should continue to favour him nothing could save the English from a complete disaster. The weather, up to the 4th November, did favour him; he made preparations such as could scarcely fail; and on the 5th November he despatched a force, overwhelming in numbers, to effect the SURPRISE which should prove to his master that his boast had not been vain.

The Russian force prepared to effect Menschikoff's purpose was, in number and position, as follows:—

On the extreme left was the corps of Prince Gortschakoff, who had taken the place of Liprandi, twenty-two thousand strong, with eighty-eight guns. It faced westward, and occupied a position which, beginning near Kamara, terminated on the Tchernaya, not far from the Inkerman ruins. To his right, beyond the Tchernaya, on the hill on which were the Inkerman ruins, facing the south and south-east, was General Dannenberg, with the corps of Soimonoff and Pauloff, forty thousand men, and one hundred and thirty-five guns, and giving touch with his right rear to the left of the force, sixty-two thousand strong, still within the lines of Sebastopol. They likewise watched the road from Bachtchi-sarai with three thousand five hundred men, and kept a small force likewise on the north side of the beleaguered fortress. Their design was, that whilst a sortie from Sebastopol should retain the French left on the ground it occupied, Dannenberg should order the corps of Soimonoff and Pauloff to ascend the northern face of the plateau of Inkerman, overwhelm the three thousand English defending that face, and pursue its victorious career in the same direction. Gortschakoff, meanwhile, was to keep Bosquet in a state of continuous alarm, and to prevent him from assisting the English until Dannenberg's success should be pronounced. Then he, too, was to join Dannenberg on the ground which he would have gained. The immediate effect, it was calculated, of the secure lodgment of an army sixty thousand strong on the eastern part of the Kherson would be not only to make the allies desist from any further attacks upon Sebastopol, but even to destroy them.* The plan was well-laid. Its success, considering the enormous superiority of the troops told off to carry it out, depended mainly upon the manner in which it should be executed.

Rain had fallen almost incessantly since 10 o'clock on the morning of the 4th November; the night that followed was damp and cold; the morning of the 5th broke foggy. But it was neither the dampness, nor the cold, nor the fog, that made that morning memorable to the English soldiers before Sebastopol. During the night the Russian soldiers had to march to occupy the positions I have named, and at dawn of day they made their rush upon the advanced posts of the still sleeping enemy. The surprise was complete. The English outposts were driven in, and the crest of the hill looking down into the valley was at once occupied by the Russian artillery and guns of position. The camp of the 2nd Division was twelve hundred yards from the crest. There, on the first sound of the Russian volleys, there had been the sudden spring from the bed, the dashing on of clothes, the mounting in hot haste, the rush to the regiment or the battery, which characterises a surprise. The Russians had, so far, done their work promptly and well. The English batteries, as they hurried to join the 2nd Division, were exposed to the fire of the guns already posted on the conquered

* "The defeat of that slender division on its ridge would have carried with it consequences absolutely tremendous. The Russians arriving on the plateau where the ground was bare, and the slopes no longer against them, would have interposed an army in order of battle between our trenches and the French lines looking on the valley. As they moved on, disposing by their mere impetus of any disjointed attempts to oppose them, they would have reached a hand to Gortschakoff on the one side, to the garrison of Sebastopol on the other, till the re-united Russian army, extended across the Chersonese, would have found on those wide plains a fair field for its great masses of cavalry and artillery. To the allies, having behind them only sea-cliffs, or the declivities leading to their narrow harbours, defeat would have been absolute and ruinous; and behind such defeat stood national degradation."—EDINBURGH REVIEW, April 1875.

crest, and the muzzles of which, only, were visible. And that fire was continuous and effective. It reached the English position, killed many men and horses, and struck down an officer, Captain Allix, not far from his own tent.

Meanwhile the men of the second division had formed to repel an attack, and had been joined by Buller's brigade of the Light Division. The formation was completed just in time. Already were the Russians creeping up the ravine. They were, however, charged and driven back. Only, however, for a brief period. On they came again in far greater numbers than before, compelled the 41st and 49th to yield ground; but were then again checked by the Guards, who had hurried as soon as possible to the support of their comrades.

But as minute succeeded minute the attack of the enemy developed itself, and it became clearer and clearer to every man of the defenders that upon his individual exertions the fortune of the day would greatly depend. There is no soldier in the world upon whom this conviction produces a greater effect than upon the Englishman. It intensifies all his good qualities. He displays the keen observation, the quickness to comprehend his orders, the fulness of resource under difficulties, the calmness and resolution in danger, which the late William Napier recorded as characterising him during the Peninsular War. On this memorable 5th of November he displayed to the full the qualities which lie, as it were, in reserve in the British character. When the Russians, covered by crowds of skirmishers, advanced on the British position, forced their artillery to limber up and retire, and pushed on in overwhelming numbers, the British soldier, whether with his regiment or separated from it in the confusion, still calmly and fearlessly looked them in the face. It is true that he fell back; that no long time elapsed before the Russians succeeded in occuping three fourths of the Inkerman plateau, and that their guns searched out the heart of the camp of the 2nd Division; but the spirit of the defenders never faltered. Pennefather, who commanded that division, set his men an example which will live for ever in history. Not even when his position was overborne in front and turned in flank, not when he saw three British guns in the hands of the assailants, did he falter or despair of the day. He was always as cool and collected as on an ordinary parade-ground.

The great features of the attack in the early morn of the 5th November were Pennefather's brilliant defence; the splendid advance of the 77th Regiment, under Colonel Egerton, and that of the 41st, led by General Adams. Not less important even than these, as diverting the attack of Soimonoff from its true point, was the attitude, on the Victoria ridge, of General Codrington. It was from dread of the flanking fire of that gallant officer's division that Soimonoff abandoned his forward march, to concentrate his attacks, as we have seen, on the Kitspur. At half-past 7 o'clock the first Russian attack, made by fifteen thousand men, of whom six thousand were actively engaged, had been repulsed; the lost guns had been recovered.

But fifteen thousand men formed but a small portion of the assailing army. Hardly had the repulse been made clear than General Dannenberg sent ten thousand fresh troops, in addition to the nine thousand reserves who had not fired a shot, to renew the attack. To meet this attack Pennefather had available not more than fourteen hundred men. Reinforcements of English troops were, it is true, approaching, which would raise that number to four thousand seven hundred, and it was believed that Bosquet would also aid him. Still neither Pennefather, nor Lord Raglan, nor Canrobert had yet realised that the attack had attained the dimensions of a great battle. The main attack of the Russians was upon one point only; and the mist, which had continued since the early morning, hid from the view of the defenders the extent of the preparations made by the assailants. The danger which threatened the former was increased by the fact that they believed the point named the Kitspur, on which was placed the Sandbag battery, to be the main object of the Russian attack. That it was one of their objects is true, and the British could not allow it to be occupied by the enemy; but, in devising means to retain it, the defenders did not sufficiently bear in mind that they left, undefended, a gap of seven hundred yards between the apex of that position and the Home Ridge!

A little after half-past seven the attack of the thousands of fresh troops upon the hundreds who had already repulsed one attack made itself felt, in front, and round the flanks. The natural effect of these attacks was to cause, on the part of groups of the defenders, a constant change of front, now to repel a flank attack, then to turn at bay to the front; and the inevitable consequence, after a time, was to cause them to lose ground. At length, as the fighting grew closer, and the weight of numbers of the assailants made itself more and more felt, the English, who had long since ceased to show any trace of formation, were forced back; the Sandbag battery was captured by the Russians; the commander of the troops on the Kitspur, General Adams, was mortally wounded; and the ridge which he had so well defended was abandoned. It was at this conjuncture that Captain Hamley* came up with three guns. These he placed in battery so skilfully that the fire from them forced the assailants who had driven the English from the Kitspur to break, turn to their left, and drop hurriedly into the shelter of the ravine. An attempt of the same troops to creep up the right bank of the ravine was repressed by the same cool and skilful officer. This "happy use of three guns"—to use the expression of the great historian of the war†—gave the English the opportunity to repair the error of the morning and to concentrate their energies of defence on the Home Ridge. But, with insular pertinacity, they still hankered after the Kitspur, now more dear to them because the post upon it which they had defended—the Sandbag battery—was in the hands of the enemy!

This battery became, then, the scene of some desperate struggles.

The troops of the Division commanded by the Duke of Cambridge attacked and took it, sweeping the enemy down the ravine. Four times was the post lost and retaken, but it was no sooner

* Now Lieutenant-General Sir Edward Hamley, K.C.B., K.C.S.I., who commanded the division which stormed Tel-el-Kebir.

† Kinglake's *Invasion of the Crimea.*

retaken than its worthlessness to the defenders became apparent; for, the parapet rising nine or ten feet from the ground, and there being no banquette, the defenders could deliver their fire only through two embrasures, or from the shoulder of the work.

The fight continued around it, however, with great fury, Colonel Walker and the Scots Fusiliers greatly distinguishing themselves. No one, however, on the English side realised that they were fighting for a point which, if held, still left a gap between its defenders and the defenders on the Home Ridge; and that it was necessary to provide equally for the defence of that gap. So little, indeed, was this realised that when the weight of numbers of the assailants began to make itself felt again on the Kitspur, the Duke of Cambridge, who had displayed all the courage and daring of his royal House, rode to the Home Ridge to endeavour to obtain from thence reinforcements. Pennefather, to whom this request was made, though hard-pressed himself, parted with half the troops which then remained to him, retaining only a force which, by the despatch from it shortly afterwards of two more wings, was speedily reduced to four hundred effective men.

The Home ridge thus defended by Pennefather, and the Kitspur by the troops already referred to, there remains the gap between the two to be accounted for. I shall deal with this presently.

It was at this critical moment of the battle that there arrived on the field two French battalions, and though, waiting for orders from their general, who had not arrived, they did not immediately take a prominent part in the action, their presence between the two extremities of the English force, afforded considerable moral support to the defenders. For, though they could not advance without orders, neither could they retreat without fighting.

Matters were in this position when Dannenberg made the attack which he hoped would prove decisive. Again was the fight renewed about the worthless Sandbag battery. A sixth time was it lost by the English. A seventh time was it retaken. Finally, the chiefest efforts of the English being concentrated on the defence of this point, the Russian left wing was driven back, and the

English pursued it down the slopes. It was, as Kinglake truly affirms, " a false, unwholesome victory," for, to gain it, the victors had to quit their advantageous position on the high ground!

This success, in fact, nearly allowed the Russians to gain all they had been fighting for. There remained but two hundred men on the Kitspur when fresh Russian troops were observed descending the heights opposite, with the evident intention to take that position in flank. At the same time, another body of the enemy had penetrated through the still undefended gap and had assailed from the height the division of Sir George Cathcart, who, against positive orders, had descended into the valley to follow up the false victory. Cathcart made a desperate effort to cut his way through the force that attacked him, but with but scanty success, and in the effort he was killed. It had been in Dannenberg's power then, by strongly reinforcing the battalion in the gap, to have gained the battle. But he had not the eye of a great commander, and he allowed the great opportunity to pass.

But the moment was critical. The Russian battalion barred the way to the junction with the force on the Home Ridge of the troops under the Duke of Cambridge, reduced now to a hundred and fifty men, still on the Kitspur, at the same time that two Russian battalions were moving against his left front. The Duke and some of his men succeeded in escaping past the Russians at the gap; but the bulk of them were not so fortunate. There were two officers who, on this occasion, covered themselves with eternal glory: the one Dr. Wolseley, a surgeon attached to the force, who, cool, clear-headed yet impassioned, pointed out to the men who clustered round him that the one road to life, if they wished to live, was to charge up the hill; the other, the ever to be lamented Fred Burnaby, who, with twenty or thirty men whom he had rallied round him near the Sandbag battery, covered the retreat of his comrades, and then dashed through the serried masses of the enemy. They failed, indeed, for the moment; and were still on the side of the Russian force furthest from the English, when there occurred that intervention which decided this period of the battle.

That intervention was the advance of the French 6th Regiment of the Line. General Bourbaki, hurrying up, saw at a glance the nature of the crisis, and gave his orders accordingly. This advance caused the Russians to retreat and rescued Burnaby and his comrades, not, however, before the splendid action of these had secured the safe arrival on the Home Ridge of the men who had charged up the hill. The French then advanced and re-captured the Sandbag battery—which had once more fallen into the hands of the enemy—but, instead of retaining it, they wisely moved on in a westerly direction.

Meanwhile, on the Home Ridge, Pennefather had, with difficulty, maintained with his handful of men a successful defence against the enormous masses of the enemy. The success was in this respect complete, in so far that, when the attack momentarily ceased, Dannenberg, with all his efforts, had failed to drive in even the English main picket.

But at half-past eight in the morning, the English had lost one half of the four thousand seven hundred men who stood under their colours at half-past seven, and, although they were now assisted by sixteen hundred French troops, and their commanders had begun to realise the real value of the several points of the position, the strength of the enemy was, for all practical purposes, undiminished; and it was clear that he had not renounced the great enterprise, the success of which Menschikoff had pre-announced to the Czar.

Deducting their losses, the Russians had at this time some seventeen thousand men on Mount Inkerman, of whom nine thousand had not, as yet, been brought under fire; they had also one hundred guns in battery; the allies had about five thousand men, all of whom had engaged; and, including a small artillery reinforcement, then close at hand, forty-eight guns.

Dannenberg now launched the third attack. Covering its advance with strong bodies of skirmishers in its front and on its flanks, he sent a massive column against the Home Ridge. Its strength, including that of the skirmishers, amounted to about six

24

thousand men. Here Pennefather had two thousand English troops, a French battalion nine hundred strong, and a small body of Zouaves. About six hundred of the English were spread in loose order to check the Russian advance; six hundred were drawn up in line, guarding the left flank of the Home Ridge; whilst the remaining eight hundred and the French battalion were on the front face of that ridge.

In one respect the conditions of the fight had changed. The atmosphere was now clear, and the combatants could see each other clearly. This, which to the fighting class of most nations, when they were greatly overmatched in numbers, would be a disadvantage, was, for the reasons already given, not unfavourable to the English.

A heavy fire from the enemy's guns preluded, this time, his advance. That advance, made by a heavy mass, was not much delayed by the British troops, who endeavoured, in detached bodies, to obstruct it. The assailants reached the ridge, captured Boothby's half-battery at its western extremity, and, though the Zouaves recaptured these, they pushed on against the position defended by the French battalion, and forced that battalion to recede. The danger to the defenders was imminent. Then it was that a hundred men of the 55th Regiment, who had been surprised and driven back from the crest on the right of the ridge, suddenly rallied, poured their fire into the Russian skirmishers standing on the ridge it had all but gained, charged them, and, after a very desperate hand-to-hand encounter, forced them to relinquish their hold. The centre body of the enemy's skirmishers was similarly forced to relinquish its grasp by an opportune charge made by two hundred men of the 77th, under Colonel Egerton, and by the rallied French battalion. Still, however, the Russian phalanx advanced till it reached the ridge at the point where the French battalion was posted, drawn up in column. Gallantly, in the face of the enemy, did that column deploy. Under its well-directed fire the Russians fell in numbers. Had it followed up its success by a bayonet charge, the Russian column must have been borne backwards. But, for some reason or other, it did not charge. In battle the opportune is decisive.

The Russians, noticing the fault committed by their enemy, were rushing forward, with the instinct of soldiers, to take advantage of it, when Colonel Daubeny, with only thirty men of the 55th, effected a diversion which gave the allies facing the Russian column the time they required. That gallant officer had the inspiring audacity to charge with his thirty men the left flank of the second battalion, that is the battalion immediately in rear of the front battalion, of the Russian column. So great was his daring, and so completely did he dominate the spirits of the enemy by his audacity, that he cleft a way through the column from flank to flank and came out unwounded with the great bulk of his followers on its right. This opportune charge did much more than astonish the Russians. It stopped the rush they were about to make upon the French in front of them. These, meanwhile, animated by the presence of General Pennefather, and by their own and English officers, responded gallantly to the order to charge, and drove the Russian mass headlong down the slope! The Russian skirmishers on the right were similarly driven back by the six hundred men who were guarding the left front of the ridge.

It was now a quarter past nine. Three attacks, all meaning mischief, had been repulsed by the allies; but the Russian masses were still practically intact, and the day was yet young. Had, indeed, the French troops at whose head Bosquet was marching to the aid of the English, been on the ground at the time of the last repulse, the day might have been won; but they were yet at a distance; and the allies on the ridge, unable effectively to follow up their success, had to feel that they would yet have to continue to defend the position they had so bravely held.

In the pursuit which had been directed from the left front of the ridge Colonel F. P. Haines,* an officer of remarkable coolness, readiness of resource, and gallantry, had, with some forty men of his own regiment, the 21st, and a few of the 63rd, pushed on till he had reached a part of the post-road across which a trench had

* Now General Sir Frederic Haines, K.C.B., late Commander-in-Chief in India.

been cut. There he had halted, and, sheltered by the steep hill-side on his left from the Russian fire, had plied with his musketry the troops below him on the post-road, whilst from over the parapet of the causeway, he dropped a plunging fire on the Russian battalions in the bed of the quarry ravine. Perceiving, however, that his position was not tenable against an advance of the enemy in force, he fell back to the main picquet wall, and prepared there to renew the defence of the barrier.

Upon Haines, then, aided by General Goldie, who commanded in the fore-central part of the field, devolved the defence of the English position against the fourth Russian attack. The number of men at the disposal of those two officers was few, but their hearts were strong, and their leaders knew that no enemy would make them budge. On came the Russians in force, only, in their first attack, to be repulsed. Both officers felt that, unless they were strengthened, this repulse was a Pyrrhœan victory; and both made earnest efforts to obtain reinforcements. At last Haines received a company of the 77th, under Lieutenant Acton, with a supply of much-wanted cartridges; and, a little later, a company of the 49th, and a few score men of the Rifles. Shortly afterwards General Goldie fell, mortally wounded; and the undivided command devolved upon Haines. Most gallantly did he exercise it, repulsing attack after attack, the enemy occupying against him a strong position and pouring in an all but overwhelming artillery fire. This fire was continuing, when suddenly the attention of the enemy's gunners was diverted by a fire upon themselves, opened under Lord Raglan's orders, by Collingwood Dickson, from two 18-pounders which had been brought to a bend of the heights which united the Home Ridge to the Fore Ridge. The fire from these guns quickly gained an ascendancy over the fire of the Russian guns; their artillery power on Shell Hill was soon broken; a fresh fire from twelve heavy French guns planted on the crest of the Fore Ridge completed their discomfiture. Then, just as this was accomplished, when the base upon which the Russians rested their aggressive power was shaken, the arrival of Bosquet with four hundred men, closely followed by fifteen hundred more, seemed to transfer the advantage, for the first time during the day, to the side of the allies. That this had become possible was owing mainly to the splendid defence of the barrier made by Haines.

But, owing to an error of judgment committed by Bosquet, the Russians had yet another chance. I omitted to mention that the French 6th Regiment of the Line, posted on the right bank of the Quarry ravine, had been assailed, at the time Haines was moving his position on the post-road to the barrier, by a strong Russian column and had been driven back; that the French 7th, the same which we have witnessed repelling and then charging the main Russian column, had marched across Haines's front at the barrier to its support; and that both had sent urgent entreaties to Bosquet for assistance. It had been the original intention of Bosquet to act in concert with Lord Raglan and Pennefather; but the despairing cry of the children of his own soil led him to ignore the English altogether and to march straight to the point where they were still drawn up—the spur on which was the Sandbag battery. This error was nearly causing a disaster.

In that direction Bosquet assumed at once the offensive. At first he made no head against the two Russian columns which had been slowly pressing upon a detachment of the 20th Regiment (English). But soon his force was augmented to three thousand men. Shortly afterwards there joined him two more regiments, Algerines and Zouaves, bounding to the front like tigers. Bosquet, however, still without communicating with the English leaders, took up a position on the extreme left, powerless for effective attack, and which left a wide gap between his troops and the English. It resulted from this, that, whilst he was riding to the right to reconnoitre, the Russian crept up, very nearly captured him, and did actually get possession of one of the twelve guns which we have seen rendering such good service on the Fore Ridge, and forming one of the six which Bosquet had recalled from that position.

Bosquet was now in great danger; the Russians were gathering on his flank. The fire of newly placed Russian batteries was so dominating the fire of the six guns still on the Fore Ridge that these had to be withdrawn. The arrival of Canrobert, at this critical moment, at the head of the Chasseurs, did not, as we shall see, improve matters. For, not knowing that the English still held their own, he believed that the battle was lost; and, had the Russians displayed the energy and activity they had shown in the earlier part of the day, it might still have been lost. But, fortunately, at this particular epoch of the day, and when a forward movement on their part might have redeemed past misfortunes, the Russians stood still, plying only their artillery. This fact, and the assurance that the English still held the Home Ridge, gave fresh hope to our allies. Bosquet, too, received a fresh reinforcement, and, animated by the hope of which I have spoken, he dashed the Zouaves and Algerian regiments, and a small band of the Coldstream Guards which had come up, against the two columns which have been referred to as slowly pressing on the 20th. The effect was electric; the Russian columns were not only overthrown, but so greatly slaughtered that remnants only succeeded in escaping from the battle-field.

It was now 11 o'clock. The French were safe, indeed; but they were separated by the gap from the English on the Home Ridge. Haines still held the barrier. The English officers, from their commander-in-chief, Lord Raglan, downwards, were for assuming the offensive by a joint movement. But Canrobert, who was, in this respect, the very reverse of Masséna—of whom it was said that on the battle-field his thoughts became clear and forcible, and his mental energy redoubled—did not comprehend the situation, and feared risking his troops, of whom he had on the ground eight thousand seven hundred beside artillery, in a forward movement. He adhered to this timid resolve during the remainder of the day.

Meanwhile Dannenberg was waiting for the result of an attack which, he hoped, the troops under Gortschakoff would make on the flank and right rear of the allies. To him, then, the decision arrived at by Canrobert was, of all the decisions, the most welcome. For, were the allies to remain inactive till Gortschakoff should make his attack felt, success seemed assured. But, fortunately for the common cause, the islanders who had borne the brunt of the battle from its commencement, who had already repulsed five attacks, and who had so shaken the enemy during the last as to obtain for themselves and allies a position of vantage, did not concur with Canrobert. Their conduct in this respect, now about to be recorded, was no part of a planned-out purpose. Rather was it the instinct of the soldier alive in the heart of every man, from Lord Raglan downwards, which prompted each to that daring aggression which thus took the form of a general movement to the front.

We have already seen how tenaciously the gallant Haines, the type of the calm and resolute soldier full of resources in danger, had held the barrier. During that defence there had been thrown out in the brushwood on the left of that post a few skirmishers who had maintained a constant fire on the gunners in the Russian batteries. Their execution had been excellent. Gradually, as it became evident that the Russian attack was shaken, these skirmishers were joined by others from behind the barrier, and, without any preconcerted design, the English began to assume here an offensive attitude. In this, doubtless, they were assisted by the still continuing fire of the two 18-pounders which, I have already shown, were posted on the crest of the hill between the Home Ridge and the Fore Ridge, and which, playing upon batteries which they dominated, and upon infantry who had been repeatedly repulsed, made great havoc alike on the numbers and *morale* of the enemy. Every moment shook the faith of Dannenberg in the possibility of another attack. For long, indeed, had he been sustained by the hope that Gortschakoff would at last make himself felt on the flank and rear of the allies. But that hope was waning when the splendid action of Haines forced him to a decision which, an hour earlier, he would have scorned.

Carefully noting the gradual change on the left front of his position from defensive to offensive action, and the increasing weakness of the enemy under its effects, Haines deemed that the hour had now arrived when that offensive action might be made still more pronounced. He accordingly threw forward a few riflemen, and assigned to them the task of taking a position which would enable them to deliver such a fire on the batteries on the hill opposite as would render it impossible to serve them. Nearly at the same time Lord West, who belonged to Haines's regiment, the 21st, and who, with that officer, had carefully examined the ground some days before the fight, directed Lieutenant Acton, with three companies, to advance against the Russian batteries. Acton was as gallant and daring an officer as ever lived. He belonged to the 77th, a regiment which on this day covered itself with glory. The other two companies did not belong to that regiment, and they, their officers as well as their men, demurred to an action which they probably deemed an act of madness. The men of the 77th, checked by this refusal of their comrades, did not, then, at once respond to the order of their officer to advance. What followed I can only tell in the words of the historian.* "Acton said: 'Then I 'll go by myself,' and moved forward accordingly; but he soon found himself quite alone, at a distance of some thirty or forty yards in front of his men. Presently, however, James Tyrrell, a private of the 77th, ran out of the ranks and placed himself by the side of his captain, saying: 'Sir, I'll stand by you.' Then a soldier sprang out from the company which was on the right of the 77th men, and placed himself close abreast of the captain, whilst Tyrrell continued to stand on the other side of him. The officer and the two soldiers moved forward towards the battery, and they compassed a few yards without being followed; but then suddenly, to Acton's infinite joy, the whole of the 77th men rushed forward after their captain, and formed up behind him." Such are our island officers, such our island soldiers! Never will men such as these yield an inch of that great Empire which their fathers and their fathers' fathers have created!

Acton then worked his way up to the battery. As he advanced, a shot from one of the two 18-pounders smashed into it. The officer who commanded it, seeing the double danger, hastened

* Kinglake's *Invasion of the Crimea*, vol. v., p. 425–6.

to limber up and retreat. With such expedition did he accomplish his task that, when Acton reached the site, his only spoil was a gun-carriage and two tumbrils!

But he, and the battery on the crest of the hill, had been the two immediate causes of the first retrograde action of the Russians. Instinctively the forward movement continued; by few men, indeed, for, scattered in small batches, their total number scarcely exceeded three hundred. But the fact of the forward movement was sufficient. It was the last feather in the scale that decided Dannenberg to retreat. Shaken by the "murderous fire," as he called it, of the 18-pounders, despairing of the promised movement of Gortschakoff, and seeing a handful of the islanders, whom he had tried to overwhelm with numbers, toiling resolutely in pursuit of his men dispirited by defeat and broken in their *morale*, the Russian Commander-in-Chief gave orders, about 1 o'clock, for a general retreat. To cover it he ordered the Vladimir regiment to the front, and, despatching from the field his batteries which had the most suffered, ordered his light field batteries, supported by the twelve battalions which had not engaged, to hold their actual positions as long as possible.

Obeying his orders, the four battalions of the Vladimir regiment, two thousand strong, advanced down the hill. So advancing, however, they came under the fire of the two 18-pounders which had rendered such splendid service, and were literally smashed. Then there fell into the hands of the allied commanders a great opportunity. A repulse goes for comparatively little unless it be converted into a ruinous defeat. The English, alone, worn out by eight hours' hard and continuous fighting, and still few in numbers, were unable to make that conversion. But by their side stood their French allies, comparatively fresh, and numbering eight thousand infantry, seven hundred cavalry, and a powerful artillery. Had they joined in the pursuit the destruction of the Russian army was assured. For some reason, however, unexplained, and on rational grounds inexplicable, Canrobert refused to associate himself in this easy enterprise, an enterprise which might have brought about the fall of Sebastopol. The most probable solution of the refusal is to be found in the character of the man. Canrobert has been, throughout his career, wanting in the decision and enterprise requisite for a commander-in-chief. It was this failing on his part which caused his removal, at a later period, from the command of the French army in the Crimea; subsequently it manifested itself, in a marked manner, at the battle of Solferino. It was a misfortune. Had Pelissier commanded the French army on the 5th November 1854, the Russian army would inevitably have been destroyed.

As it was, Lord Raglan was unable to launch his tired troops in pursuit of an enemy who, though beaten, was still at least five times more numerous. He had to content himself with the fact that, with a force at first much less than, and at no time exceeding, eight thousand, he had, with but slight assistance from the French, repulsed six attacks made by a general who had at his disposal more than 60,000!*

As long as England lives will this famous battle be remembered. It was a battle which brought into prominence all the virtues of the national character. Coolness, endurance, calm courage, presence of mind, the possession of resources under difficulties, were conspicuous in every phase of it. Never did the unreformed officer of the British army, not yet worried and emasculated by worthless book examinations, show to a greater extent his fitness to command the men with whom the admirable system of long service had filled the regiments! It was indeed a soldier's battle; a battle which only men could have fought to a successful issue. The officers, gentlemen who had obtained their commissions by purchase, were worthy of the men; the men were worthy of the officers. Fortunate indeed was it for England, when she became involved in that famous contest, that the age of Military Harlequinade had not superseded the age of Military Common Sense; that the system of appointing to high places Acrobats who had

* In the battle the Russians lost in killed, wounded, and prisoners, 11,959; the English lost 2,573, of whom 635 were killed. The French lost 1,800, of whom 175 were killed.—Kinglake.

risen by depreciating soldiers of solid worth and capacity, had not even been dreamt of; that the filching by one officer of the honours achieved by another, and the suborning of newspaper correspondents to assail the reputation of subordinate officers, were crimes which would have ensured universal reprobation. With other times have come other manners. Still there remains to us, and to the sound-hearted men who constitute the majority of the military profession, the great remembrance of Inkerman, a battle which "so fought"—alas! that we cannot say, "so followed," —"and so fairly won,"—

> Came not, till now, to dignify the times,
> Since Cæsar's fortunes.*

* Henry IV.

CHAPTER X.

ÁRAH AND ÁZAMGARH.*

No epoch of British history has been more fruitful of heroic deeds than the epoch known as the Indian Mutiny. Those deeds knew no distinction of sex, of profession, of early training. In the presence of a sudden and unexpected danger, the undaunted courage which has made England great was equally conspicuous in the tenderly-nurtured woman, in the merchant who toiled at his desk, in the member of the Civil Service who, sitting alone in his court surrounded by men of the same blood as the mutineers, continued still to deal out justice, and in the soldier by profession. In all, when the emergency came, the island blood spoke out. History gives no brighter example of stern determination under difficulties, of cool and calm courage, alike in reverses and when success seemed to loom in the future, of the mind equal to all circumstances, than that which illustrated this splendid epoch of Great Britain's career!

Not, indeed, that the conduct of all men, though all were inspired by the same noble instincts, led to one uniform result. Conduct depends upon character and upon intellectual power. A man may be brave, may be careless of life, and may yet be wanting in the ability to lead and direct his fellow men. In military life it is often experienced how a soldier who has proved

* For this chapter I have used the same sources of information which formed the basis of my *History of the Indian Mutiny*.

an excellent second-in-command, becomes a poor commander-in-chief. No one supposes that Collingwood would have equalled Nelson. The power of commanding is a gift vouchsafed to few. It is not only that a man must be able to bear responsibility; that he must be so self-reliant as to act without hesitation and with decision in a sudden emergency; he must possess a judgment which is calm and cool under all circumstances, however unexpected and startling; he must be able to risk much in order to gain all; he must know how to forego the temptation to gain a small advantage, if he sees that, by foregoing, a greater opportunity will offer. Above all, he will give as few chances as possible to his enemy. Whilst watching for a weak point in his order of battle, he will expose none which he cannot strengthen in his own array. There are few men so naturally gifted as to be able to fulfil these conditions. For it is a natural gift. Hannibal, as we have seen in the first chapter of this volume, possessed it to perfection. It constituted the strength of Turenne. It was conspicuous in Marlborough and Eugene, in Loudon and in Frederic. Napoleon displayed it to perfection in that wonderful campaign of 1796. The reader has noticed in these pages how richly, in this respect, Moreau, Masséna, and Lecourbe were endowed; how many officers who fought at Inkerman rivalled even these. The Indian Mutiny brought to the front several who possessed those qualities to perfection, some likewise from whose composition they were absolutely wanting. Eminent amongst the former were, among many others, Lord Napier of Magdala, Lord Strathnairn, the late Vincent Eyre, and Lord Mark Kerr. Of the latter class I propose to give in this chapter a single illustration, and, in contrasting the AMBUSH AND SURPRISE of A'rah with the AMBUSH AND SURPRISE of A'zamgarh, to show how, in the first case, the absence of the qualities I have enumerated led to defeat, whilst in the second they enabled their possessor to gain a great and decisive victory!

The mutiny, on the 10th May 1857, of the native troops stationed at Mírath, and the seizure by those troops of the imperial city of Dihlí, found no greater sympathy throughout India that in the city of Patná. The province of western Bihár, of which that city was the capital, is famous, first, as having been the cradle of Buddhism, and secondly, as having formed, when the Mughuls ruled India, a powerful province under the sway of a Muhammadan Governor. It lost none of its importance on its complete transfer from the Muhammadans to the English in 1764. The district of Tirhút, between Nipál and the Ganges, gradually became the seat of an industry which attracted to it from England men who were prepared to devote a great portion of their lives to the cultivation of its soil that they might return with an independence to pass their declining years in their own country. Of this district Muzaffarpúr was made the chief station for the English officials who should dispense equal justice to the inhabitants, European and native alike. Tirhút contained in 1857 an area of 6,343 square miles occupied by a population of 4,349,000. Besides indigo, it produced cereals, sugar, tobacco, opium, and saltpetre. The other towns in it are Hajípúr, on the confluence of the Ghandak and the Ganges, opposite Patná; Sonpúr, famous for its fair, and Darbanghá, the seat of the Tirhút rájahs. The other districts, forming together western Bihár, were (2) Champáran, between the Ghandak, Nipál, and the Bágmátí, with an area of 3,531 square miles, and a population of 1,439,000, and having as its chief station Motíhárí, as its largest town Bhetiá; (3) Sáran, between the Ghághrá, the Ghandak, and the Ganges, with an area of 2,654 square miles, a population of 2,063,000 souls, producing in abundance cereals, indigo, opium, and sugar: its civil station was Chaprá; (4) Sháhábád, between the Ganges, the Karamnásá, and the Són, with an area of 4,385 square miles, a population numbering 1,726,000, and having, besides its chief station, A'rah, the important towns of Domráon, the seat of the Ujjainiah Bhojpúr rájahs; Baksar, the seat of the victory which gave Bihár to the English in 1764; Chausá, at the confluence of the Karamnásá and the Ganges; Sahasrám, an important post on the grand trunk road leading from Calcutta to the north-west; and Rohtás, near the Són, a renowned fortress. The fifth district was that of Gayá, with an area of 4,718 square miles and a population of 1,944,000. Its chief station, Gayá, on the Phalgú, was famous for its places of pilgrimage and Buddhistic remains. Its other principal towns were Shergháti to the south, on the road from Hazáribagh to Banáras; Dáúdnagar, so called from Dáúd Khán, Governor of Bihár under Aurangzíb; and Arwal, famous for its manufacture of paper. Last of all came the sixth and, in many respects, the most important district, the district of Patná, possessing an area of 2,101 square miles and a population of 1,559,000 souls, and producing rice, cereals, oilseeds, opium, tobacco, sugar, and grapes. Its capital was the famous city of Patná, known to the Sanscrit student as Palibothra, capital of Magadha, re-named by the Muhammadans Azímábád, but still universally called Patná. It extended a mile and a half, from east to west, along the right bank of the Ganges and three-quarters of a mile backwards from it. It contained many large buildings, several mosques, now but little used, and a population of about 285,000. The European quarter, called Bákípúr, corrupted into Bankipúr, lies on the western side of the city. Ten miles by road, and twelve by water, further to the west of Patná, is the military station of Dánápúr.

It is significant of the prestige attaching to the English that this large province, consisting, we have seen, of 23,732 square miles, and bearing upon its surface a native population of over thirteen millions, should have been administered by a few of their countrymen scattered at isolated towns in the several districts, and that order in it should have been ensured by the presence at one point near the capital of a small military force. The civil servants at the isolated towns in the several districts were subordinate to an officer at Bákípúr, holding the rank of Commissioner of the Patná division. Dánápúr was the head-quarters of a military division. The force, then, was thus under the immediate command of a Major-General.

In 1857 the Commissioner of Patná was Mr. William Tayler. In describing the character of Mr. Tayler I prefer to use the language of a gentleman totally unconnected with India, and, therefore, not subject to the party feeling which, however unconsciously, often sways the minds of men who have passed their lives in that country—a gentleman who, influenced by his love of truth alone, examined carefully every document bearing upon the controversy between Mr. Tayler and Sir F. Halliday, and who recorded his views in a history of the Indian Mutiny which is, in all respects, admirable. "William Tayler," writes Mr. T. R. E. Holmes,* "was a man of culture, keen sense of humour, and wide sympathies. His spirits were marvellously elastic and buoyant for his years; and withal he was by nature so combative that he could not always bring himself to live submissively under a superior whom he did not respect. This temper, however, though it was injurious to his prospects of official success, did not weaken his efficiency as a public officer. Deploring the want of sympathy which prevented the average English official, in spite of the conscientious industry with which he fulfilled his duties, from becoming familiar with the habits of thought of the natives and their real feelings towards British rule, he had not contented himself with working for the material prosperity of his people, but had tried, like Henry Lawrence, to reach their hearts as well. But the tenderness which moved him to make allowance for their weaknesses was balanced by a stern resolution which would never allow them to dispute his supremacy. He was not a man of iron, however, but a man of tempered steel. The sympathy and kindliness of his nature were allied with a keen sensitiveness. He felt that the duty which lay before him was a grave one, that his responsibility was appalling."

I have said that the action of the mutineers on the 10th May found more sympathy in the city of Patná than in any other city in India. For Patná was the head-quarters of a sect of Muhammadan reformers, called the Wahábís, who tenets inspired them with the most fanatical hatred of all who do not acknowledge the pure

* *A History of the Indian Mutiny, and of the Disturbances which accompanied it among the Civil Population.* London: W. H. Allen & Co.

faith of Islám. At the head of the Patná section of these reformers, were three Maulawís, who, like the Council of Three in the Venetian Government, secretly directed all the movements of the body. The danger, therefore, was pressing and urgent.

It is due to Mr. Tayler to record that he met the crisis with a courage and firmness worthy of the highest praise. How he, unsupported at Calcutta, made head against the storm; how he paralysed the action of the Wahábís by placing their three leaders in confinement; how he repressed a dangerous rising in the city itself; how he cowed the disaffected, won over the wavering, and, by his high example, encouraged the loyal, has been told by the historians of the Indian Mutiny. The difficulties that beset him arose not less from the action of the disaffected than from the half-hearted measures respecting the troops at Dánápúr which alone the Government of India could be prevailed upon to authorise, and which culminated in a disaster such as, but for the vigorous and decisive action of two brave men—William Tayler and Vincent Eyre—and the courage of the English stationed at Árah, would have been irremediable.

The garrison of Dánápúr consisted of the 10th Foot, the 7th, 8th, and 40th regiments N. I., one company of European and one of native artillery. The officer commanding was Major-General Lloyd, an officer who had rendered excellent service in his day, and who, only four years before, had successfully repressed the rising in Santháls.

When the Mutiny broke out in May 1857, the only Englishmen in India who did not at once comprehend its full significance were the gentlemen who constituted the Government of India and the Government of Bengal. These gentlemen refused to see that a crisis had come upon them which affected the entire country to the east and north-west of Calcutta. An illustrious French writer once stated that the great fault of English administration was that though it punished, it made no attempt to prevent, crime. This fault constituted the besetting sin of the administration of Lord Canning in the early days of the Mutiny. With rare exceptions, and then only when the measure was forced upon them, the Government of India waited till a regiment had actually mutinied before they attempted to deprive it of the weapons which made its mutiny formidable. They deliberately refused to recognise the fact that the leaven of revolt had permeated the whole native army. They preferred to wait until each in its turn should seize its own opportunity for outbreak and slaughter.

Thus it was, that although, before the reinforcements for which they had sent to Persia, Ceylon, and Burmá, had reached Calcutta, the 10th Foot was the only English regiment occupying the long line between the capital and Kánhpúr, the Government of India steadily refused to accede to the requests pressed upon them that the three native regiments stationed at Dánápúr should, as a precautionary measure, be disarmed. It can easily be realised how enormously this refusal added to the difficulties of Mr. Tayler. With the entire province of western Bihár upon his hands, having at his disposal only a few Sikhs belonging to a regiment raised by Major Rattray and commanded by that gallant officer, he could not rest, as under ordinary circumstances he would have rested, upon the support of the English regiment at Dánápúr. That regiment had enough to do to watch the three native regiments, to hold itself in readiness in case these should rise. Nay, more. It was confidently anticipated by the disaffected that the three native regiments would seize an opportunity to steal a march on the English regiment, and to surprise Patná. So far, then, from the garrison at Dánápúr being a support to the Commissioner of Patná it constituted a very real danger.

Unmoved and obstinate, the Governments of India and of Bengal watched Mr. Tayler, as he steered his charge safely through the storm during the months of May and June and during a part of July. In the last-named month the expected reinforcements had arrived: the Government was preparing to execute the plans they had formed for the relief of Lakhnao; fresh English regiments were to pass by Dánápúr. To all sensible Englishmen in India, then, the moment seemed opportune to effect by those regiments, on the journey, that disarming of the native regiments at Dánápúr which would ensure, in the first place, the safety of western Bihár in the second, the security of Calcutta itself.

For, when once the fresh regiments should have reached Kánhpúr there would be no English force available for the preservation of the capital, except a wing which held Fort William. In the always possible contingency, then, of the three native regiments at Dánápúr evading the 10th Foot, and causing by their action a rising in western Bihár, it was certain that eastern Bihár would follow, and, in that case, the safety of Calcutta would be seriously imperilled. The danger then of leaving the three native regiments in possession of their arms was a very real danger.

The representations on this subject were so urgent that the Government met at the end of the first week of July to consider them. After a brief discussion they arrived at a decision fatal to their prescience as statesmen, worthy of the lay-figures which, in the early days of the revolt, they actually were. They declined to take upon their shoulders the responsibility of giving any order whatever; they would neither direct that the regiments should retain their arms nor that they should be deprived of them. They left the decision absolutely to Major-General Lloyd.

This decision, though not communicated to the public, was privately conveyed to the mercantile community of Calcutta. That community had large interests in western Bihár, and its members were disinclined to see those interests, and possibly interests nearer to them, sacrificed by the abdication by the Government of their authority. They accordingly made a respectful request to the Government to receive a deputation from their body on the subject. Lord Canning, his mind already made up, received the deputation, listened to what they had to say, and then, very curtly, informed them that he declined to listen to their prayer!

Meanwhile, the regiment which was to pass by Dánápúr on its way to Allahabad, the 5th Fusiliers, arrived, the 22nd July, off the former station. This was the regiment which General Lloyd was to use, in conjunction with the 10th Foot, to disarm, should he decide to disarm, the three native regiments. Apparently he had made up his mind not to disarm, for he allowed the 5th Fusiliers to proceed on their way the following morning.

But no sooner had the 5th Fusiliers quitted Dánápúr than General Lloyd began to doubt whether he had acted rightly. He could not indeed call them back; but when, two days later, two companies of the 37th Foot arrived off the station, he ordered them to disembark. The next morning he directed them, in junction with the 10th Foot and the European company of artillery, not, indeed, to disarm the native troops, but to render their muskets powerless by carting away all their percussion caps!

Had General Lloyd directed that the native troops should be disarmed, and had he carried out the order in a soldierly manner, the result would have been obtained, as it was obtained at Láhor, at Barrakpúr, and at other stations, without disorder and without bloodshed. But, in attempting the half-measure he had devised, without taking the necessary precautions, he induced a very great calamity. The English troops were dining when the sipáhís were asked by their own officers to surrender the percussion-caps about their persons. The demand, resented as an insult, became the signal for mutiny, and when the English troops were in a position to act, the three native regiments, unpursued, were marching at full speed for A'rah!

Why did not General Lloyd pursue them? He had still time to repair his error of the morning, for the European troops were drawn up on the parade-ground. But General Lloyd had taken up his post on board the steamer which had carried the two companies of the 37th. None of the officers below him cared to take the responsibility of acting in his absence. Not one of them even took the precaution to ascertain the route which the mutineers had taken. They burned, indeed, the lines of the sipáhís, and then remained quiescent, confused, almost terrified. They had allowed the three native regiments to decamp with arms and ammunition, with bag and baggage, no one knew, no one inquired, whither!

The following morning, indeed, General Lloyd made a faint

effort to obtain some information by despatching a steamer carrying a few riflemen to the river Són. But the effort produced no result, for the draught of water in the Són was insufficient for the steamer.

The sipáhís, meanwhile, greatly exasperated against their late masters, had reached the left bank of the Són river early on the morning of the 26th July. Here, had General Lloyd displayed the smallest energy, he might have caught and destroyed them, for the one ferry-boat there was insufficient for so large a number, and the men spent the whole day on the right bank whilst boats were being collected whereupon to cross. In this service they were greatly aided by the adherents of a large and influential land-owner, whose estates lay in the vicinity of A'rah, named Kunwár Singh. This nobleman, a Rájpút chief of ancient lineage, had been so cruelly treated by the Bengal Board of Revenue, that his feelings, naturally loyal towards the British, had undergone a complete revulsion, and, old as he was—and he had seen eighty summers—he was only anxious now to take revenge for insults which he deemed premeditated. Not unobservant of the storm that was brewing at Dánápúr, he had foreseen that the timid action of the Government would inevitably produce the result they desired to avoid.

No sooner, then, had Kunwár Singh heard of the successful mutiny of the three regiments than he sent messengers to assure them of his sympathy, and to urge them to march on A'rah. The fact that his retainers assisted them to collect boats on the Són during the 26th is thus explained. Before night had fallen every man of them had crossed. Kunwár Singh himself was on the spot, and, acting under his advice, the sipáhís decided to march on A'rah, treat the Europeans there as enemies, and plunder the treasury. Success there would give all western Bihár, excepting the city of Patná, into their hands!

The station of A'rah is twenty-five miles distant from Dánápúr. The Europeans and Eurasians there numbered fifteen, but these had at their disposal fifty Sikhs of Rattray's regiment, whom, with the prescience which characterised all his actions during this terrible crisis, Mr. W. Tayler had sent thither from Patná. The residents of A'rah had been neither blind nor indifferent to the signs of the times. One of them, Mr. Vicars Boyle, a civil engineer connected with the railway, had, from a very early period, regarded an attack upon the station as coming within the range of possibility. He had, therefore, despite the jeers of some, and the covert ridicule of others, fortified the smaller of the two houses in his compound in a manner which would enable its defenders to resist a sudden attack. This house was a small, two-storied, detached building, about fifty feet square, surmounted by a flat roof. In it Boyle had stored supplies of all kinds. As soon as the information of the approach of the rebels justified Boyle's precautions, the residents, accompanied by one Muhammadan gentleman and the fifty Sikhs, hurried thither for protection.

They were well sheltered behind its walls, and had begun the demolition of the larger house, to prevent it serving as a protection to the rebels, when these, inspired by Kunwár Singh, entered the station, plundered the treasury, and proceeded to search for the Europeans. They had not gone far when they found their progress checked by the one small house which Vicars Boyle had armed and provisioned.

During that evening, the following day, and the day after, the 29th July, the sipáhís did all they knew to expel the defenders from that solitary bungalow. They brought guns to bear upon it, they poured into it a strong musketry fire, they attempted to gain over the Sikhs, they offered terms. In vain. The persistent skill and gallantry of the little garrison kept them off; the Sikhs remained true to their salt; the terms were refused.

At length at midnight on the 29th the garrison were roused by a sound of musketry not very far from them. The sound, however, gradually faded away in the direction of Dánápúr. It then became clear to all that an attempt had been made to relieve them, and that the attempt had failed.

We left General Lloyd content with having made one feeble

effort to reconnoitre the mutineers in the direction of the river Sŏn. The next day, information having reached him that Kunwár Singh had joined the revolted sipáhis, he resolved to intrench himself at Dánápúr. But for the urgent entreaties of the Commissioner of Patná, he probably would have done so. But Mr. Tayler, alive to the exigencies of the crisis, implored him, with all the fervour of his energetic nature, to be up and doing. There was still time to smite the rebels hip and thigh, still a chance of retrieving the lost opportunity. Possibly not even the eloquence of Mr. Tayler would have produced the desired effect, but that it was backed by the information that the rebels were besieging the residents of A'rah in a house into which these had crowded. Upon this the general determined, the evening of the 27th, to despatch 193 men of the 37th in a steamer, to be landed at a point on the banks of the Són where the road to A'rah touches that river. The steamer started that night, but ran aground and stuck fast. General Lloyd, disheartened, then gave up the enterprise, but the Commissioner of Patná, strongly urging, persuaded him to renew it. The General then ordered that 250 men of the 10th Foot, with 70 Sikhs and some volunteers, under Colonel Fenwick of the 10th, should steam to the Són, on the morning of the 29th, in a vessel which had arrived the previous day. Subsequently the number was reduced by 100 men, and the command was given to Captain Dunbar of the 10th. The detachment left Dánápúr that morning, amid the cheers of the European soldiers and traders, picked up the 193 men of the 37th on the way, and began to disembark at the nearest point to A'rah about 2 P.M. They numbered 343 Europeans, 70 Sikhs, and some twenty volunteers.

Captain Dunbar had had no experience of command. He had been paymaster to his regiment, and his knowledge of mere routine military work was small. General Lloyd, in a letter justifying his own action, stated his suspicion that Colonel Fenwick was aware of Dunbar's unfitness for command. That he was incompetent for such a duty the story of his conduct I am now about to tell proves most fully.

The disembarkation began, I have said, about 2 P.M. by the mouth of a nálá the course of which led to a point much nearer to A'rah. But to reach this it was necessary to proceed in country boats. Some time was spent in procuring these, but at length a sufficient number was obtained to carry the troops for some miles. From these they disembarked about 7 P.M. They were fasting, for, although some attempt had been made to cook the dinners whilst search was being made for the boats, it had not, owing to want of proper arrangements, succeeded. At 7 o'clock, when the men disembarked for the second time, the night was so fine and bright with moonlight, that Dunbar, encouraged by news brought him by some natives that the garrison of A'rah was holding out, gave the order to move forward at once. The Sikhs marched in front in skirmishing order; then followed the Europeans. No other precaution was taken against surprise. In this formation they pushed on till the troops reached a bridge across a stream about two miles from A'rah. Here some of the officers about him advised Dunbar to halt for the night and serve out biscuits and rum to the men. But Dunbar, anxious about the besieged garrison, determined to press on.

He had but just crossed the bridge when some horsemen were sighted in front: these, however, galloped away before they could be fired upon. Dunbar then called in the skirmishers, and pushed on in close order. Just then the moon went down. A few minutes later, as the column, marching along the road, gave its right flank to a dense mango-grove, a tremendous musketry-fire opened suddenly from the grove and caused considerable loss. The sound of the volley from the grove had not died away when a similar fire was opened from a group of men immediately in front of the column. Dunbar himself and several of his officers were shot down; the deadly fire continued: the white uniform of the Europeans presented a sure mark, while the dusky bodies of the sipáhís were invisible. Officers and men fell fast; the English troops became bewildered; no great spirit was on the spot to rally them; confusion almost immediately supervened; the men clustered together in groups and began to fire wildly; every moment increased the number of casualties; when suddenly one officer, cooler than his comrades, who had discovered an empty tank in the vicinity, proceeded thither and sounded the "assembly." The men responded to the well-known call; rallied round their officer, and, from the somewhat sheltered spot, recommenced, under better auspices, the musketry combat. Still, however, their white clothing told terribly against them. For some time they stood, uncertain as to their next movement. After a brief interval, however, the surviving officers resolved to hold out where they were until daylight, and then to retreat by the road along which they had advanced.

Daylight came. The survivors of the British force formed up and set out on the road leading towards Dánápúr. But the enemy had calculated their movement. They had occupied every point of vantage in the retrograde route. The British soldiers, suffering from hunger, from fatigue, from exhaustion, pushed on without order, exposed to a galling fire, intent only on reaching the nálá which they had crossed but a few hours earlier. Before they reached it their retreat had become a rout. The men pressed doggedly forward, scarcely returning the fire which long rained upon them. Fortunately for them, some time before the nálá was reached the ammunition of the pursuers ran short. But for that, hardly a man would have escaped. At length the nálá was sighted. It was realised that the boats were still there. Then there was a tumultuous rush forward, the men crowding and hustling one another, all bent on putting the water between themselves and their pursuers.

The confusion of the embarkation beats description. The defeated soldiers rushed at the stranded boats to push them into the stream, whilst those of the sipáhís who still had ammunition crowded near them, and poured upon them a galling fire. Some of these, more daring than their comrades, even set fire to some of the boats. Many of the British were shot here; some were burnt; some were drowned. The commands and entreaties of their officers were of no avail. At length a start was made, and the survivors landed on the opposite bank. They were now safe from pursuit. But their losses had been terrible. Of the four hundred and thirteen Europeans and Sikhs who had crossed that nálá the previous evening only fifty re-crossed it unwounded. Of fifteen officers twelve were killed or wounded. The survivors made their way without further incident to the steamer, which conveyed them back to Dánápúr.

Such was the catastrophe of A'rah—a catastrophe due solely to bad leadership. It was repaired, it is true, by the courage of the garrison of the isolated house at that station, and especially by the splendid achievement of Vincent Eyre, leading a smaller number of men against the same enemy. But, though repaired, the memory of it remained. A British force had been led into an ambush, had been surprised and totally defeated, by an enemy whom it affected to despise.

Contrasting strongly in its results with the SURPRISE and AMBUSH of A'rah was the SURPRISE and AMBUSH of A'zamgarh. The first was disastrous to the surprised, because their leader was incompetent; the second was disastrous to the surprisers, because the commander of the surprised was a real soldier, a man who thoroughly understood his craft, whose mental energies were never better directed than when he was in the presence of danger!

Kunwár Singh, after Eyre had defeated the rebel sipáhís and relieved the noble garrison of A'rah, had retired to Jagdispúr, a town or village of about five thousand inhabitants, near his ancestral estates. Here Eyre attacked and defeated him on the 11th August. So complete was the discouragement attending the overthrow, that Kunwár Singh quitted the Sháhábád district, and during the remainder of the year contented himself with the career of a freebooter nearer to the Oudh frontier, plundering in every direction, whilst he trained and augmented his levies. In the early part of 1858, however, an opportunity presented itself to him of dealing a severe blow against his old masters. The necessity of concentrating as many troops as possible before Lakhnao had seriously weakened the British garrisons in the dis-

districts joining the eastern border of Awadh. Noticing this, Kunwár Singh entered that province, rallied to his standard many sipáhís belonging to the force which had been then but recently defeated by General Franks on his march to Lakhnao, and, crossing the border, occupied Atraolia, a fortified village surrounded by a wall fifteen feet high, in the district of A'zamgarh, twenty-five miles from Sultánpúr, and the same from the station of A'zamgarh.

The A'zamgarh district was then guarded by a small British force consisting of two hundred and six men of the 37th Regiment—the very regiment a detachment from which had, it will be remembered, formed part of the surprised force under Dunbar—sixty native cavalry, and two light guns, under the command of Colonel Milman, of the 37th Foot. When Kunwár Singh seized Atraolia, Milman was encamped at Koelsa, near A'zamgarh. On the 21st March he received information from Mr. Davies, magistrate of A'zamgarh, of the vicinity of Kunwár Singh and his following. Milman at once broke up his camp, marched all night, and at daybreak of the 22nd came upon the advanced outposts of the enemy occupying some mango-groves. He at once drove them from their position, then halted in the groves he had gained, and ordered his men to cook their breakfasts. The arms were consequently piled, the fires were lighted, and the coffee was yet simmering in the cans, when, before it could be tasted, information came to him that the enemy in full force was close at hand resolved to deliver a counter attack!

It was indeed true. At last the opportunity for which Kunwár Singh had been longing since Eyre expelled him from Jagdíspúr had come to him. He had before him a force which, though mainly composed of Europeans, was yet small in comparison with his own; those Europeans, too, the regimental comrades of men whom his men—for he still had with him the remnants of the three Dánápúr regiments—had surprised and slaughtered; now, too, as on the last occasion, they were tired after a long march, they were fasting, they were miles from their base, and they had no supports! Every point was in his favour. He believed he was marching to an assured victory!

Milman, meanwhile, on receiving the information to which I have adverted, had made his men fall in, and had ridden forward to assure himself of its truth. It was true. He saw the enemy stretched out, some halted, covered by a mud wall, others advancing through fields of sugar-cane and mango groves. For a moment he hoped that a forward demonstration on his part might check them. He tried it accordingly, but soon recognised that whilst he could make no impression on the stolid centre, he was outflanked on both wings. He could only, then, fall back on Koelsa. This movement he effected in good order, pursued all the way. That he was able to reach his destination safely was due very much to the conduct of sixty men of the native cavalry (4th Madras L.C.), whose constant charges did more than anything else to keep the enemy at a respectful distance.

His loss in killed and wounded had been very severe. Nor, indeed, when he reached Koelsa, did he find there the refuge he had expected. The rumour of his mishap had preceded him. As was natural under such circumstances, the greater number of his camp-followers and cartmen had fled, the latter taking their bullocks with them. The enemy was near and was still advancing; the supplies in camp had been all but exhausted, nor had Milman any means of procuring more at Koelsa. Finding it thus impossible to remain at that place, Milman resolved to continue his retreat to A'zamgarh. This movement forced him to abandon his camp-equipage. There was no help for it; so he fell back, reached A'zamgarh the same day, and despatched at once messengers to Banáras, to Alláhábád, and to Lakhnao for succour. Without any delay, he proceeded to intrench himself within the jail, leaving the town to its fate.

The distances between Milman and the three places named were considerable. A'zamgarh is separated from Banáras by eighty-one miles of good road passing through Jaunpúr; from Alláhábád by a hundred and nine miles, and from Lakhnao by a hundred and seventy-one. The town, founded in 1649 by A'zam Khán, an officer of the Emperor Sháh Jahán, lies on the river Tons, which, navigable downwards to its confluence with the Surjú, is here crossed by a bridge of boats. Its population numbered then nearly fourteen thousand souls.

Kunwár Singh had followed up Milman, but, from some cause which has never been explained, not very closely. Probably he was waiting for reinforcements, feeling certain, as he must have felt, that the news that he was driving the English before him would bring hundreds if not thousands to his standard. And, undoubtedly, the news did produce that result. For, on the 26th March, he advanced and seized the town of A'zamgarh, and threatened the jail. This, fortunately for the defenders, was surrounded by a deep ditch.

Meanwhile responses from two quarters had been made to Milman's appeal. On the 25th a hundred and fifty men, and on the 26th, before Kunwár Singh attacked the town, a hundred and thirty, of the 37th, led by Colonel Dames, and forty-six Madras sipáhís, marched through the town and entered the jail. Colonel Dames, as senior officer, then assumed the command. The day following, the 27th, Dames attempted a sortie at the head of two hundred Europeans and the sixty Madras cavalry, taking with him also two guns. At first it seemed as though he would succeed, but the enemy, recovering almost immediately, drove him back with the loss of one officer and eleven men killed and wounded. He resolved, then, to act thenceforward on the defensive, until a force should come to relieve him.

That very day, the 27th, Milman's express reached Alláhábád. There, the Governor-General of India, Lord Canning, held his court. The news caused him great anxiety. Knowing what sort of a man Kunwár Singh was—that he possessed audacity and courage, and that hatred of the British had become the passion of his life—he realised at once the danger of the situation. He saw how Kunwár Singh, should he succeed in overwhelming the British at A'zamgarh, might make a raid upon Banáras, then almost unprovided with troops, and, by seizing that capital of Hinduism, sever the communications between Calcutta and the North-West.

An able commander, flushed with victory, and commanding a very large force, might even do more; for every available British soldier, the very slender garrison of Calcutta excepted, was at the moment in, or in progress to, the North-West Provinces.

The danger then was, unless it could be conjured away, a very real danger. How to meet it became the question of the moment. The Commander-in-Chief and the British army were at Lakhnao. Thence there was no time to procure aid. At Alláhábád there were two wings of two regiments, the 13th Light Infantry and the 54th. To move the latter, however, was impossible, for the men were struck down by sickness. Practically, then, but one of the wings, that of the 13th Light Infantry, was available. Its commander was Colonel Lord Mark Kerr. For the moment that officer and the wing he commanded were the props upon which the Government of India rested to rescue them from a great difficulty. Lord Canning, then, sent for Lord Mark and explained to him the situation. Lord Mark comprehended it on the instant. The same evening he set out for Banáras, with orders to pick up whatever troops he might find there, and push on with all haste to A'zamgarh.

For such a service there was not an officer in the British service more competent than Lord Mark Kerr. Spare of body, active, a splendid horseman, inured to fatigue, cool under all circumstances, and becoming even cooler in the presence of danger, Lord Mark was the type of the resolute British soldier. He had already shown in the Crimea, on the field of Tchernaya and before Sebastopol, when in command of the same regiment, the 13th Light Infantry, that he knew how to lead his men to victory. In India he had excited the wonder and admiration of his comrades, especially those of the Indian Service, by the immunity with which, bareheaded, he was able to dare the rays of an Indian sun. A stern disciplinarian, he was just, and, being just as well as stern, he was beloved by his men. So deeply, indeed, was the principle of justice implanted in his character that the fact of his being employed to suppress a mutiny of the Indian soldiers and a partial rising of the Indian people did not blind him to the many good qualities of the races he was called upon to combat. These, too, instinctively appreciative, recognised in their enemy a loyal soldier, to whom they might appeal with due confidence that their case was not prejudged. The name of Lord Mark Kerr lives, even now, in the recollection, and is enshrined in the hearts, of the natives of India, especially of those natives of the Western Presidency amongst whom his later service was passed. The reader will not be surprised when I add that such a man, so clear-headed, so resolute, so daring, so proof against the scares and alarms which overwhelm feebler natures, was pre-eminently the man to conduct to a safe issue the dangerous mission upon which the Governor-General of India despatched Lord Mark Kerr!

Lord Mark Kerr and the wing of his regiment, three hundred and ninety-one strong, inclusive of nineteen officers, set out that same night, reached Banáras on the 31st, picked up there a troop—two officers and fifty-five men—of the Queen's Bays; one artillery officer, seventeen gunners manning two 6-pounder guns and two 5½-inch mortars, and, thus strengthened, started for A'zamgarh at ten o'clock on the night of the 2nd April—Good Friday. His total force, numbering twenty-two officers and four hundred and forty-four men, was but little, if at all, in excess of that which followed the ill-fated Dunbar;* whilst the rebels, led by the same Kunwár Singh, were very much more numerous than the force which surprised and defeated that officer.

Pushing on with all speed, Lord Mark reached Sarsána, eight miles from A'zamgarh, the evening of the 5th August. There he received, and during the night he continued to receive, most urgent letters from the staff officer with Colonel Dames's force, setting forth the necessity of hastening on without a moment's delay. But, hurriedly to march a small force of four hundred and forty-four Europeans, tired from a long march, in the height of the rainy season, across a country utterly unknown to any one of them,

* Dunbar had with him many volunteers, in addition to his four hundred and thirteen soldiers.

and with which he himself had no personal acquaintance, to relieve a place besieged it might be by fifteen thousand, it could not be by less than five thousand, men, was an idea not to be entertained by such a commander as Lord Mark Kerr. Defeat would but precipitate the evil he had been sent to avert. Wisely, then, did Lord Mark permit his soldiers to enjoy their well-earned slumber, determined to march on when the dawn of day should approach.

At four o'clock in the morning, then, he formed his men, refreshed and re-invigorated by their sleep, in marching order. Lord Mark led the way, accompanied by a party of the Bays, well ahead of the infantry. After marching about two hours, Lord Mark, whose fine powers of observation had been all the time keenly on the alert, noticed some buildings and a mango grove to the left of the road a little beyond him. Halting to examine these attentively, he observed that both the building and the grove were occupied by sipáhís. Sweeping then his glass suddenly round to the right, he noticed that the banked ditches which lined the road in that direction were likewise crowded with armed men. His mind was instantly made up. He remained where he was till the infantry came up, and kept them there halted till the train of elephants, camels, and carts had closed in. When he had his force in hand, in compact order, he despatched a company to the right front to clear the ditches of the enemy. In this he succeeded; but the success had hardly been pronounced, when, from the buildings and mango grove of which I have spoken there opened a heavy fire upon his column. Instantly he threw out his infantry in skirmishing order and, bringing the guns to the front, opened fire at a distance of about five hundred yards on the buildings and their occupants. The enemy's foot-men, however, were so numerous that they spread out on both sides of him, threatening his rear as well as his flanks. The position was full of danger, because, with a small body of men, Lord Mark had to defend his baggage-train, the drivers and animals composing which had at the very first shot shown symptoms of terror and who soon became uncontrollable. For a whole hour, however, Lord Mark maintained, though with difficulty, his position, covering his baggage, and repulsing the enemy alike from his front, from his rear, and from his flanks. Upon the enemy's position, however, he had made no impression.

It was just at this moment, when he was meditating offensive action, that he discerned the reserves of the enemy forming up in quarter-distance column, whilst from their right a compact column was being detached, for the evident purpose of penetrating between him and his baggage-train. The situation was now more critical than ever. The detached column of the enemy, waxing bolder as they beheld their comrades holding their own against the British, did succeed in setting fire to many of the carts. Meanwhile, Lord Mark had pushed the two 6-pounders to within sixty yards of the main building, but their fire failed for some time to produce much effect upon it. He had been anxious to try the effect of shelling, but the two mortars had been halted in a very disadvantageous position for the purpose, and to move them so that they could be brought to bear with effect it was necessary to draw them back a short distance. When he had, in the earlier stage of the fight, ordered such a movement, he noticed that it had been interpreted by the enemy to signify a retreat, and had caused them to rise to their feet and advance with loud shouts. He felt now, however, that at all cost the main building must be carried. As soon, then, as the two 6-pounders had effected a small breach in the main building, he called for volunteers to storm it. Some thirty or forty men rushed at once to the front. They found the breach not quite practicable, but, like true British soldiers, they set to work with their bayonets to enlarge it. Their labours, however, disclosed an inner wall yet uninjured. Upon this Lord Mark ordered them to set fire to the roof and wooden parts of the building; then to fall back. They obeyed the order with alacrity. It was a splendid piece of work, for, shortly after they had fallen back and the firing had recommenced from one gun—the other being used to prevent the enemy from pressing too closely on the rear—the flames from the fire the men had kindled forced the enemy to evacuate the building. Instantly Lord Mark—who had but just arranged for another infantry attack on the building—sent the Bays to the front. The rebels did not stay to meet the shock of their charge, and space was at once cleared for a further advance.

Just then, however, the attack on his rear had become very pronounced. The rebels had seized an embankment which there lined the road and from it poured in a heavy and continuous fire. To stop it, Captain Wilson Jones, commanding the company of the 13th which formed the rear-guard, at once faced about, charged and drove the enemy from the embankment. Unfortunately Jones fell in the moment of victory.

Lord Mark's position was now peculiar. He had pierced the enemy's centre: the way to A'zamgarh lay open to him: on his left the rebels, terrified by the catastrophe at the building, were rapidly falling back: but on the right they were still menacing, whilst, in the rear, the carmen and drivers having run away, the baggage remained exposed. Under these circumstances, bearing in mind the object of his mission, Lord Mark resolved to leave a sufficient force to his right, whilst with the main body he should push on to A'zamgarh, rally to himself, and return with, any loyal sipáhís he might find there, in the belief that these, on a pinch, might drive the carts. He probably argued that the enemy, believing themselves threatened by the movement, would be glad to retreat while they could. Detailing then Major Tyler of his regiment, a cool and capable officer, for the duty of holding the enemy's left in check, Lord Mark pushed on to A'zamgarh.

His anticipations were realised sooner even than he had expected. The enemy's left wing, frightened at his forward movement, beat a hasty retreat. No sooner had it disappeared than the carmen and drivers emerged from their hiding-places and resumed their seats on their carts and on their elephants. Then Major Tyler, pushing rapidly after his chief, rejoined him, and Lord Mark led the united force to A'zamgarh.

He encountered no further opposition. The enemy, terrified by their defeat, had even abandoned a village which had to be traversed and which a few men might easily have defended. At eleven o'clock Lord Mark reached the bridge of boats leading across the river to the jail occupied by Colonel Dames. The rebels had so injured the bridge with their guns as to render it for the moment impassable, and they were still maintaining a heavy fire upon it. Under this fire it was, however, speedily repaired by Lieutenant Colomb, R.A., acting under Lord Mark's orders. Then, in accordance with the programme he had laid down, he summoned the Madras sipáhís from the intrenchment and despatched them to aid in escorting the carts and elephants. They accomplished this task without opposition.

This splendid victory was not gained without loss. Eight officers and men were killed and thirty-four severely or dangerously wounded, nearly ten per cent. of the whole force. The total casualties of the enemy were never ascertained.

The battle of A'zamgarh stands out as the only instance in history in which an army surprised by an enemy lying in ambush for it succeeded in defeating the surprisers. If we look at other instances we shall note that Flaminius succumbed at Lake Trasimenus, and Varus in the German forest; that Braddock was crushed by the Indians; that Van Rautzau was completely beaten at Tuttlingen; that even the great Turenne was overwhelmed at Mergentheim;* that Mosel and Ziethen were all but destroyed at Domstädtl.† It is true that Lecourbe got the better of Souvoroff, but Lecourbe was not tempted into an ambush, neither was the surprise complete, for, as pointed out in the chapter referring to the action, Lecourbe was expecting the attack which has been called, perhaps not quite correctly, a surprise. The same remark applies to Inkerman. There there was no question of ambush. Our troops were, it is true, surprised in the grey of the morning and they succeeded, with the aid of the French, in repulsing the assailants. The case of the battle of A'zamgarh is, then, unique in history. It is, in very deed, I repeat, the only instance on record

* *Battlefields of Germany*, pp. 158–67, and 187–93. W. H. Allen & Co.
† *Loudon*, pp. 68–72. Chapman & Hall.

in which an army, surprised by an enemy in ambush, succeeded in inflicting a crushing defeat upon that enemy.

That it did so was due to its leader. Lord Mark Kerr at the head of four hundred and forty-four men defeated an army led by the same chieftain who, with far fewer numbers, had destroyed the almost equal force of Dunbar. It is true that Lord Mark had at his side Colonel Longden of the 10th Foot and Mr. Venables, the famous indigo-planter of A'zamgarh, and that Mr. Venables knew the country thoroughly. But Lord Mark was the leader. Upon his shoulders rested the entire responsibility. It was for him to say whether the force should advance, should halt, or should retire. He had to make all the arrangements, to do all the reconnoitring. And he did not flinch from the task. Manfully did he fight that up-hill battle—calm, cool, steadfast, daring, and collected in every phase of it. When, at the period when his men were hardly pressed and the enemy were making great efforts to seize his baggage it was suggested to him that he should abandon the convoy and cut his way into the intrenchment, he merely replied: "Wait a bit; we'll win yet." And, persevering, he did win. His mind embraced every section of the battle-field, and his coolness enabled him to act exactly as the exigencies of the moment required.

In another point of view his conduct is not less to be commended. Dunbar hurried into a trap, because in his desire to relieve the garrison of A'rah he scorned every precaution. But the garrison of A'zamgarh believed themselves to be as hardly pressed as was the garrison of A'rah. Nothing could surpass the imploring tone of the letters received by Lord Mark on the night of the 5th at Sarsána. A less prudent commander would have hurried in impetuously, and would have been destroyed in the darkness of the night. That Lord Mark Kerr was most anxious to relieve his countrymen is unquestionable. A warm-hearted man, he felt the responsibility of halting only a few hours whilst the garrison of A'zamgarh might be in imminent danger. It required a greater courage to resist the pressure to move forward than many men are endowed with. Yet he did resist it, and, resisting it, succeeded.

It subsequently transpired that on that memorable day Kunwár Singh had at his disposal twelve thousand men. That he did not bring all these into the field is possible, but he brought at least a sufficient number, had his execution been equal to his mental power, to have crushed a force so small as that which Lord Mark Kerr commanded. By his timely action, then, Lord Mark Kerr did more than relieve the garrison of A′zamgarh. He defeated projects so formidable that, unchecked, they would have greatly neutralised the effects of Sir Colin Campbell's triumph at Lakhnao, and have forced that commander to send a considerable force to restore the prestige of England in a part of India which had been untrodden by the rebels. It remained untrodden, thanks to Lord Mark Kerr, to the very end!

INDEX.

A.

B.

C.

D.

E.

H.

I.

J.

K.

L.

M.

N.

O.

P.

R.

S.

T.

W.

Y.

Z.

www.ingramcontent.com/pod-product-compliance
Lightning Source LLC
LaVergne TN
LVHW011631200926
848484LV00011B/2554/J
* 9 7 8 1 8 4 7 3 4 2 2 7 0 *